AGING AND STABILIZATION OF POLYMERS

STARENIE I STABILIZATSIYA POLIMEROV

СТАРЕНИЕ И СТАБИЛИЗАЦИЯ ПОЛИМЕРОВ

AGING
AND STABILIZATION
OF POLYMERS

Edited by

M. B. Neiman

Translated from Russian

CONSULTANTS BUREAU
NEW YORK
1965

First Printing – March 1965

Second Printing – November 1968

ISBN-13: 978-1-4615-8551-0   e-ISBN-13: 978-1-4615-8549-7
DOI: 10.1007/978-1-4615-8549-7

The original Russian text was published for the Institute of Chemical Physics of the Academy of Sciences of the USSR by Nauka, the Academy of Sciences Press, in Moscow in 1964.

Library of Congress Catalog Card Number 64-23249

CONTENTS

PREFACE

The need for a broad development of the production of polymer materials has become evident. All these materials are subject to various types of aging (destruction); hence, stabilizers which permit the storage, reprocessing, and use of polymer materials without any appreciable change in their properties must be introduced into them. In recent years, this problem of stabilizing polymers has attracted the attention of many scientists and technologists, both in the USSR and abroad. The scientific basis of the foreign studies will be found in a number of theoretical premises, but chiefly the theory of chain reactions with unbranched chains.

In the Soviet Union, the concepts of Academician N. N. Semenov on chain reactions with degenerate branches have become the starting point of theoretical studies of the stabilization and destruction of polymers. Soviet scientists have developed a theory of critical concentrations of antioxidants and have shown that the processes of stabilization have a very complex chemical character. The nature of the polymers themselves greatly affects these processes and consequently, different stabilizers are required for polymers of different structures. In addition, it has been shown that the antioxidants used thus far can not only cause chain termination, but can also initiate oxidation and give rise to degenerate branches.

All these studies were designed to create a scientific basis for the purposeful selection of effective stabilizers. They were published, over a period of years, in various journals, and by 1962 the need was felt for a monograph on the aging and stabilization of polymers in which the results of investigations of Soviet scientists would be systematically presented, together with a critical survey of the studies of foreign researchers.

To speed up the publication of this monograph, a large group of authors — specialists in polymer stabilization — was enlisted in its writing. Some problems in which Soviet scientists have been interested to a lesser degree, for example, thermal and photooxidative destruction, have been treated in less detail in the monograph.

The introduction to the book was written by Academician V. A. Kargin; the individual chapters were written by the following: Chapter I by Professor M. B. Neiman; Chapter II by Candidate of Chemical Sciences A. L. Buchachenko; Chapter III by V. V. Mikhailov and I. P. Maslova; Chapter IV by Candidate of Chemical Sciences A. F. Lukovnikov and E. N. Matveeva; Chapter V by Candidate of Chemical Sciences B. M. Kovarskaya and Professor M. B. Neiman; Chapter VI by Professor A. A. Berlin, Candidate of Chemical Sciences D. M. Yanovskii, and Candidate of Chemical Sciences Z. V. Popova; Chapter VII by Candidate of Chemical Sciences I. I. Levantovskaya; Chapter VIII by Candidate of Chemical Sciences B. M. Kovarskaya; Chapter IX by Academician Professor K. A. Andrianov; Chapter X by Professor A. S. Kuz'minskii; and Chapter XI by Professor G. L. Slonimskii.

Professor M. B. Neiman

Introduction

SIGNIFICANCE AND PROSPECTS OF THE STABILIZATION
OF POLYMERS AND ARTICLES MADE FROM THEM

The problem of stabilization in the broad meaning of the word is that of the conservation of the initial properties of polymer substances and materials under the most varied influences: heat, radiation, mechanical and chemical influences, primarily oxidation and hydrolysis, and, finally, the influence of microorganisms. It is quite natural that in this effort we come up against all the variety of chemical and structural processes that occur during the breakdown of polymers, and we arrive at the conclusion that stabilization cannot be accomplished for all polymers by any single method. Hence, the particular problem that forms the basis for polymer breakdown must be defined in each case.

The need for stabilizers arises already during reprocessing of the polymers. It is known that the polymer substance should be converted to the fluid state for reprocessing, and that the region of reprocessing is the temperature region between the melting point and the temperature of thermal decomposition of the polymer. The broader this region, and the greater the excess of the reprocessing temperature over the melting point, the more easily and completely the process of reprocessing will occur. Of course, it is more favorable to expand this region by raising the decomposition temperature of the polymer by stabilization than by lowering the melting point, since lowering the melting point also entails a reduction of the thermal stability of the material. At present the overwhelming majority of polymers are reprocessed with a stabilizer additive.

Stabilization also plays as important a role in applications of polymer materials, especially stabilization against the combined action of light and oxygen and the effects of continuous mechanical action. Even though the variety of mechanisms of polymer breakdown is great, chain processes of oxidation and chain decomposition play a basic role. Hence investigations related to the processes of the appearance of free radicals, the elucidation of their role as initiators of various chain processes of oxidation and destruction, and, finally, a search for inhibitors of these processes are the proper basic directions for studies in the field of stabilization. These lines of study have been very fully explored, and we shall therefore skip over them, and instead discuss less popular trends.

The action of inhibitors on free radicals formed at the ends of decomposing chains stops the development of further chain processes; it cannot, however, prevent a decrease in molecular weight due to the initial cleavage of the chain molecule as a result of either chemical action or mechanical force. This undesirable effect may be reduced by using nonvolatile monomers as inhibitors and plasticizers; terminating the chain decomposition processes, these monomers simultaneously increase the length of the parts of the chain molecules cleaved. Probably the use of polymer inhibitors, in which the inhibiting groups are contained in the molecules of such polymers themselves, would be still better. In this case the event of inhibition would represent chemical combination of the decomposing molecule with the polymer inhibitor molecule. By regulating the molecular weight and number of active groups in the polymer inhibitor molecule, we can gain the possibility of not only maintaining the molecular weight, but of increasing it as well as the degree of branching of the polymer to be stabilized.

In addition, the stabilization of polymers by solid particles of colloidal dimensions is also possible. It would seem that the use of truly soluble inhibitors should give a far greater probability that the inhibitor molecules will encounter the torn-off ends of the chains. However, we now know that practically all polymers exhibit well-defined structures and that extraneous substances, including soluble ones, are arranged among the elements of these structures. It is thus possible that the advantages of inhibitors soluble in the polymer are not as great as it would seem at first glance. At the same time, inhibition by reaction with solid surfaces possesses a number of advantages, both in the selection of substances and in improving the bond between the polymer and the pigment or filler particles. The appearance of structuring in this case compensates for the drop in molecular weight. Dispersed inhibitors may give satisfactory stabilization results, as well as the application of inhibitors deposited on filler particles and pigments.

We should note one further category of phenomena that occur in polymers – especially in crystalline polymers – during their use and storage. Investigations of the structure of polymers in recent years have shown an unexpected richness of structural forms and a close connection between these forms and the mechanical properties of the polymers. It is also known that under mechanical action or during storage change occur in the structure of the polymers that are analogous to the phenomena of aging of metals. The problem arises then of not only creating but also of stabilizing the most favorable structures in polymer materials. This problem will most likely have to be solved in entirely different ways from the problem of chemical stabilization. Of course, in this case also the use of polymer additives that prevent rapid recrystallization of the polymers and at the same time possess the properties of inhibitors is possible. Probably it is always more profitable to use polymer substances in the form of mixtures or grafted compositions to regulate almost any processes in polymers.

Chapter I

MECHANISM OF THE THERMOOXIDATIVE DESTRUCTION AND STABILIZATION OF POLYMERS

1. OXIDATION IN THE GAS AND LIQUID PHASES

After A. N. Bakh [1] and K. O. Engler [2] formulated the peroxide theory of oxidation, a large number of studies appeared in which the oxidation of a number of hydrocarbons and various other organic substances was studied. It was found that many oxidation reactions are autocatalytically accelerated and are characterized by well-defined induction periods.

The development of studies along this line entered a new phase after N. N. Semenov formulated the theory of chain reactions with branches [3], and especially after the creation of the theory of degenerate explosions [4].

The results of the basic studies on oxidation were generalized from the viewpoint of chain theory [5]. Under the influence of the studies of the schools of Semenov and Hinschelwood, a large number of investigations of the mechanism of the oxidation of organic compounds in the gas and liquid phases appeared. The basic problem of research in this field was soon formulated — determining the constants of the elementary reactions of the complex radical-chain oxidation process in order to calculate the rate of the oxidation reaction according to its hypothetical mechanism. A comparison of the theoretical calculation with the experimental data would make it possible to judge the correctness of the mechanism proposed.

The successes of studies along this line were summarized by N. N. Semenov [6] and V. N. Kondrat'ev [7] in their recently issued monographs.

The most important results were obtained in an investigation of the reactions of oxidation in the gas phase. A. B. Nalbandyan and V. V. Voevodskii, using the literature data and the results of their own studies, substantiated the radical-chain scheme of the oxidation of hydrogen and showed that all the conclusions that follow from this scheme are well justified experimentally [8]. The values of many of the constants used in the study mentioned are still being verified and refined in a number of investigations [9].

In view of the recent development of the method of electron paramagnetic resonance (EPR), it has been possible to detect experimentally a number of free radicals and atoms formed during the oxidation, and thus once again to confirm the correctness of the scheme of the oxidation of hydrogen and some other substances adopted.

The formation of atoms of hydrogen [10, 11], oxygen [12], and the radicals HO_2 [13] in the oxidation of hydrogen has been demonstrated.

N. N. Semenov and associates, constructing a scheme for the oxidation of methane and using the values of the constants of the elementary reactions available in the literature, calculated the maximum concentration of formaldehyde formed in this process [6]. The calculated value differs by less than an order of magnitude from the value found experimentally, which should be considered good coincidence, since only approximate values of many of the constants used for the calculations are known.

In the investigation of the mechanism of oxidation in the liquid phase, matters are considerably worse than for oxidation in the gas phase. In spite of the large number of studies conducted in this field, the results of which are presented in a number of surveys [14-18], it has not yet been possible to resolve the problem of constructing a mechanism even of the simplest processes of oxidation of organic substances in the liquid phase from the elementary reactions. Hence studies investigating the elementary processes that occur in the liquid phase and determining their rate constants [19, 20] are acquiring great significance.

In recent years the development of studies in this field has been substantially accelerated in connection with the use of new methods of investigation. For example, much new information has been successfully obtained using the method of labelled atoms, especially using the isotopic kinetic method [21].

The sequence of the formation of various intermediate products in the oxidation of cyclohexane and other hydrocarbons, as well as the rate of formation and dissociation of the intermediate products, can be determined by the isotopic kinetic method [22].

The use of labelled atoms has made it possible to determine the ratio of the rates of a number of radical reactions in the liquid phase [23-26].

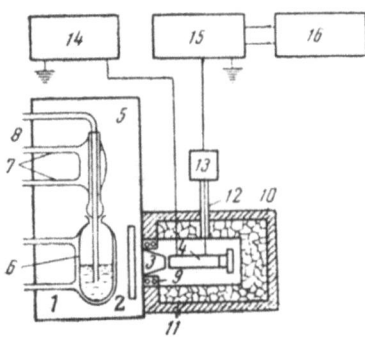

Fig. 1. Scheme of a setup for measuring the intensity of chemiluminescence. 1) Reaction vessel with sleeve; 2) light filters; 3) light pipe; 4) photomultiplier; 5) light-impermeable casing; 6) thermostat; 7) inlet and outlet of reflux condenser; 8) inlet of gas to the reaction vessel; 9) furnace for heating the light pipe; 10) casing of plastic foam for cooling the photomultiplier with dry ice; 11) casing of photomultiplier; 12) shielded inlet to casing of photomultiplier; 13) outside cascade of electrometric amplifiers; 14) high-voltage stabilizer; 15) electrometric amplifier; 16) automatic recording electronic potentiometer.

An extremely promising method has been developed for investigating the rates of certain radical reactions in the liquid phase – the method of chemiluminescence [27], which was recently used successfully to demonstrate the participation of the radicals $R^{\cdot}$ and $RO_2^{\cdot}$ in the processes of hydrocarbon oxidation in the liquid phase [28]. The latter work studied the intensity of chemiluminescence in the oxidation of methylbenzene at 60°C in the presence of an initiator.

A scheme of the setup used is depicted in Fig. 1. In the reaction vessel 1, where the temperature is kept constant, the hydrocarbon to be oxidized, in which oxygen is dissolved, and the initiator, which decomposes slowly at the temperature of the experiment, can be placed. In the recombination of radicals $RO_2^{\cdot}$, chemiluminescence arises; its intensity is measured by a photomultiplier 4 with a system of amplification, the intensified photo current being recorded by means of an electronic potentiometer 16.

One of the chemiluminescence curves in the case when an oxygen-saturated solution of ethylbenzene was placed in the reaction vessel, is cited in Fig. 2. After introduction of the initiator, the chemiluminescence intensity rose rapidly, after which constant intensity of the luminescence was observed for a long time (curve 1). When the oxygen was almost entirely consumed, the concentration of radical $RO_2^{\cdot}$ began to decrease rapidly, which led to a decrease in the rate of recombination of these radicals and produced a decrease in the chemiluminescence intensity. At the lowest portion of curve 1, recombination of radicals $R^{\cdot}$ was observed, related to the smaller luminescence yield.

We succeeded in calculating the ratio $k_1/\sqrt{k_2}$ in the case of the oxidation of ethylbenzene at 60°C by means of an electronic computing machine for the reactions

$$R^{\cdot} + O_2 \overset{k_1}{=} RO_2^{\cdot}$$

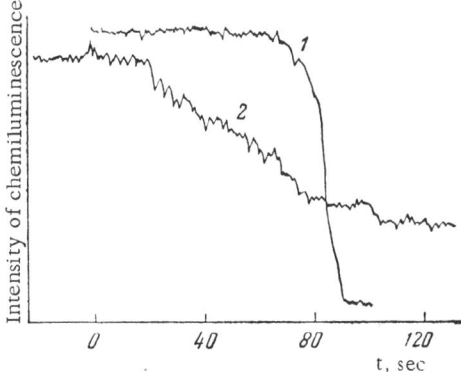

Fig. 2. Curve of the chemiluminescence of ethylbenzene
(1) and methyloleate (2) at 60°C. Rate of initiation w_i
$= 7.5 \times 10^{-7}$ mole · liter^{-1} · sec^{-1}.

and

$$R^{\cdot} + R^{\cdot} \overset{k_2}{=} RR$$

This ratio proved to be equal to 2.7×10^4 liters$^{0.5}$ (mole · sec)$^{-0.5}$.
Curve 2 shows the analogous drop of the chemiluminescence intensity as
oxygen was consumed in the oxidation of methyloleate (at 60°C).

The absolute rate of a number of oxidation reactions, the effectiveness of various initiators, and the rate constants of the reactions of the
radicals with certain antioxidants have been successfully measured using
the chemiluminescence method [29]. It is easy to imagine the enormous
prospects of the development of this method for investigating the mechanism of oxidation in the liquid and solid phases.

Of the other methods, we should mention the method of polarography,
which was first used to determine peroxides and aldehydes formed in the
oxidation of hydrocarbons in [30], as well as the chromatographic method,
widely used in the investigation of the kinetics of reactions in the gas,
liquid, and solid phases [31].

The so-called inhibitor method, based on a measurement of the rate
of consumption of an inhibitor [32], is frequently used to determine the
rates of initiation and development of chains during reactions of liquid-
phase oxidation. The weak aspects of this method have been indicated
in [33, 34].

The mechanism of the radical-chain oxidation of normal decane has
been investigated in very great detail by various methods in [35-39]. On
the basis of these studies and those of other authors, the following
scheme of oxidation of the hydrocarbon RH in the liquid phase can now
be considered most probable:

$$RH + O_2 \overset{w_i}{=} R^{\cdot} + HO_2^{\cdot}$$

A trimolecular initiation reaction is also possible [40], since it is less endothermic:

$$2RH + O_2 = 2R^{\cdot} + H_2O_2$$

Then the chain reaction develops in the following way:

$$R^{\cdot} + O_2 \overset{k_1}{=} RO_2^{\cdot}$$

$$RO_2^{\cdot} + RH \overset{k_2}{=} ROOH + R^{\cdot}$$

Chain termination occurs by recombination of radicals:

$$\left. \begin{array}{l} RO_2^{\cdot} + RO_2^{\cdot} \overset{k_3}{=} \\[4pt] R^{\cdot} + R^{\cdot} \overset{k_3'}{=} \\[4pt] R^{\cdot} + RO_2^{\cdot} \overset{k_3''}{=} \end{array} \right\} \begin{array}{l} \text{inactive} \\ \text{products} \end{array}$$

Degenerate branching occurs as a result of decomposition of the hydroperoxide:

$$ROOH \overset{k_4}{=} RO^{\cdot} + {}^{\cdot}OH$$

In a number of works using the sector method [41], the values of the constants k_1-k_4 and the initiation rate w_i have been successfully determined, which made it possible to calculate the rate of the complex reaction at the first stages of oxidation, when the oxidation products, except for hydroperoxide, play no vital role.

Since the values of the constants depend on the structure of the radical $R^{\cdot}$ and the mechanism of the process changes as the active reaction products accumulate, it is clear how much work must be done before we succeed in solving the problem of constructing a mechanism for liquid-phase oxidation from the elementary reactions.

2. OXIDATION IN THE SOLID PHASE

The mechanism of solid-phase oxidation, in particular, the oxidation of polymers has been still less investigated. Work in this field has begun comparatively recently.

Since the known polymers are obtained from monomers belonging to various classes of organic compounds, they differ not only in the mechanisms of oxidation, but even in the character of the gross processes.

In the oxidation of polyvinyl chloride, for example, a vital role is played by splitting out of HCl molecules and the formation of double bonds. The oxidation of polyamides proceeds without any appreciable induction period and differs sharply in character from the oxidation of hydrocarbons. The thermooxidative destruction of polysiloxanes also proceeds uniquely. Condensation polymers – epoxide resins, polyarylates, and polycarbonates begin to be oxidized at comparatively high temperatures, and thus far no methods of stabilizing them are known. The oxidation of raw and cured rubbers also proceeds extremely uniquely. The available material on the oxidation of the enumerated polymers is described in Chapters IV-IX of this book.

In this chapter we shall consider only the results of investigations of the oxidation of polypropylene, since the process of oxidation of this polymer has been investigated in the greatest detail; moreover, it has been found that its mechanism is extremely close to the mechanism of the oxidation of hydrocarbons in the liquid phase.

First of all, we should mention that the EPR method has made it possible to detect the formation of two types of radicals in polymers under irradiation by γ rays or fast electrons – alkyl $RCH_2\dot{C}HCH_2R'$ and allyl $RCH_2\dot{C}HCH = CHCH_2R'$. We might assume that radicals of the allyl type are formed in the reaction of alkyl radicals with macromolecules with double bonds [42].

It has also been shown by the EPR method that radicals $R^{\cdot}$ in polymers can add oxygen, forming peroxide radicals $RO_2^{\cdot}$ [43].

In [44, 45], conducted using the sector method, the oxidation of specially synthesized 2, 4, 6-trimethylheptane, simulating a segment of the polypropylene molecules, was studied. In these works the light-initiated oxidation was completed according to a radical chain scheme:

$$RH + h\nu \xrightarrow{w_i} R^{\cdot}$$
$$R^{\cdot} + O_2 = RO_2^{\cdot}$$
$$RO_2^{\cdot} + RH \xrightarrow{k_2} ROOH + R^{\cdot}$$
$$RO_2^{\cdot} + RO_2^{\cdot} \xrightarrow{k_3} \text{inactive product}$$

Chain termination occurred according to the following reaction in the case of the addition of diphenylamine:

$$RO_2^{\cdot} + \langle \rangle - N - \langle \rangle \xrightarrow{k_5} ROOH + \langle \rangle - \dot{N} - \langle \rangle$$
$$\qquad\qquad | \qquad\qquad$$
$$\qquad\qquad H \qquad\qquad$$

$$\langle\!\bigcirc\!\rangle\!-\!\dot{N}\!-\!\langle\!\bigcirc\!\rangle + RO_2^{\cdot} = \langle\!\bigcirc\!\rangle\!-\!\underset{\underset{O}{|}}{N}\!-\!\langle\!\bigcirc\!\rangle + RO^{\cdot}$$

The experiments were conducted in the temperature interval 60–80°C; moreover, the values of all the constants of the elementary reactions were determined. In addition, the formation of a stable diphenyl-azotoxide radical was demonstrated.

Ethylbenzene was oxidized in [46], the authors of which demonstrated the formation of a peroxide radical

$$C_6H_5 - \underset{\underset{O-O^{\cdot}}{|}}{CH} - CH_3$$

in the oxidation reaction by the EPR method.

3. OXIDATION OF POLYPROPYLENE

Isotactic polypropylene, just like other polyolefins, is oxidized according to a radical chain mechanism with degenerate branches. Evidence of this is the presence of a well-defined induction period and auto-acceleration of the reaction, which occurs according to the law $w = Ae^{\varphi t}$. Hydroperoxides are the branching products [47]. The introduction of peroxide into the polymer reduces the induction period in proportion to $\sqrt{Per}$, which gives evidence of the presence of quadratic chain termination.

The accumulation of peroxides in the oxidation of atactic polypropylene was studied in [48, 49], and the effective activation energy of the formation of some of its oxidation products was determined.

A more detailed investigation of the kinetics of the oxidation of isotactic polypropylene was conducted in [50]. The experiments were conducted on a circulation setup, the scheme of which is depicted in Fig. 3.

The temperature of the reaction vessel was maintained by means of a liquid thermostat with an accuracy of ±0.2°. The volatile destruction products were frozen out in traps cooled with dry ice. The system was filled with oxygen at the selected pressure.

In the case of a pressure drop in the system, a relay connected to the contact manometer 5 was turned on, and the circuit of an induction coil in one of the arms of the valve 6 was closed. An iron rod was drawn into the magnetic field of the coil, the mercury in the other arm of the manometer dropped, and oxygen from vessel 7 entered the system through a glass tube with a Schott filter. The amount of absorbed oxygen was calculated according to the pressure drop in this vessel.

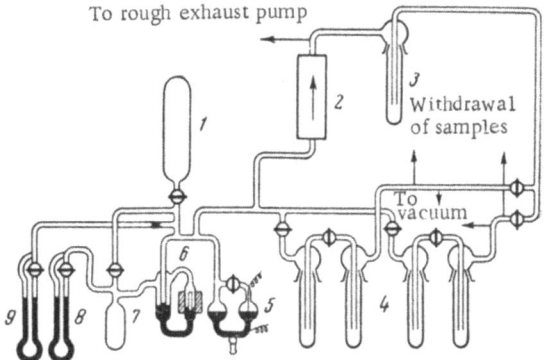

Fig. 3. Scheme of the circulation setup. 1) Vessel with oxygen; 2) circulation pump; 3) reaction vessel; 4) traps; 5) contact manometer; 6) valve; 7) vessel for measuring the absorption of oxygen; 8 and 9) mercury manometers.

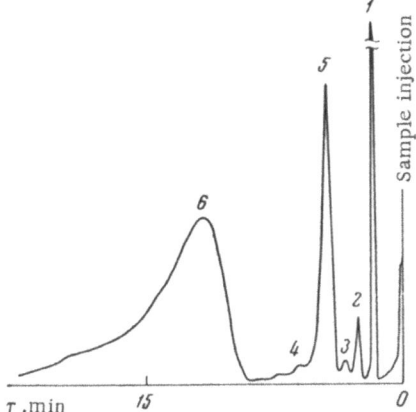

Fig. 4. Chromatogram of the volatile prod-ucts obtained in the oxidation of isotactic polypropylene. 1) $H_2 + CO$; 2) CH_3CHO; 3 and 4) unidentified products; 5) CH_3COCH_3; 6) H_2O.

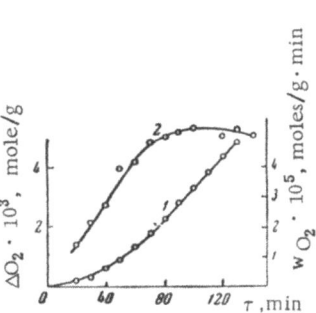

Fig. 5. Oxidation of polypropylene at 130°C and p = 400 mm Hg. 1) Absorption of oxygen; 2) rate of ab-sorption of oxygen.

Isotactic polypropylene was introduced into the reaction vessel in the form of films 40 μ thick, which were pressed in the absence of oxygen.

Water, formaldehyde, acetaldehyde, acetone, methanol, hydrogen peroxide, CO, CO_2, and H_2 were detected among the volatile products. The aldehydes and hydrogen peroxide were determined by a polarograph-ic method. Gas and gas-liquid chromatography were used to analyze the other products. Coal and fine white sand with dipropionitrile ester as the stationary phase were used as the adsorbants. The gas carrier was helium.

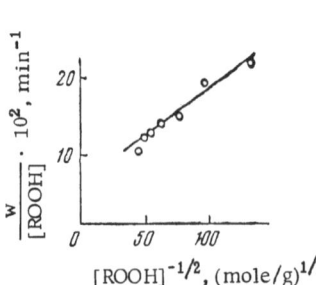
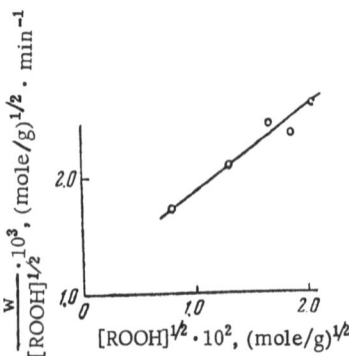

Fig. 6. Dependence of the rate of oxygen absorption on the hydroperoxide concentration.

Fig. 7. Dependence of the rate of oxygen absorption on the hydroperoxide concentration.

Figure 4 presents one of the chromatograms as an example illustrating the separation of the oxidation products.

The hydroperoxide formed in the polypropylene was determined by an iodometric method. The accumulation of C = O and OH groups was monitored by means of the infrared absorption spectra.

Figure 5 depicts the absorption curve of oxygen in one of the experiments, conducted at 130°C and an oxygen pressure of 400 mm Hg. Curve 2 presents the variation of the oxidation rate during the process. The maximum of the rate of oxygen absorption corresponds to the oxidation of 15% of the monomer units in the polymer. In evaluating the amount of polymer oxidized, it was considered that the rate of oxygen absorption is equal to the rate of oxidation of the monomer units, i.e., that one monomer unit adds one molecule of oxygen.

A quantitative consideration of the results showed that the rate of oxygen absorption w_{O_2} can be written in the form

$$w_{O_2} = a[ROOH]^{1/2} + b[ROOH]$$

According to this expression, the graph of the dependence of the rate of oxygen absorption on the peroxide concentration is a straight line in two systems of coordinates:

$$\frac{w_{O_2}}{[ROOH]}, \quad [ROOH]^{-1/2} \quad \text{and} \quad \frac{w_{O_2}}{[ROOH]^{1/2}}, \quad [ROOH]^{1/2}$$

This conclusion is confirmed by Figs. 6 and 7, on which the experimental results are presented in the coordinates indicated above.

The rates of formation of H_2O, CH_2O, CH_3CHO, $(CH_3)_2CO$ were also determined. Figure 8 shows as an example the curves of the rate of formation of H_2O and $(CH_3)_2CO$. The curves of the rates of formation of the

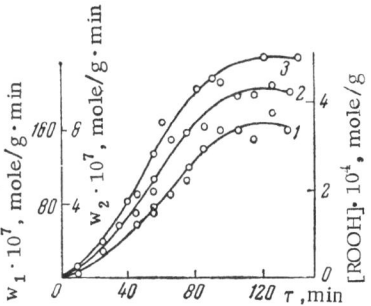

Fig. 8. Formation of oxidation prod-ucts of polypropylene at 130°C and p = 400 mm Hg. 1) Rate of formation of acetone; 2) rate of formation of water; 3) hydroperoxide concentration.

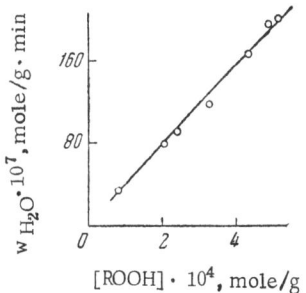

Fig. 9. Dependence of the rate of for-mation of water on the peroxide con-centration at 130°C and p = 400 mm Hg.

other products possess the same form. Curve 3 represents the variation of the peroxide concentration during the pro-cess. As we can see, the rate of for-mation of the products varies in accord with the peroxide concentration. We might assume that all the secondary products are formed as a result of de-composition of the primary oxidation product – the hydroperoxide. It fol-lows from the experimental data that the rate of formation of the products is proportional to the peroxide con-centration (Fig. 9). The experimental points fit well on a straight line. Anal-ogous results have also been obtained for other products. This fact con-firms the assumption that all the prod-ucts are obtained as a result of de-composition of hydroperoxides.

In addition, the available experi-mental data [50] indicate that hydro-peroxide decomposes during the reac-tion with the polymer according to the equation

$$ROOH + RH = RO^{\cdot} + H_2O + R^{\cdot}$$

i.e., according to a bimolecular law, and not a monomolecular law, as was assumed above. The possibility of such a mechanism of peroxide de-composition was indicated earlier in [51, 52, 53], since the endothermic effect of this reaction (–15 kcal/mole) is 25 kcal/mole smaller than the endothermic effect of monomolecular decomposition. The decomposition of the hydroperoxide would occur even more rapidly if it reacted with a compound in which the bond energy of hydrogen were smaller than that in RH.

As is well known, the energy of the N – H bond in antioxidants is sub-stantially smaller than the energy of the C – H bond in hydrocarbons. Hence, [54, 55] compared the rate of decomposition of hydroperoxide formed in the oxidation of isotactic polypropylene, in the absence and in the presence of antioxidants. It was found that the rate of decomposition of the hydroperoxide in the presence of polypropylene at temperatures from 90 to 130°C is substantially smaller than the rate of decomposition

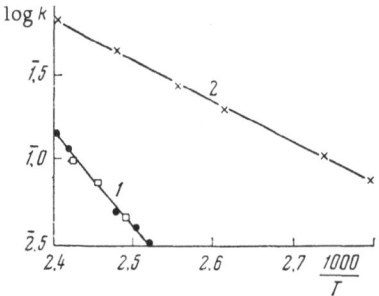

Fig. 10. Temperature dependence of the rate of decomposition of the hydroperoxide of pure polypropylene (1) and in the presence of N-cyclohexyl-N'-phenyl-p-phenylenediamine (2). The squares on curve 1 represent the results obtained in iodometric titration; the points correspond to the data obtained from the rate of formation of the products.

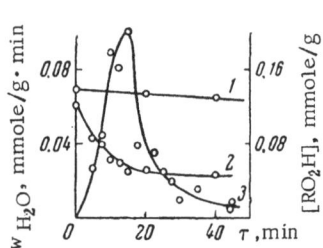

Fig. 11. Kinetics of the decomposition of the hydroperoxide of pure polypropylene at 110°C (1) and in the presence of 0.29 millimole/g dihexadecyl sulfide (2) and the rate of formation of water in the decomposition of the hydroperoxide in the presence of the sulfide (3).

of the same hydroperoxide when 0.38 millimole/g of N-cyclohexyl-N'-phenyl-p-phenylenediamine is introduced into polypropylene (Fig. 10).

The activation energy of the decomposition of polypropylene hydroperoxide proved to be equal to 25 kcal/mole, while in the presence of the indicated antioxidants it was 13 kcal/mole.

K. Ingold has shown that a number of antioxidants of the class of substituted phenols and aromatic amines also accelerate the decomposition of the hydroperoxides formed in the oxidation of white oil [56]. The same result was obtained by V. V. Dudorov, A. L. Samvelyan, A. F. Lukovnikov, and P. I. Levin in an investigation of the decomposition of polypropylene hydroperoxide in the presence of antioxidants.

Dialkyl sulfides accelerate the decomposition of polypropylene hydroperoxide according to an entirely different mechanism. Data on the influence of an admixture of dihexadecyl sulfide on the decomposition of polypropylene hydroperoxide at 110°C are cited in [55]. As can be seen from Fig. 11, the concentration of the hydroperoxide in the absence of the sulfide drops extremely slowly, while in the presence of the sulfide it drops considerably more rapidly, especially at the beginning of the reaction. Correspondingly, the rate of formation of water during the decomposition of the hydroperoxide in the presence of the sulfide is characterized by a sharply pronounced maximum.

Such reaction kinetics can be explained by the formation of an addition product of the sulfide to the hydroperoxide, which decomposes to the

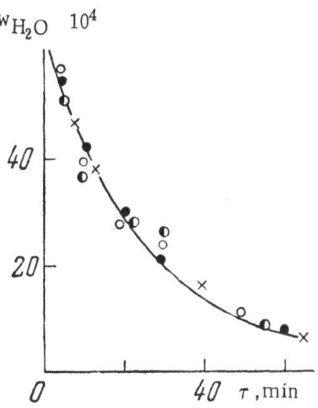

w_{H_2O} 10^4

40

20

0   40 τ, min

Fig. 12. Variation of the rate w of formation of volatile products in the decomposition of polypropylene hydroperoxide at 130°C. O) Rate of formation of acetone; ◐) water; ×) acetaldehyde; ●) formaldehyde.

sulfoxide, water, and olefin according to the scheme:

$$\begin{array}{ccc} O-H & R_2SO & H_2O \\ | & & \\ R_2S...O \quad H & \rightarrow & \\ | \quad | & & | \quad | \\ \sim C-C\sim & \sim C=C\sim \\ | \quad | & | \quad | \\ CH_3 \; H & CH_3 \; H \end{array}$$

In the bimolecular decomposition of hydroperoxides, water, alkyl and alkoxyl radicals are always formed. When alkoxyl radicals decompose, depending on their structure, formaldehyde, acetone, acetaldehyde, and other volatile products may be formed according to the reactions:

$$\begin{array}{c} CH_3 \\ | \\ R-C-CH_2O^{\cdot} \rightarrow R\dot{C}HCH_3 + CH_2O \\ | \\ H \end{array}$$

$$\begin{array}{c} CH_3 \\ | \\ R-C-CH_3 \rightarrow R^{\cdot} + CH_3COCH_3 \\ | \\ O \\ \cdot \end{array}$$

$$\begin{array}{c} CH_3 \\ | \\ R-C-O^{\cdot} \rightarrow R^{\cdot} + CH_3CHO \\ | \\ H \end{array}$$

In addition to the volatile products, the decomposition of alkoxyl radicals produces nonvolatile alcohols, ketones, and aldehydes, the presence of which in the oxidized polymer has been demonstrated by the method of infrared spectroscopy.

Such nonvolatile products contain about 70% of the absorbed oxygen.

The observed facts are well explained by the following scheme of oxidation of polypropylene:

$$RH + O_2 \overset{w_i}{=} R^{\cdot} + HO_2^{\cdot}$$

$$R^{\cdot} + O_2 \overset{k_1}{=} ROO^{\cdot}$$

$$ROO^{\cdot} + RH \overset{k_2}{=} ROOH + R^{\cdot}$$

$$\text{ROO}^\cdot + \text{ROO}^\cdot \overset{k_3}{=} \text{inactive products,}$$

$$\text{ROOH} + \text{RH} \overset{k_4}{=} \text{RO}^\cdot + \text{H}_2\text{O} + \text{R}^\cdot \to 2\,\text{R}^\cdot$$

The differential equations for the variation of the concentrations of the radicals $\text{R}^\cdot$ and $\text{RO}_2^\cdot$ follow from this scheme:

$$-\frac{d[\text{R}^\cdot]}{dt} = w_i + k_2\,[\text{RH}]\,[\text{RO}_2^\cdot] + 2k_4\,[\text{RH}]\,[\text{ROOH}] - k_1\,[\text{R}^\cdot]\,[\text{O}_2] \qquad (1)$$

$$\frac{d\,[\text{RO}_2^\cdot]}{dt} = k_1\,[\text{R}^\cdot]\,[\text{O}_2] - k_2\,[\text{RH}]\,[\text{RO}_2^\cdot] - k_3\,[\text{RO}_2^\cdot]^2 \qquad (2)$$

Using the Bodenstein-Semenov method and adding the equations cited, we obtain:

$$w_i + 2\,k_4\,[\text{RH}]\,[\text{ROOH}] - k_3\,[\text{RO}_2^\cdot]^2 = 0 \qquad (3)$$

From this we obtain the following expression for the concentration of the radicals $\text{RO}_2^\cdot$:

$$[\text{RO}_2^\cdot] = \sqrt{(w_i + 2k_4\,[\text{RH}]\,[\text{ROOH}])/k_3} \qquad (4)$$

Considering the fact that $k_1\,[\text{R}^\cdot][\text{O}_2]$ is equal to the rate of absorption of oxygen w_{O_2}, substituting the value of $[\text{RO}_2^\cdot]$ into equation (1), and neglecting the value of w_i, we obtain:

$$w_{\text{O}_2} = k_2\sqrt{2k_4/k_3}\,[\text{RH}]^{3/2}\,[\text{ROOH}]^{1/2} + 2k_4\,[\text{RH}]\,[\text{ROOH}] \qquad (5)$$

Let us transform equation (5) by dividing both parts by $[\text{RH}]\cdot[\text{ROOH}]$, to the form

$$\frac{w_{\text{O}_2}}{[\text{RH}]\,[\text{ROOH}]} = 2k_4 + k_2\sqrt{\frac{2k_4}{k_3}\frac{[\text{RH}]}{[\text{ROOH}]}} \qquad (6)$$

If the oxidation scheme cited above corresponds to the mechanism of the oxidation of polypropylene, then the dependence of the rate of absorption of oxygen on the hydroperoxide concentration should be depicted by a straight line in the coordinates $\dfrac{w_{\text{O}_2}}{[\text{ROOH}]}$ and $\sqrt{\dfrac{1}{[\text{ROOH}]}}$. We can see from Fig. 6 that the experimental data are in good agreement with equation (6).

If water and the other volatile products are formed as a result of the decomposition of hydroperoxides, then the rate of formation w of the volatile products in the thermal decomposition of hydroperoxides should decrease with time in proportion to the decrease in the peroxide con-

TABLE 1. Relative Rates of Formation of Volatile
Products in the Oxidation of Polypropylene
and Decomposition of Hydroperoxides

Product	Oxidation of polypropylene	Decomposition of hydroperoxides
H_2O	1	1
CH_3COCH_3	0.025	0.025
CH_2O	0.017	0.014
CH_3CHO	0.008	0.007

centration. Figure 12 depicts the results of experimental determinations
of the rates of formation of water, acetone, and aldehydes. When the
scales are selected suitably, all these rates fit well on one curve and
are described by the equation according to which the concentration of the
decomposing hydroperoxide varies. The scales of the rates of formation of water, acetone, formaldehyde, and acetaldehyde in Fig. 12 are
related as 1 : 40 : 72 : 138.

It is important to mention that the relative rates of formation of
volatile products in the oxidation of polypropylene and in the decomposition of hydroperoxides are practically equal (Table 1).

The data cited give evidence that the primary oxidation product of
polypropylene represents hydroperoxide, in the decomposition of which
a number of volatile and nonvolatile secondary products are formed.

4. AUTOINHIBITION IN THE OXIDATION OF POLYMERS

According to N. N. Semenov's theory of degenerate explosions [4],
products that can give rise to chain branching are formed in the oxidation of many substances. If the drop in the oxygen pressure x in the reaction vessel at the first stages of oxidation is related to the formation
of a branched product, then the kinetics of oxidation can be described by
the differential equation:

$$\frac{dx}{dt} = w_i + \phi x \qquad (7)$$

where ϕ usually bears the name of the autocatalysis factor. When this
equation is integrated, N. N. Semenov's approximate law

$$x = A e^{\phi t} \qquad (8)$$

which gives a good description of the beginning of the oxidation process,
is obtained. The products of branching are hydroperoxides, aldehydes,

and other readily decomposed or oxidized products. The kinetic curve
is exponential in character at the first stages. However, in the oxidation of most polymer substances, the oxidation rate reaches a maximum
value, then decreases gradually, and finally the reaction practically stops.

Thus, the kinetics of the oxidation of the overwhelming majority of
polymers is described by S-shaped curves, as is the case, for example,
in the thermooxidative destruction of polyolefins, polyethylene oxide,
polyamides, polycarbonates, polyarylates, epoxide resins, rubbers, etc.

Experiments have shown that the decrease in the oxidation rate of
a number of polymers is related to the formation of various substances
that inhibit oxidation and are capable of terminating the reaction chains.
In the thermooxidative destruction of polycarbonate, diphenylolpropane
is such a reaction-inhibiting substance. If the products obtained in the
oxidation of polyamides are added to the polyamide in small amounts,
then the oxidation of the latter is greatly decelerated.

It can be assumed that the accumulation of an inhibiting product
leads to a decrease in the autocatalysis factor ϕ, the kinetics of the
oxidation being described by the equation

$$\frac{dx}{dt} = w_i + (\phi - ky)\,x \tag{9}$$

If we assume that the concentration of the inhibiting product y varies in
proportion to the amount of absorbed oxygen, i.e., according to the law
$y = k_i x$, then equation (9) approaches the form

$$\frac{dx}{dt} = w_i + \phi x - g x^2 \text{, where } g = kk_1 \tag{10}$$

Introducing new parameters

$$\alpha = \frac{f}{2g} + \sqrt{\frac{f^2}{4g^2} + \frac{w_i}{g}}$$

$$\beta = -\frac{f}{2g} + \sqrt{\frac{f^2}{4g^2} + \frac{w_i}{g}}$$

$$\gamma = \sqrt{f^2 + 4gw_i}$$

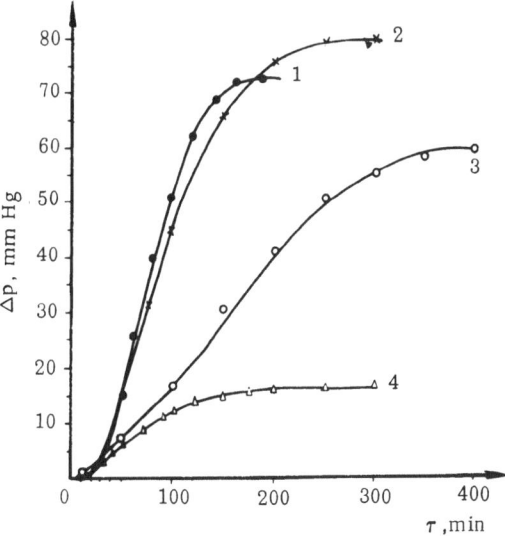

Fig. 13. Calculated kinetic curves of the oxidation of polyarylate TD at 300° and $P_{O_2} = 300$ mm (1); hardened epoxide resin ED-6 at 200° and $P_{O_2} = 300$ mm (2); poly-carbonate at 300° and $P_{O_2} = 200$ mm (3); and poly-carpoamide at 180° and $P_{O_2} = 200$ mm (4). Experimental data depicted by points.

and integrating equation (10), we obtain

$$\frac{a-x}{\beta+x} \cdot \frac{\beta}{a} = e^{-\gamma t} \tag{11}$$

The parameter α is equal to the value of the pressure drop x at which the reaction rate is practically equal to 0. If we know two points of the kinetic curve x_1, t_1, and x_2, t_2, we can calculate the values of the parameters β and γ. Using the quantities α, β, and γ we can easily construct the entire kinetic curve according to equation (11).

The results of calculation of a number of kinetic curves for the thermooxidative destruction of polyarylate, epoxide resin, polycarbonate, and polycaproamide according to formula (11) are cited in Fig. 13. As can be seen from the figure, the experimental points in all cases fit well on curves calculated according to the semiempirical formula (11).

In deriving this formula we assumed that substances that accelerate and inhibit oxidation accumulate in concentrations proportional to the amount of absorbed oxygen. Actually, the accumulation of the branching agent and reaction-inhibiting substance on account of their consumption occurs more slowly.

The changes in the acceleration and inhibition of the oxidation process in comparison with the hypotheses adopted are apparently approximately compensated, which follows from the good coincidence of the experimental data with the curves calculated according to formula (11) for a number of polymers of various structures.

It is easy to calculate the quantities w_i, f, and g, contained in equation (10), according to the values of the parameters α, β, and γ:

$$w_i = \frac{a_1 \beta \gamma}{\alpha + \beta} \tag{12}$$

$$f = \frac{a - \beta}{\alpha + \beta} \cdot \gamma \tag{13}$$

$$g = \frac{\gamma}{\alpha + \beta} \tag{14}$$

5. CRITICAL CONCENTRATION OF ANTIOXIDANTS

About 40 years ago, D. Christiansen [57] and H. Bäkstrom [58] explained the inhibition of the oxidation reaction by small impurities, assuming that oxidation is a chain process, and impurities terminate the reaction chains. This theory was developed by N. N. Semenov in his monograph "Chain Reactions" [5].

It is customarily believed that carbon chain polymers are oxidized according to the mechanism of chain reactions with degenerate branches, proposed by Semenov [3]. Initiation occurs as a result of attack on the RH molecule by oxygen according to the reaction

$$RH + O_2 \overset{w_i}{=\!=} R^{\cdot} + HO_2^{\cdot}$$

The chain reaction develops through the radicals $R^{\cdot}$ and $RO_2^{\cdot}$:

$$ROO^{\cdot} + RH \overset{k_2}{=\!=} ROOH + R^{\cdot}$$

$$R^{\cdot} + O_2 \overset{k_1}{=\!=} ROO^{\cdot}$$

Moreover, molecules of the hydroperoxide ROOH are formed as the primary reaction product; they are capable of decomposing, thus giving rise to branching:

$$ROOH \overset{k_4}{\rightarrow} RO_2^{\cdot}$$

It is assumed that the antioxidants JH react with active $RO_2^{\cdot}$ radicals, forming inactive products:

$$ROO \cdot + JH \overset{k_5}{=} \text{inactive products}$$

Semenov showed that the presence of a "critical" concentration of the antioxidant [6] follows from the mechanism cited, described by the differential equations

$$\frac{d[RO_2^{\cdot}]}{dt} = w_i - k_5[JH][RO_2^{\cdot}] + k_4[ROOH] \tag{15}$$

$$\frac{d[ROOH]}{dt} = k_2[RH][RO_2^{\cdot}] - k_4[ROOH] \tag{16}$$

Actually, the solution of the system of linear equations cited above is written in the form

$$[RO_2^{\cdot}] = \sum A_i e^{\lambda_i t} + B_i \tag{17}$$

where λ represents the roots of the secular equation

$$\begin{vmatrix} -k_5[JH] - \lambda & k_4 \\ k_2[RH] & -k_4 - \lambda \end{vmatrix} = 0 \tag{18}$$

From this we obtain the quadratic equation

$$\lambda^2 + (k_4 + k_5[JH])\lambda + k_4 k_5[JH] - k_4 k_2[RH] = 0 \tag{19}$$

Since the coefficient with λ is positive, the sum of the roots of the equation is negative. This means that either both roots are negative, or one root is positive, while the other is negative, the absolute value of the latter exceeding the absolute value of the positive root. If both roots are negative, then the values of $RO_2^{\cdot}$ and ROOH do not increase to infinity, but approach some constant value, i.e., we are dealing with a slow steady-state reaction. If one of the roots is positive, then the concentration of the radicals constantly increases, and we are dealing with a nonsteady-state accelerating process.

We can answer the question of whether the process is steady-state or nonsteady-state by determining the sign of the free factor in equation (19). If

$$[JH] > \frac{k_2}{k_5}[RH] \tag{20}$$

then the free factor is positive, and, consequently, both roots of equation (19) are negative. This means that when inequality (20) is observed, the oxidation process is steady-state. Consequently, the antioxidant hinders the acceleration of oxidation. If

$$[JH] < \frac{k_2}{k_5}[RH] \tag{21}$$

then the free factor of equation (19) is negative, and, consequently, one of the roots of this equation is positive. This means that the concentrations of the radicals, in accord with equation (17), will increase to large values, and the oxidation reaction will be accelerated, i.e., will be nonsteady-state.

The concentration of the antioxidant, determined by the equation

$$[JH]_{cr} = \frac{k_2}{k_5} [RH] \qquad (22)$$

is called the "critical" concentration. At smaller concentrations of the antioxidant, the oxidation process proceeds as a nonsteady-state reaction, while at larger concentrations it proceeds on a steady-state basis.

An analogous result was obtained by a somewhat different method in [59]. If the number of radicals is equal to n, the rate of initiation w_i, autocatalysis factor φ, and the concentration of the antioxidant x, then

$$\frac{dn}{dt} = w_i + \varphi n - knx \qquad (23)$$

and

$$-\frac{dx}{dt} = knx \qquad (24)$$

Here k is the rate constant of the reaction of the radicals with the antioxidant.

It is clear that under the condition

$$kx < \varphi \qquad (25)$$

the reaction will be autoaccelerated, while under the condition

$$kx > \varphi \qquad (26)$$

we shall have a steady-state reaction.

Thus, the critical concentration of the antioxidant is determined by the equation

$$x = \varphi/k \qquad (27)$$

If the condition (26) is observed at the beginning of the reaction, then the concentration of radicals (Fig. 14, curve 1) reaches some steady-state value soon after the beginning of the reaction, which will be preserved for a time from zero to τ. During this time the concentration x will drop comparatively slowly to a value x_{cr} (curve 2). Then the reaction begins to be accelerated, whereupon the concentration of radicals will increase sharply, while the concentration of the autooxidants rapidly drops to zero.

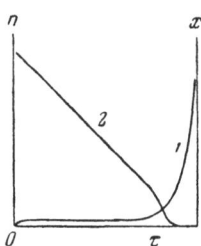

Fig. 14. Variation of the concentration of radicals n (1) and the antioxidant x (2) in a chain oxidation reaction.

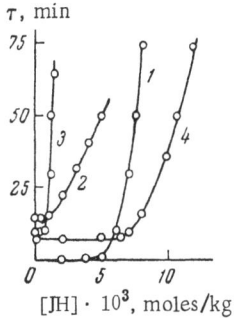

[JH] · 10^3, moles/kg

Fig. 15. Dependence of the induction period during oxidation on the concentration of the antioxidant. 1) Polyethylene oxide with stabilizer 22-46, 145°C, P_{O_2} = 340 mm Hg; 2) tetralin and tri-tert-butylphenol, 140°C, P_{O_2} = 300 mm Hg; 3) cetane and diphenylamine,170°C, P_{O_2} = 300 mm Hg; 4) polypropylene and phenyl-β-naphthylamine, 200°C, P_{O_2} = 300 mm Hg.

The experiments of a number of authors [59 - 64] have shown that when the oxidation of liquid hydrocarbons and polymers is inhibited by various antioxidants, the phenomenon of the critical concentration is actually observed (Fig. 15).

In the book by N. M. Emanuél' and Yu. I. Lyaskovskaya [65], the scheme of inhibition of oxidation is corrected, namely: in place of the equation ROO˙ + JH = inactive products, ROO˙ + JH = ROOH + J˙ is correctly written. However, Semenov's conclusion is repeated unchanged, the secular equation (18) and the formula for the critical inhibitor concentration (22) being cited in unchanged form.

However, if we assume Emanuél''s correction as reasonable, then instead of equation (16) we should write:

$$\frac{d[ROOH]}{dt} = k_2[RH][RO_2˙]$$

$$+ k_5[JH][RO_2˙] - k_4[ROOH] \qquad (28)$$

Then the secular equation will take the form

$$\begin{vmatrix} k_5[JH] - \lambda & k_4 \\ k_2[RH] + k_5[JH] - k_4 - \lambda \end{vmatrix} = 0 \qquad (29)$$

Under this condition the free factor of the quadratic equation for λ will not depend on [JH] and will always be negative. This means that one of the roots of the quadratic equation will be positive, and the oxidation reaction at all concentrations of the antioxidant will be non-steady-state. Thus, the scheme adopted does not provide the possibility of explaining the presence of a critical concentration of antioxidants.

The following hypotheses are placed at the basis of the generally accepted scheme of inhibited oxidation:

1) the inhibitor terminates the reaction chain, but does not influence the process of chain generation;

2) the hydroperoxide decomposes, forming two active radicals.

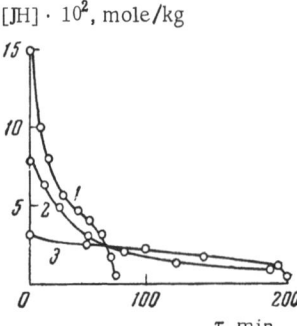

[JH] · 10^2, mole/kg

τ ,min

Fig. 16. Consumption of antioxidants — 2,6-di-tert-octyl-4-
methylphenol (1), di-tert-butyl-
p-cresylmethane (2) and phenyl-
α-naphthylamine (3) in polypropylene at 200°C and p_{O_2}
= 300 mm Hg.

It follows from these premises that the inhibitor is consumed according to a linear law, if its concentration exceeds the critical value. The rate of consumption of the inhibitor in this case is equal to the rate of initiation, and the induction period increases in proportion to the inhibitor concentration.

However, the experimental investigations conducted have shown that the premises cited above actually are not justified. In [33] it was shown that the antioxidants usually are not consumed according to a linear law. Thus, 2,6-di-tert-octyl-4-methylphenol, when introduced into polypropylene, is consumed first rapidly, then more slowly, and then again more rapidly. The antioxidant di-tert-butyl-p-cresylmethane in polypropylene is consumed according to a first order law until its concentration is reduced to the critical value, after which it disappears rapidly. If polypropylene is stabilized by phenyl-α-naphthylamine, then the latter is consumed first slowly, and then considerably more rapidly during oxidation. The curves of the consumption of the antioxidants mentioned are depicted in Fig. 16. Anallogous results have been obtained in other works in which the kinetics of the consumption of diphenylamine in the inhibition of the oxidation of polypropylene, cetane, and other hydrocarbons was investigated.

6. INITIATION OF OXIDATION BY ANTIOXIDANTS

The enumerated experiments give evidence of a complex mechanism of the action of antioxidants. We might think that the latter not only terminate the reaction chains, but can also initiate oxidation. If an antioxidant only terminates the chains, then its introduction into the polymer cannot lead to a reduction of the induction period. If it possesses the ability to terminate and initiate the chain, on the other hand, then its addition under certain conditions can lead to a reduction of the induction period.

In order to realize the latter case, we introduced 0.02 mole/kg di-tert-butyl-p-cresylmethane into polypropylene, which increased the induction period to 180 min at 200°C and p_{O_2} = 300 mm Hg. If another antioxidant, 2,6-di-tert-octyl-4-methylphenol, is added to polypropylene thus stabilized, the induction period is reduced from 180 to 50 min [33]

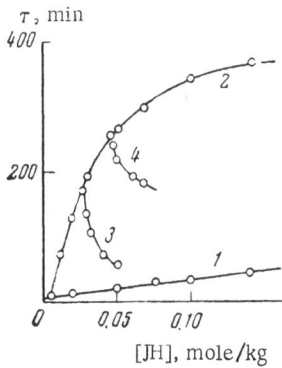

Fig. 17. Induction period in the oxidation of polypropylene in the presence of antioxidants, 2,6-di-tert-octyl-4-methylphenol (1), di-tert-butyl-p-cresylmethane (2) and mixtures of the two antioxidants (3 and 4); T = 200°C, P_{O_2} = 300 mm Hg.

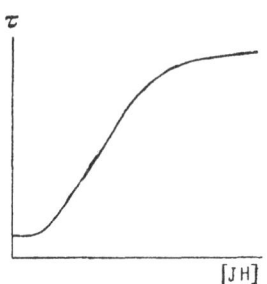

Fig. 18. Typical curve of the dependence of the induction period on the concentration of the antioxidant.

(Fig. 17). Ingold observed a similar effect, which he called antagonistic, when two antioxidants were introduced into hydrocarbon fuel: 2,4,6-trimethylphenol and N-phenyl-α-naphthylamine or di-tert-butyl-p-cresylmethane and N,N'-di-sec-butyl-p-phenylenediamine [56].

The presence of initiation of oxidation by antioxidants also permits us to explain the fact that the induction period does not increase linearly, but more slowly when the concentration of the inhibitor increases (Fig. 18).

It is clear that the effectiveness of an antioxidant depends on the ratio of the rates of chain termination and initiation. Since the latter process is characterized by a greater activation energy, when the temperature increases, the rate of initiation increases more rapidly than the rate of chain termination. Hence the effectiveness of an antioxidant should decrease with increasing temperature, which is actually observed experimentally.

1. Initiation can occur as a result of oxidation of the inhibitor according to the reaction

$$JH + O_2 = HO_2^{\cdot} + J^{\cdot}$$

In this case a comparatively active radical $HO_2^{\cdot}$ is formed, the presence of which during oxidation was recently demonstrated by the EPR method [13].

2. Relatively inactive inhibitor radicals $J^{\cdot}$ can sometimes be encountered in the reaction with molecules of the polymer

$$J^{\cdot} + RH = JH + R^{\cdot}$$

In this case active radicals are formed. The possibility of such a reaction has been demonstrated by the EPR method [66].

3. Antioxidants can accelerate oxidation, by reacting with hydroperoxides according to the scheme:

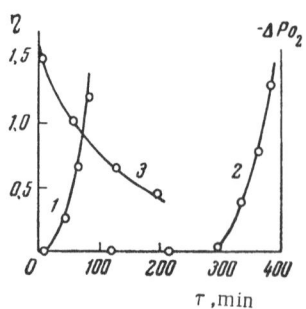

$$ROOH + JH = RO^{\cdot} + H_2O + J^{\cdot}$$

This reaction was described in [55].

An important fact indicating the initiation of a reaction, connected to consumption of the antioxidant, is the drop in the molecular weight of the polymer during the induction period. In this case the number of resorbed bonds is proportional to the consumption of the antioxidant [67].

Fig. 19. Absorption of oxygen in the oxidation of pure polypropylene (1) and stabilized by di-tert-butyl-p-cresyl disulfide and pyrocatechol phosphite naphthol (2); (3) shows the drop in viscosity in the latter case.

This phenomenon is especially evident for the example of the thermooxidative destruction of polypropylene with an admixture of di-tert-butyl-p-cresyl disulfide, investigated by A. F. Lukovnikov, P. I. Levin, and A. G. Vasil'eva [68].

In this case the induction period increases greatly, but the molecular weight of the polymer drops catastrophically (Fig. 19).

Thus, the combination of the known facts leads to the conclusion that the first assumption that antioxidants do not influence initiation is erroneous. The second premise also is doubtful – the formation of two active radicals in the decomposition of the hydroperoxide. It has been shown in a number of studies [69, 70] that hydroperoxides can decompose in solution without forming radicals, giving rise to no chain generation.

As was shown by Yu. A. Shlyapnikov, the introduction into the scheme of polymer oxidation of the hypothesis that δ-active radicals $RO_2^{\cdot}$ are ultimately formed in the decomposition of the hydroperoxide makes it possible to determine the conditions, under the observance of which a "critical" antioxidant concentration is possible [71].

The mechanism of the oxidation of a polymer RH with an admixture of an impurity JH can be represented by the scheme:

$$JH + O_2 \overset{w_i}{=} J^{\cdot} + HO_2^{\cdot} \rightarrow RO_2^{\cdot}$$

$$RO_2^{\cdot} + RH \overset{k_2}{=} ROOH + R^{\cdot}$$

$$ROOH + RH \overset{k_4}{=} \delta RO_2^{\cdot}$$

$$RO_2^{\cdot} + JH \overset{k_5}{=} ROOH + J^{\cdot}$$

If a sulfide R_2S, capable of decomposing hydroperoxide without the formation of active products [72], is introduced into the polymer, then the following equation is added to the scheme cited above:

$$ROOH + R_2S \xrightarrow{k_6} \text{inactive products}$$

In the case under consideration, equations (15, 16) take the form

$$\frac{d\,[ROOH]}{dt} = k_2\,[RH]\,[R\dot{O}_2] + k_5\,[JH]\,[R\dot{O}_2] - k_4\,[RH]\,[ROOH] - k_6\,[R_2S]\,[ROOH] \quad (30)$$

$$\frac{d\,[R\dot{O}_2]}{dt} = w_i - k_5\,[JH][R\dot{O}_2] + \delta k_4\,[RH]\,[ROOH] \quad (31)$$

Let us write the secular equation

$$\begin{vmatrix} -k_4\,[RH] - k_6\,[R_2S] - \lambda & k_2\,[RH] + k_5\,[JH] \\ \delta k_4\,[RH] & -k_5\,[JH] - \lambda \end{vmatrix} = 0 \quad (32)$$

After development, it will take the form

$$\lambda^2 + \{k_4\,[RH] + k_5\,[JH] + k_6\,[R_2S]\}\,\lambda + k_4k_5\,[RH]\,[JH] + k_5k_6\,[JH]\,[R_2S] - $$
$$- k_2k_4\delta\,[RH]^2 - k_4k_5\,[RH]\,[JH] = 0 \quad (33)$$

An analysis analogous to the analysis of equation (19) permits us to calculate the value of the critical antioxidant concentration:

$$[JH]_{cr} = \frac{\delta k_2 k_4\,[RH]^2}{(1-\delta)\,k_4k_5\,[RH] + k_5k_6\,[R_2S]} \quad (34)$$

In the particular case when the concentration $[R_2S]$ is equal to zero, we obtain

$$[JH]_{cr} = \frac{\delta}{1-\delta}\,\frac{k_2}{k_5}\,[RH] \quad (35)$$

We can see from (35) that the steady-state reaction and critical concentration of the antioxidant can occur only in the case when $\delta < 1$, i.e., when less than one active radical on the average is obtained in the decomposition of the hydroperoxide. If $\delta \geq 1$, then at any concentration of the antioxidant the reaction proceeds in a nonsteady-state basis and no critical antioxidant concentration exists.

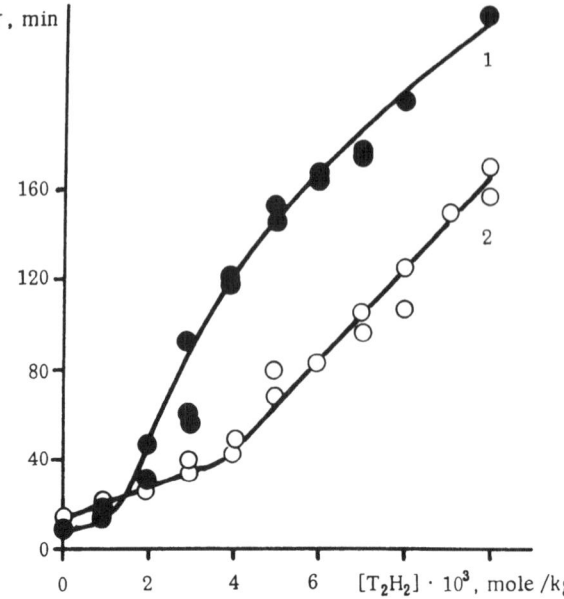

Fig. 20. Dependence of the induction period in the oxidation of polycaproamide on the concentration of antioxidant
I in the absence (1) and in the presence (2) of antioxidant
II. T = 200°C; p_{O_2} = 200 mm Hg. (1) 2,2'-methylene-bis-
(4-methyl-6-tert-butylphenol); (2) 2,2'-methylene-bis(4-
methyl-6-tert-butylphenol) + 0.01 mole/kg 2,6-di-(1,1-
dimethylhexyl)-4-methylphenol.

In addition to the initiation of oxidation considered above, the antioxidant can also increase the probability of degenerate branching, for example, by accelerating the decomposition of hydroperoxides, forming
an active radical, as was indicated above.

To verify the correctness of this hypothesis, we conducted experiments on the inhibition of the oxidation of polypropylene by a mixture of
two antioxidants – 2,2-methylene-bis-(4-methyl-6-tert-butylphenol) (I)
and 2,6-di-(1,1-dimethylhexyl)paracresol (II). When impurities of the
two antioxidants are present in concentrations x_1 and x_2, the increase in
the concentration of active centers n as against equation (23) is described
in the equation

$$\frac{dn}{dt} = w_n + \phi n - k_1 x_1 - k_2 x_2 \tag{36}$$

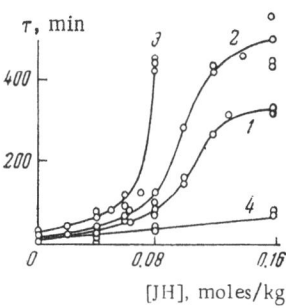

Fig. 21. Dependence of the induction period in the oxidation of polypropylene on the concentration of 2,6-di-tert-octyl-4-methylphenol (4) and with admixtures of disulfide in concentrations: 1) 0.04 mole/kg; 2) 0.08 mole/kg; 3) 0.12 mole/kg. T = 200°C; p_{O_2} = 300 mm Hg.

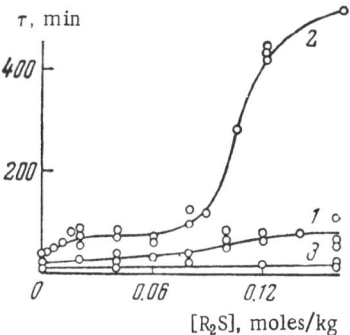

Fig. 22. Dependence of the induction period in the oxidation of polypropylene on the didecyl sulfide concentration (3). 1) The same in the presence of 0.04 mole/kg, 2,6-di-tert-octyl-4-methyl-phenol; 2) the same in the presence of 0.08 mole/kg 2,6-di-tert-octyl-4-methylphenol. T = 200°C; p_{O_2} = 300 mm Hg.

The critical concentration of the first antioxidant, by analogy with equation (27), is expressed in the following way:

$$(x_1)_{cr} = \frac{\phi}{k_1} - \frac{k_2}{k_1} x_2 \qquad (37)$$

Taking the derivative with respect to x_2, we obtain:

$$\frac{\partial (x_1)_{cr}}{\partial x_2} = \frac{1}{k_1} \frac{\partial \phi}{\partial x_2} - \frac{k_2}{k_1} \qquad (38)$$

As can be seen from equation (38), an increase in $(x_1)_{cr}$ with increasing concentration of the second antioxidant is possible only in the case when

$$\frac{\partial \phi}{\partial x_2} > 0 \qquad (39)$$

The latter inequality indicates an increase in the autocatalysis factor with increasing concentration x_2.

Figure 20 shows the results of experiments on the determination of the induction periods in the oxidation of polypropylene as a function of the concentration of the antioxidant I in the absence and in the presence of the antioxidant II. As can be seen from the figure, the critical concentration of the antioxidant I increases from 1.3×10^{-3} mole/kg to 4×10^{-3} mole/kg under the influence of an addition of antioxidant II. As it follows from formula (39), this means that the addition of antioxidant II produces an increase in the autocatalysis factor, i.e., increases the probability of degenerate branching.

7. THEORY OF SYNERGISM

The introduction of the sulfide R_2S into the polymer in sufficient concentration can make the denominator in the right-hand portion of equation (26) positive even if $\delta > 1$, which makes it possible for a critical antioxidant concentration to be realized. The correctness of this conclusion was confirmed by the experiments described in [71].

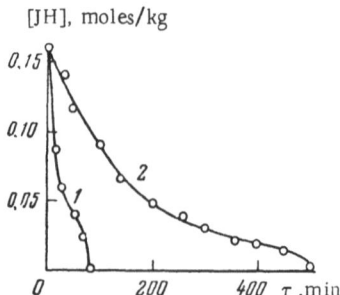

[JH], moles/kg

Fig. 23. Consumption of 2,6-di-tert-octylcresol in the oxidation of polypropylene (1) and in the presence of didecyl sulfide (2). T = 200°C; p_{O_2} = 300 mm Hg.

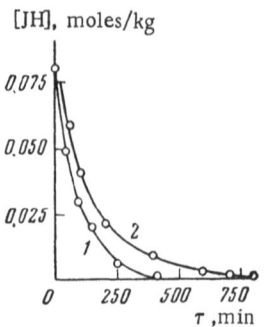

[JH], moles/kg

Fig. 24. Consumption of di-tert-butyl-p-cresylmethane in the oxidation of polypropylene (1) and in the presence of didecyl sulfide (2). T = 200°C; p_{O_2} = 300 mm Hg.

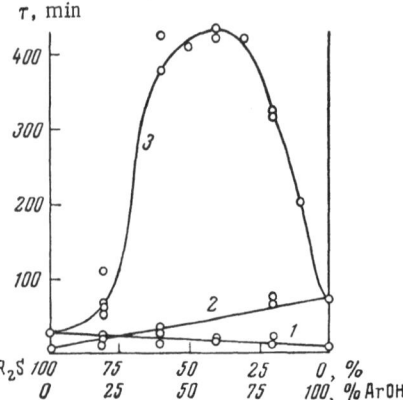

τ, min

Fig. 25. Dependence of the induction period in the oxidation of polypropylene on the molecular fraction of didecyl sulfide and 2,6-di-tert-octylcresol. Total addition of stabilizer 0.2 mole/kg; T = 200°C; p_{O_2} = 300 mm Hg.

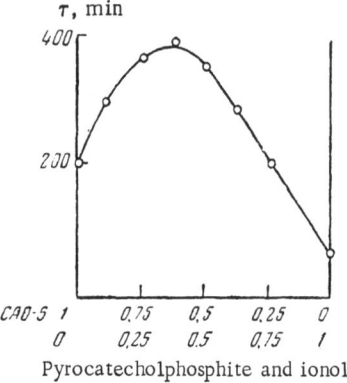

τ, min

Fig. 26. Induction period of the thermo-oxidative destruction of polypropylene stabilized by mixtures of SAO-6 and pyrocatechol phosphite with ionol. Summary concentration of the stabilizers 0.015 mole/kg; T = 200°C; p_{O_2} = 200 mm Hg.

If the antioxidant 2,6-di-tert-octyl-4-methylphenol is introduced into polypropylene in various concentrations, then at 200°C and p_{O_2} = 300 mm Hg, the induction period increases linearly from 8 to 65 min, i.e., no critical antioxidant concentration is observed. If, in addition to the indicated phenol, didecyl sulfide is introduced into the polymer, a critical concentration will be distinctly observed, and will decrease with increasing sulfide concentration, in accord with the formula of [34] (Fig. 21).

Didecyl sulfide itself is a relatively ineffective antioxidant, and its introduction into polypropylene produces an extremely slight increase in the induction period. However, the introduction of 2,6-di-tert-octyl-4-

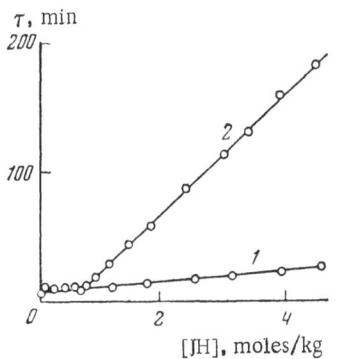

τ, min

[JH], moles/kg

Fig. 27. Dependence of the induction period of thermooxidative destruction of stabilized polypropylene on the concentration of diphenylamine (1) and on the diphenylamine concentration in the presence of 3% polyphenylene (2). $T = 172°C$; $p_{O_2} = 200$ mm Hg.

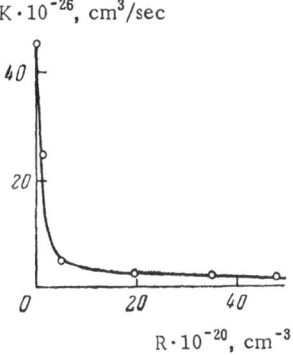

$K \cdot 10^{-26}$, cm^3/sec

$R \cdot 10^{-20}$, cm^{-3}

Fig. 28. Dependence of the rate constant of consumption of tri-tert-butyl-phenoxyl on the concentration of diethylphosphorous ester in benzene solution ($T = 200°C$).

methylphenol into the polymer together with didecyl sulfide leads to a sharp increase in the induction period and to the appearance of a critical concentration (Fig. 22).

Experiments have shown that the introduction of didecyl sulfide into polypropylene sharply reduces the rate of consumption of 2,6-di-tert-octyl-4-methylphenol (Fig. 23). The rate of consumption of the antioxidant di-tert-butyl-p-cresylmethane in polypropylene is analogously reduced, but to a lesser degree, if didecyl sulfide is introduced into the polymer, as can be seen from the curves presented in Fig. 24. These facts are in full agreement with the theory developed in [7].

It is precisely by decomposition of the hydroperoxide and deceleration of the antioxidant consumption that the effect of mutual intensification of the effectiveness, observed in mixtures of 2,6-di-tert-octyl-4-methylphenol with didecyl sulfide is explained. Each of these inhibitors of oxidation, taken individually, only negligibly lengthens the induction period. In a mixture, on the other hand, they are capable of lengthening the induction period in the oxidation of polypropylene to 450 min at 200°C and an oxygen pressure of 300 mm Hg (Fig. 25).

We cannot doubt that the large number of cases of mutual intensification of the effectiveness, observed when mixtures of antioxidants with sulfides are introduced into polymers [73, 74], is explained by the theory formulated above. Of all the mixtures investigated up to this time, a mixture of pyrocatechol phosphite ionol with di-tert-butyl-p-cresol sulfide (SAO-6) has proved the most effective. At a summary concentration of 0.015 mole/kg, this mixture lengthened the induction period of the oxidation of polypropylene to 400 min at 200°C and an oxygen pressure of 200 mm Hg (Fig. 26).

Intensification of the effectiveness of the antioxidant in those cases when the radical $J^{\cdot}$ of the antioxidant produces initiation of the oxidation can be achieved by adding a second substance, which captures $J^{\cdot}$ radicals.

As Likhtenshtein has shown [75], such a case is realized when polypropylene is stabilized with diphenylamine, if polyphenylene or diphenylpicrylhydrazyl is introduced into the polymer as the second substance.

The data obtained in the latter work are depicted in Fig. 27. When the diphenylamine concentration is increased, the induction period τ slowly rises (curve 1). If diphenylamine is introduced into polypropylene together with polyphenylene, then the dependence of τ on the concentration of diphenylamine is depicted by curve 2, i.e., a well-defined intensification effect is observed. Moreover, the appearance of a critical antioxidant concentration is distinctly visible.

The phenomenon of the effect of mutual intensification of antioxidants, frequently called synergism, has been developed in detail in a number of monographs and surveys [65, 66, 76, 77].

A study recently appeared [78], in which the effect of mutual intensification of the inhibition of the oxidation of white oil by mixtures of substituted phenols with disubstituted esters of phosphorous acid was investigated. The authors of this study believe that in the reaction of phenol with $RO_2^{\cdot}$ radicals, the relatively inactive phenoxyl radical $C_6H_5O^{\cdot}$ is formed, then reacting not with the phosphorous ester, but with the product of its isomerization, which occurs according to the scheme

$$(RO)_2 P - OH \rightleftarrows (RO)_2 P \diagup\!\!\!\!{}^{O}_{\diagdown H}$$

The reaction of phenoxyl according to the equation

$$C_6H_5O^{\cdot} + (RO)_2 P \diagup\!\!\!\!{}^{O}_{\diagdown H} \rightarrow C_6H_5OH + (RO)_2P^{\cdot}{=}O$$

leads to the formation of an active radical, which adds an alkyl radical

$$R_1^{\cdot} + (RO)_2P^{\cdot}{=} O \rightarrow (RO)_2 P \diagup\!\!\!\!{}^{O}_{\diagdown R_1}$$

To verify this hypothesis by EPR, we investigated the kinetics of the reaction of the tri-tert-butylphenoxyl radical with the ester $(C_2H_5O)_2POH$ in benzene solution [91]. We found that the rate constant of the consumption of the phenoxyl radical drops rapidly when the concentration of the phosphorous ester is increased (Fig. 28).

It follows from the experimental results that the phenoxyl radical reacts with hydrogen of the monomer form of the phosphorous ester.

The dimer formed by means of hydrogen bonds does not react with free radicals. When the concentration of the ester is increased, the number of dimer molecules increases, and the rate constant of the reaction of di-tert-phenoxyl with the phosphorous ester decreases.

From the material cited above, the negative significance of the fact that antioxidants not only terminate chains, but also take part in the initiation of oxidation is clear. Moreover, an important role is played by the formation of hydroperoxides in the reaction of peroxide radicals $RO_2^{\cdot}$ with antioxidants, the molecules of which contain weakly bonded hydrogen. Such antioxidants include widely used derivatives of phenols and aromatic amines.

It seems to us that one of the most important problems in the field of inhibition of oxidation is the problem of the search for new classes of antioxidants which do not contain weakly bonded hydrogen in their molecules. Of undoubted interest in this respect are trisubstituted esters of phosphorous acid [79], as well as polyphenylenes, polyphenylacetylenes, and polyazophenylenes [80].

8. FREE RADICALS AS STABILIZERS

In connection with this problem, we undertook an attempt to use radicals produced by replacing the active amine hydrogen with an oxygen atom according to the scheme:

$$R - N - R' \rightarrow R - N - R'$$
$$\overset{|}{H} \qquad \overset{|}{O}$$

as antioxidants in place of amine derivatives.

We synthesized [81] the following radicals:

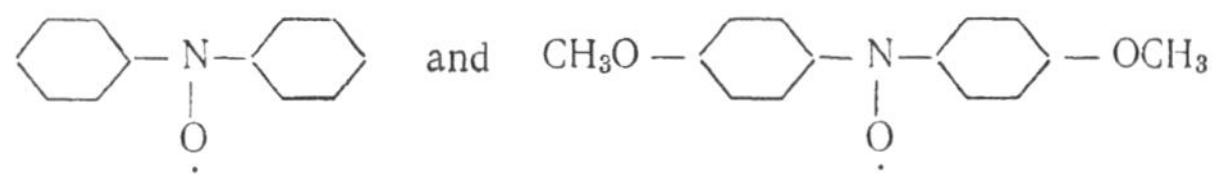

These radicals are comparatively stable and are good antioxidants of polypropylene and a number of liquid hydrocarbons.

The fact that the azotoxide radical, produced by the oxidation of diphenylamine, is itself an antioxidant was noted by Thomas, who compared the inhibition of the oxidation of white oil by additives of small amounts of diphenylamine and diphenylazotoxide [82].

G. I. Likhtenshtein [83] has shown that diphenylazotoxide is consumed very slowly during the induction period of the oxidation of cetane
at 150°C, apparently at a rate equal to the rate of initiation. Only at the
end of the induction period does the rate of consumption of diphenylazotoxide increase (Fig. 29).

A number of aliphatic azotoxides [84], the structural formulas of
which are cited below, have also been synthesized from triacetoneamine I:

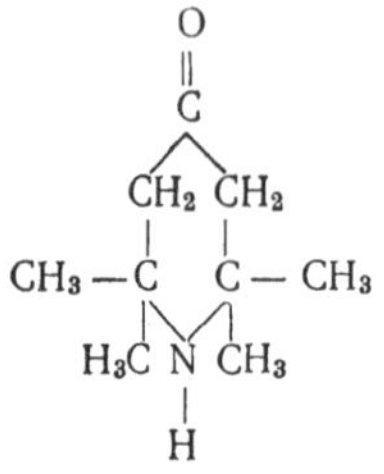

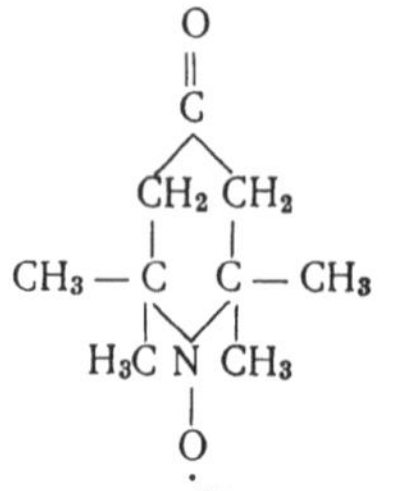

As has been shown [85], the radical II is a substantially better antioxidant, inhibiting the oxidation of polypropylene, than is triacetoneamine (Fig. 30). This effect, as we believe, is related to the fact that
triacetoneamine reacts with the radicals $RO_2^{\cdot}$ according to the scheme

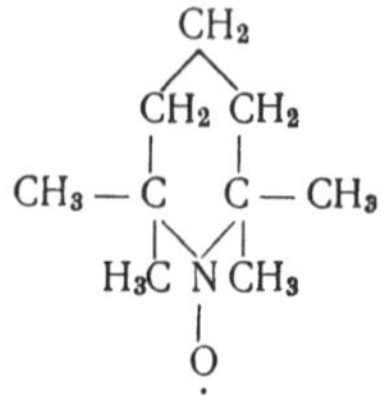

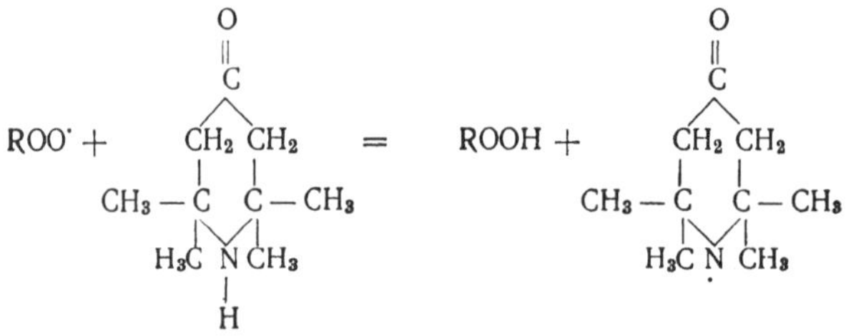

while radical II adds radicals $R^{\cdot}$ according to the scheme

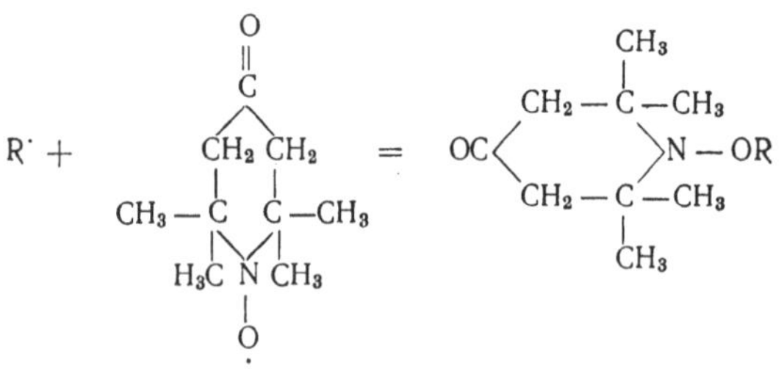

without forming peroxide molecules.

The high stability of the radicals II and III is apparently related to
the strong shielding of NO by the methyl groups.

Radical II is a good antioxidant, inhibiting the thermooxidative destruction of polyamides.

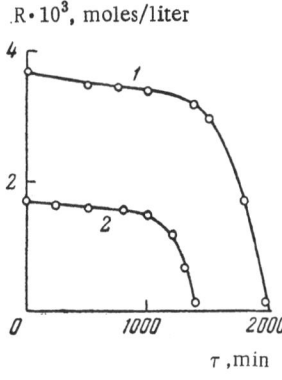

Fig. 29. Consumption of diphenylazot-oxide in the oxidation of cetane: T = 150°C, p_{O_2} = 500 mm Hg. Initial radical concentrations — 0.0038 mole/liter (1) and 0.0018 mole/liter (2).

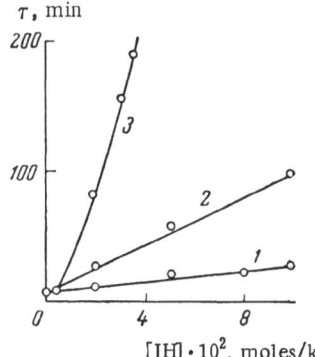

Fig. 30. Dependence of the induction period in the oxidation of polypropylene on the concentration of triacetoneamine (1), diphenylamine (2), and triacetoneazotoxide (3). T = 170°C.

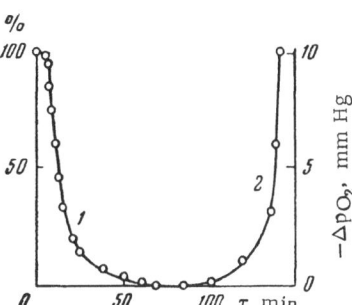

Fig. 31. Consumption of triacetonazotoxide during the induction period (1) and absorption of oxygen in the oxidation of polyamide 68 (2) (scale on the right).

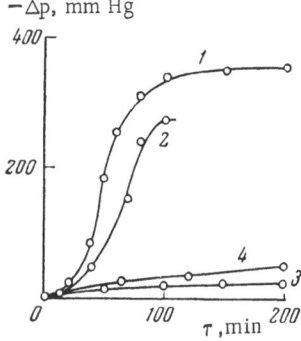

Fig. 32. Increase in pressure in the oxidation of acetylated polyformaldehyde (1), in the presence of 2% polyamide (2), 1.2% polyamide and 0.8% triacetone-azotoxide (3), and in the oxidation of delrin (4). T = 200°C, p_{O_2} = 600 mm Hg.

The consumption of radical II during the induction period has been measured by the EPR method. It has been found that only after the disappearance of the EPR signal, i.e., after the consumption of the radical, does the autocatalytic process of the oxidation of polycaproamide begin, as can be seen from the curves cited in Fig. 31.

Since radical II is a universal antioxidant in the oxidation of a number of hydrocarbons, polypropylene, and polyamides, we decided to test

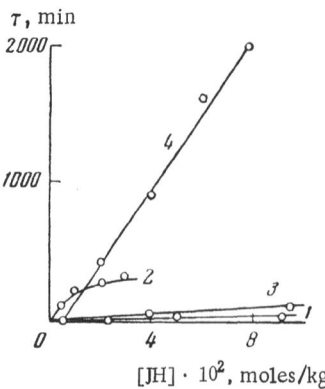

τ, min

[JH] · 10^2, moles/kg

Fig. 33. Dependence of the induction period in the oxidation of polypropylene on the concentration of 2,6-di-tert-octyl-4-methylphenol (1), stabilizer 22-46 (2), polygard (3), and the ester of SAO-6 and boric acid (4). T = 200°C; P_{O_2} = 300 mm Hg.

it as an inhibitor of the thermooxidative destruction of polyformaldehyde.

As is known from [86, 87], polyformaldehyde undergoes rapid thermooxidative destruction at 200°C. The investigations of N. S. Enikolopyan [88] have shown that the destruction of polyformaldehyde begins with the ends of the molecule, formaldehyde being liberated and then reacting with oxygen, forming formic acid. To stabilize polyformaldehyde, polyamide must be added to it to tie up the formaldehyde.

The introduction of 1.2% polyamide resin and 0.8% of radical II into acetylated polyformaldehyde exerts a good stabilizing action of polyformaldehyde (Fig. 32).

Studies in the field of radical inhibitors have begun recently, and the successes achieved in this short time permit us to consider this line as quite promising.

9. BORIC STABILIZERS

Another line of work in the field of searches for new stabilizers of polymers which has been developed recently, is the use of complete esters of boric acid, containing no weakly bonded hydrogen, as antioxidants. A number of these esters have been described in [89]. The structural formulas of some of the most effective antioxidants among the esters tested are cited below:

B-1   B-2

B-3   B-S

The antioxidants B-1, B-2, and B-3 are esters of pyrocatechol-boric acid and cyclohexanol, phenol, and α-naphthol. They are all good antioxidants, inhibiting the thermooxidative destruction of polypropylene at 200°C. B-2 and B-3 possess a well-defined critical concentration; B-3 is a strong antioxidant, while the action of B-2 and B-1 is considerably weaker.

It is interesting that the ester of boric acid and biphenol sulfide (SAO-6), which was denoted above as B-S, proved to be the most effective antioxidant of all the boric esters we investigated, in spite of the presence of hydroxyl hydrogen in the molecule. Figure 33 presents the results of experiments on the determination of the induction period in the oxidation of polypropylene stabilized by various antioxidants at a temperature of 200°C and an oxygen pressure of 200 mm Hg. As can be seen from the figure, B-S is a far more effective stabilizer than 2,6-di-tert-octyl-4-methylphenol, 22-46, and polygard. At a concentration of 0.04 mole/kg, it inhibits oxidation for 840 min, while polygard restrains oxidation for only 50 min, and 22-46 exerts its action for approximately 300 min.

We hope that the studies described above on the search for new antioxidants will serve as a starting point for the development of research in this field, of great importance for polymer chemistry.

BIBLIOGRAPHY

1. A. N. Bakh, Zhur. Russ. Fiz.-Khim. Obshchestva 29:373, 1897.
2. K. O. Engler and A. Wild, Ber. 30:1669, 1897.
3. N. N. Semenov, J. Phys. Chem. 2B:161, 1929; Chem. Revs. 6:347, 1929.
4. N. N. Semenov, J. Phys. Chem. 2B:464, 1931.
5. N. N. Semenov, Chain Reactions, Leningrad, State Press for Chemical Literature, 1934.
6. N. N. Semenov, Some Problems of Chemical Kinetics and Reactivity, Moscow, Academy of Sciences USSR Press, 1958.
7. V. N. Kondrat'ev, Kinetics of Reactions in the Gas Phase, Moscow, Academy of Sciences USSR Press, 1959.
8. A. B. Nalbandyan and V. V. Voevodskii, Mechanism of the Oxidation and Combustion of Hydrogen, Moscow, Academy of Sciences USSR Press, 1949.
9. V. V. Azatyan, V. V. Voevodskii, and A. B. Nalbandyan, Kinetika i Kataliz 2:340, 1961.
10. V. I. Panfilov, D. D. Tsvetkov, and V. V. Voevodskii, Kinetika i Kataliz 1:233, 1960.
11. V. V. Azatyan, V. I. Panfilov, and A. B. Nalbandyan, Kinetika i Kataliz 2:295, 1961.
12. V. V. Azatyan, L. A. Akopyan, A. B. Nalbandyan, and B. V. Ozherel'ev, Doklady Akad. Nauk 141:129, 1961.

13. L. I. Avramenko and R. V. Kolesnikova, Doklady Akad. Nauk SSSR 140:1100, 1961.
14. J. D. Bolland, Quart. Revs. (London) 3:1, 1949.
15. C. E. Frank, Chem. Revs. 46:155, 1950.
16. L. Bateman, Quart. Revs. (London) 8:147, 1954.
17. D. L. Knorre, Z. K. Maizus, L. K. Obukhova, and N. M. Emanuél', Uspekhi Khim. 26:416, 1957.
18. G. A. Russel, J. Chem. Educ. 36:111, 1959.
19. J. D. Bolland, Trans. Faraday Soc. 46:358, 1950.
20. T. A. Ingles and H. W. Melville, Proc. Roy. Soc. A218:163, 175, 1953.
21. M. B. Neiman, Zhur. Fiz. Khim. 28:1235, 1954.
22. I. V. Berezin, L. G. Berezkina, and T. A. Nosova, Collection: Oxidation of Hydrocarbons in the Liquid Phase, Moscow, Academy of Sciences USSR Press, 1959.
23. G. A. Razuvaev, N. S. Vyazankin, Yu. I. Dergunov, and O. S. D'yachkovskaya, Doklady Akad. Nauk 132:364, 1960.
24. I. V. Berezin, I. F. Kazanskaya, and K. Martinek, Zhur. Fiz. Khim. 35:2039, 1961.
25. I. V. Berezin and O. Dobish, Doklady Akad Nauk SSSR 142:105, 1962.
26. I. V. Berezin and Go Chu, Doklady Akad Nauk SSSR 142:383, 1962.
27. R. F. Vasil'ev, O. N. Karpukhin, and V. Ya. Shlyapintokh, Doklady Akad. Nauk SSSR 125:106, 1959.
28. R. F. Vasil'ev, Doklady Akad. Nauk SSSR 144:143, 1962.
29. O. N. Karpukhin, V. Ya. Shlyapintokh, and N. V. Zolotova, Izvest. Akad. Nauk SSSR, Otdel. Khim. Nauk, No. 7:1325, 1963.
30. M. B. Neiman, A. A. Dobrinskaya, and V. I. Gnyubkin, Investigation of the Mechanism of the Oxidation of Butane, Moscow, N. D. Zelinskii University Press, 1939.
31. M. B. Neiman and V. S. Pudov, Neftekhimiya 2(6):918, 1962.
32. Z. K. Maizus, I. P. Skibida, and N. M. Emanuél', Doklady Akad. Nauk SSSR 131:880, 1960.
33. Yu. A. Shlyapnikov, V. B. Miller, and E. S. Torsueva, Izvest Akad.Nauk SSSR, Otdel. Khim. Nauk, p. 1966, 1961.
34. Z. K. Maizus, N. M. Emanuél', and V. N. Yakovleva, Doklady Akad Nauk SSSR 143:366, 1962.
35. L. S. Vartanyan, Z. K. Maizus, and N. M. Emanuél', Zhur. Fiz. Khim. 30:862, 1956.
36. D. G. Knorre, Z. K. Maizus, and N. M. Emanuél', Doklady Akad. Nauk SSSR 112:457, 1957.
37. N. M. Emanuél, S. K. Maizus, and L. G. Privalova, Intern. J. Appl. Radiation and Isotopes 7:111, 1959.
38. Z. K. Maizus, I. P. Skibida, N. M. Emanuél', and V. N. Yakovleva, Kinetika i Kataliz 1:55, 1960.

39. L. K. Obukhova and N. M. Emanuél', Izvest. Akad. Nauk SSSR, Otdel. Khim. Nauk, p. 1544, 1960.
40. E. T. Denisov, Doklady Akad Nauk SSSR 130:1055, 1960.
41. C. H. Bamford and M. J. S. Dewar, Proc. Roy. Soc. A198:252, 1949.
42. L. T. Koritskii, Yu. N. Molin, B. N. Shamshev, N. Ya. Buben, and V. V. Voevodskii, Vysokomolekulyarnye Soedineniya 1:1182, 1959.
43. Yu. D. Tsvetkov, Ya. S. Lebedev, and V. V. Voevodskii, Vysokomolekulyarnye Soedineniya 1:1519, 1959.
44. A. L. Buchachenko, K. Ya. Kaganskaya, M. B. Neiman, and A. A. Petrov, Kinetika i Kataliz 2:44, 1961.
45. A. L. Buchachenko, K. Ya. Kaganskaya, and M. B. Neiman, Kinetika i Kataliz 2:161, 1961.
46. Ya. S. Lebedev, V. F. Tsepalov, and V. Ya. Shlyapintokh, Doklady Akad.Nauk SSSR 139:1084, 1961.
47. V. B. Miller, M. B. Neiman, V. S. Pudov, and L. I. Lafer, Vysokomolekulyarnye Soedineniya 1:1696, 1959.
48. Z. Manyasek, D. Berek, M. Michko, M. Lazar, and Yu. Pavlinets, Vysokomolekulyarnye Soedineniya 3:1104, 1961.
49. D. Ryshavy, L. Balaban, V. Slavin, and R. Ruzha, Vysokomolekulyarnye Soedineniya 3:1110, 1961.
50. V. S. Pudov, B. A. Gromov, M. B. Neiman, and E. G. Sklyarova, Neftekhimiya 3:743, 1963; V. S. Pudov and M. B. Neiman, Neftekhimiya 3:750, 1963.
51. V. Stannett and R. B. Mesrobian, J. Am. Chem. Soc. 72:4125, 1950.
52. Z. K. Maizus, I. P. Skibida, N. M. Emanuél', and V. N. Yakovleva, Kinetika i Kataliz 1:55, 1960.
53. Z. K. Maizus, I. P. Skibida, and N. M. Emanuél', Doklady Akad. Nauk SSSR 131:880, 1960.
54. M. B. Neiman and V. S. Pudov, Izvest.Akad. Nauk SSSR, Otdel. Khim. Nauk, No. 5:932, 1962.
55. V. S. Pudov and M. B. Neiman, Neftekhimiya 2(6):918, 1962.
56. K. U. Ingold, J. Inst. Petrol. 47:375, 1961.
57. J. A. Christiansen, J. Phys. Chem. 28:145, 1924.
58. H. Bäckstrom, J. Am. Chem. Soc. 49:1460, 1927.
59. M. B. Neiman, V. B. Miller, Yu. A. Shlyapnikov, and E. S. Torsueva, Doklady Akad. Nauk SSSR 136:647, 1961.
60. N. M. Emanuél', A. B. Gagarina, and Z. K. Maizus, Doklady Akad. Nauk SSSR 135:354, 1960.
61. D. Ryshavy, Vysokomolekulyarnye Soedineniya 3:464, 1961.
62. V. B. Miller, M. B. Neiman, V. S. Pudov, and L. I. Lafer, Vysokomolekulyarnye Soedineniya 1:1696, 1959.

63. R. Goglev and M. B. Neiman, Vysokomolekulyarnye Soedineniya 5:1050, 1963.

64. G. I. Likhtenshtein, M. B. Neiman, and N. Sysoeva, Tr. po Khim. i Khim. Tekhnol. p. 155, 1963.

65. N. M. Emanuél' and Yu. I. Lyaskovskaya, Inhibition of Processes of Fat Oxidation, Moscow, State Press for the Food Industry, 1961.

66. M. B. Neiman, Zhur. Vses. Khim. Obshchestva im. D. I. Mendeleeva 7(2):166, 1962.

67. Yu. A. Shlyapnikov, V. B. Miller, M. B. Neiman, and E. S. Torsueva, Vysokomolekulyarnye Soedineniya 4(8):1228, 1962.

68. A. F. Lukovnikov, P. I. Levin, and A. G. Vasil'eva, Aging and Stabilization of Polymers, Moscow, Academy of Sciences USSR Press, 1963.

69. J. L. Bolland and P. Ten Have, Trans. Faraday Soc. 43:201, 1947.

70. V. A. Shushunov, B. A. Redoshkin, and L. V. Kodintseva, Tr. po Khim. i Khim. Tekhnol. p. 463, 1961.

71. Yu. A. Shlyapnikov, V. B. Miller, M. B. Neiman, and E. S. Torsueva, Vysokomolekulyarnye Soedineniya 5(10):1507, 1963.

72. G. W. Kennerly and W. L. Patterson, Ind. Eng. Chem. 48:1917, 1956.

73. A. F. Lukovnikov, P. I. Levin, M. B. Neiman, and M. S. Khloplyankina, Vysokomolekulyarnye Soedineniya 3:1243, 1961.

74. A. F. Lukovnikov, P. I. Levin, and M. S. Khloplyankina, Summaries of the Conference on the Aging and Stabilization of Polymers, Moscow, Academy of Sciences USSR Press, 1961, p. 7.

75. G. I. Likhtenshtein, ibid., p. 5.

76. M. B. Neiman, A. S. Kuz'minskii, and L. G. Angert, Vestnik Akad. Nauk SSSR, 30(11):36, 1960.

77. K. U. Ingold, Chem. Revs. 61:563, 1961.

78. G. G. Knapp and H. D. Orloff, Ind. Eng. Chem. 53:63, 1961.

79. P. A. Kirpichnikov et al., Summaries of the Conference on the Aging and Stabilization of Polymers, Moscow, Academy of Sciences USSR Press, 1961.

80. A. A. Berlin, Z. V. Popova, and D. M. Yanovskii, Doklady Akad. Nauk SSSR 131:563, 1961.

81. E. G. Rozantsev, L. A. Kalashnikova, and M. B. Neiman, Izvest. Akad. Nauk SSSR, Otdel. Khim. Nauk, p. 501, 1962.

82. J. R. Thomas, J. Am. Chem. Soc. 82:5955, 1960.

83. G. I. Likhtenshtein, Dissertation, Institute of Chemical Physics, Academy of Sciences USSR, 1962.

84. M. B. Neiman, Yu. G. Mamedova, and Z. G. Rozantsev, Khim. Zhur. AN Azerb.SSR, No. 6:37, 1962.

85. G. I. Likhtenshtein, Zhur. Fiz. Khim. 36:750, 1962.

86. W. Kern and H. Cherdron, Makromol. Chem. 40:101, 1960.

87. W. Kern, H. Cherdron, et al., Angew. Chem. 73(6):5177, 1961.

88. N. S. Enikolopyan and M. S. Vardanyan, Zhur. Vses. Khim. Obshchestva im. D. I. Mendeleeva 7(2):194, 1962.

89. E. G. Rozantsev, L. A. Krinitskaya, and B. V. Rozynov, Plast. Mass. No. 11:48, 1963.

90. Yu. A. Shlyapnikov, V. B. Miller, M. B. Neiman, and E. S. Torsyeva, Doklady Akad. Nauk SSSR 151:148, 1963.

91. A. L. Buchachenko, E. I. Sdobnov, S. R. Rafikov, and M. B. Neiman, Izvest. Akad. Nauk SSSR, Otdel. Khim. Nauk, p. 1118, 1963.

Chapter II

STABLE RADICALS OF INHIBITORS OF OXIDATIVE PROCESSES

In the oxidation of polymers, as was shown in Chapter I, the active centers that carry the oxidation chains are radicals of the type of RO· and RO$_2$·. Such radicals can be determined in the substance being oxidized by the EPR method only in exceptional cases [1], since the rate of their destruction is very great, and hence their concentrations are small and lie below the limit of sensitivity of the modern EPR spectrometers. However, if substances capable of reacting rapidly with active radicals and thereby forming stable radicals, incapable of intensively continuing the oxidation chain, are added to the substrate being oxidized, then the concentrations of such radicals can be determined by the EPR method, and their structures can also be identified [2].

This chapter is devoted to an investigation by the EPR method of the structure of the radicals of the most widespread classes of inhibitors – phenols, naphthols, aromatic amines – their electronic structure, mechanism of formation, and their physical and chemical properties. *

1. PHENOXYL RADICALS

Alkyl-substituted phenols are the most widespread inhibitors. The phenoxyl radicals corresponding to them are readily produced by oxidizing the phenols with lead dioxide [3], in photolysis, γ and β radiolysis, as well as in their reaction with active radicals produced in the thermal or catalytic decomposition of organic peroxides and hydroperoxides [4]. The stability of the radicals formed is determined by the structure of the initial phenol. The most stable radicals are strongly shielded phenols. Thus, 2,4,6-tri-tert-butylphenol (I), when oxidized by PbO$_2$, forms phenoxyl radicals, the EPR spectrum of which is presented in Fig. 34a, in almost 100% yield:

---

* For detailed information on these problems see the monograph [15].

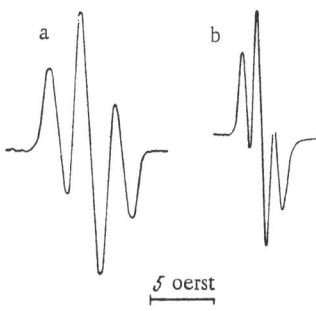

Fig. 34. EPR spectrum. a) 2,4,6-tri-tert-butylphenoxyl; b) radical obtained from the peroxide (II).

OH
(CH₃)₃C ─⟨ ⟩─ C(CH₃)₃
          │
        C(CH₃)₃
          I

The triplet with intensity ratio 1 : 2 : 1 and splitting 1.7 Oe arises when the unpaired electron interacts with two equivalent meta-protons of the benzene ring. This gives evidence of strong delocalization of the unpaired electron over the system of π-bonds of the benzene ring and σ-bonds of the substituents. However, the interaction of this electron with the protons of the methyl groups of the substituents, shielded by the potential barrier of two C – C bonds, is so small that it does not appear in the EPR spectrum.

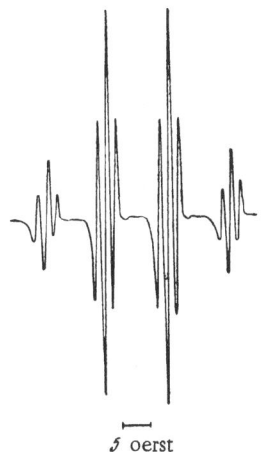

Fig. 35. EPR spectrum of 2,6-di-tert-butyl-4-methylphenoxyl.

Radicals can be obtained from I in the crystalline state by evacuating the solvent in which the radicals were synthesized. The dark blue crystals of the radical sublime readily under vacuum at temperatures of 70-100°C. With respect to its chemical properties, the radical possesses a dual nature, which is related to the delocalization of the unpaired electron and to transfer of the reaction center. On the one hand, many reactions (for example, stripping of hydrogen from the solvent molecules, re-action with sodium) occur at the oxygen atom; on the other hand, the reactions with halogens, the addition of alkyl and peroxide radicals [5, 7], and the addition of oxygen take place in the benzene ring in the para-position to the oxygen atom. In this case the peroxide II is formed with oxygen:

        C(CH₃)₃   C(CH₃)₃
          \   C(CH₃)₃   /
O ═⟨   ⟩─ O ─ O ─⟨   ⟩═ O
          /   (CH₃)₃C   \
        C(CH₃)₃   C(CH₃)₃
                    II

when heated to 80-120°C, it in turn decomposes, forming the radicals whose EPR spectrum is presented in Fig. 34b. The nature of the spec-

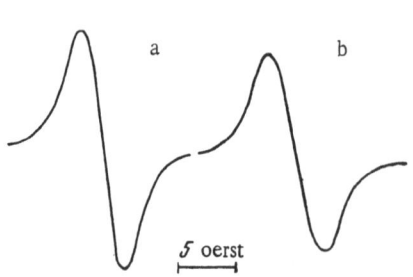

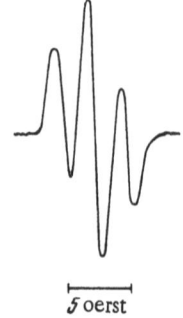

Fig. 36. EPR spectra. a) α-Naphthoxyl; b) β-naphthoxyl.

Fig. 37. EPR spectrum on the binaphthol radi-cal [formula (VI)].

trum does not change; however, the splitting between the components of the triplets decreased to 1.2 Oe. Probably the peroxide decomposes at the O – O bond, the presence of two oxygen atoms in the para-position decreasing the constant of hyperfine interaction of the unpaired electron with the meta-protons in the radical formed. The EPR spectrum of the radical obtained from the phenol III

$$\text{OH}$$
$$(CH_3)_3C - \text{⟨ring⟩} - C(CH_3)_3$$
$$C\overset{O}{\underset{H}{<}}$$
III

is of the same character. However, the splitting between the components of the triplet is increased to 2.4 Oe.

The introduction of substituents containing α-hydrogens into the para-position of phenol sharply changes the EPR spectra: splitting on these protons appears. Thus, the radicals of 2,6-di-tert-butyl-4-methyl-phenol (ionol) (IV)

$$\text{OH}$$
$$(CH_3)_3C - \text{⟨ring⟩} - C(CH_3)_3$$
$$CH_3$$
IV

are characterized by the EPR spectrum shown in Fig. 35.

The four groups of lines (quadruplet) with splitting between their centers of 10.7 Oe and intensity ratio 1 : 3 : 3 : 1 arise in the interac-

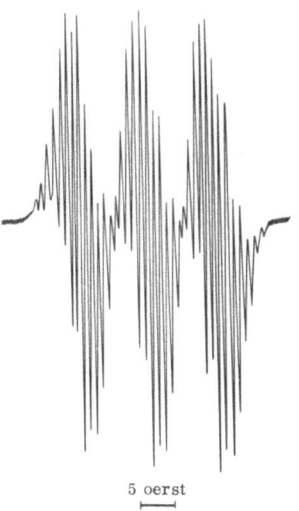

5 oerst

Fig. 38. EPR spectrum of diphenylazotoxide.

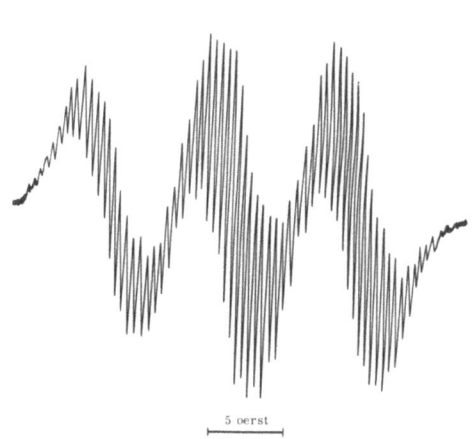

5 oerst

Fig. 39. EPR spectrum of tetramethoxydiphenyl-
azotoxide.

tion of the unpaired electron with the three equivalent protons of the
methyl groups in the para-position, while the triplet splitting of the components of the quadruplet is related primarily to the meta-hydrogens.
The same spectrum is also characteristic of the radicals formed from
the phenol (V):

$$CH_3 — (CH_2)_4 — C — \underset{CH_3}{\overset{CH_3\ OH\ \ CH_3}{\text{⬡}}} — C — (CH_2)_4 —CH_3$$

V

If a $-CH(CH_3)_2-$ or $-CH(CH_3)_2$-group stands in the para-position in
place of the methyl group, then two or three groups of lines, corresponding to the number of α-protons of the para-substituent, appear in
the EPR spectrum.

Thus, an analysis of the EPR spectra shows that phenoxyl radicals
are formed by stripping of the hydroxyl hydrogen. This is also confirmed by the infrared spectra of the radicals, where the absorption
band of the OH group disappears, and bands characteristic of a system
of quinoid bonds appear [6], as a result of delocalization of the unpaired
electron. Calculation shows that about 50% of the spin density of the unpaired π-electron is accounted for by the carbon atom in the para-position, which explains the high reactivity of this position in the radical.

TABLE 2. EPR Spectra of Phenoxyl Radicals

Radical	Number of components of the EPR spectrum	Splitting (in oersteds) on the protons of the substituents and other nuclei				Literature
		Para-	Ortho-	Meta-	Other nuclei	
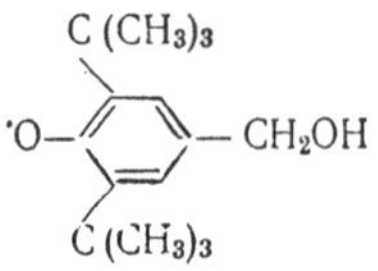 $\cdot O-$ (ring with $C(CH_3)_3$ top and $\dot{C}(CH_3)_3$ bottom) $-CH_2OH$	Triplet of triplets	12.3	—	—	—	[3]
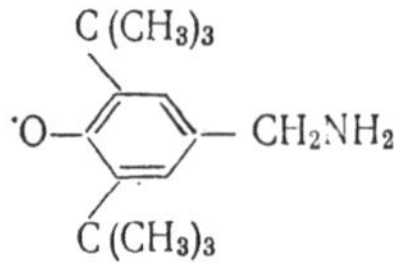 $\cdot O-$ (ring with $C(CH_3)_3$, $\dot{C}(CH_3)_3$) $-CH_2NH_2$	Triplet of quadruplets	11.5	—	1.4	1.4(N)	Data of author
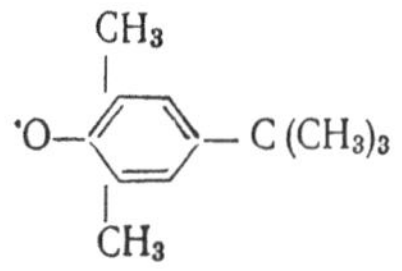 $\cdot O-$ (ring with CH_3 top and $\dot{C}H_3$ bottom) $-C(CH_3)_3$	Septet of triplets	—	6.1	1.8	—	[3]
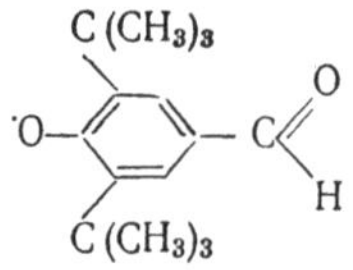 $\cdot O-$ (ring with $C(CH_3)_3$, $\dot{C}(CH_3)_3$) $-C\overset{O}{\underset{H}{\diagdown}}$	Triplet	—	—	2.5	—	Data of author
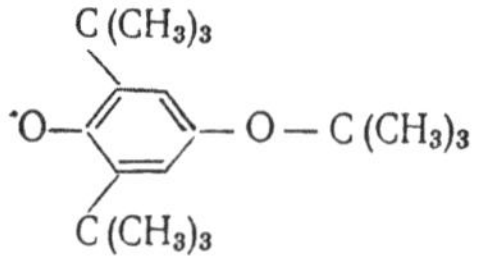 $\cdot O-$ (ring with $C(CH_3)_3$, $\dot{C}(CH_3)_3$) $-O-C(CH_3)_3$	Triplet	—	—	1.8	—	[7]
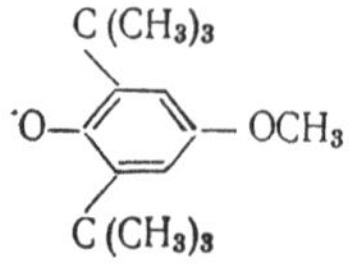 $\cdot O-$ (ring with $C(CH_3)_3$, $\dot{C}(CH_3)_3$) $-OCH_3$	Nine lines	1.6	—	—	—	[7]
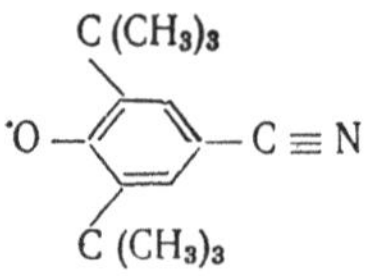 $\cdot O-$ (ring with $C(CH_3)_3$, $\dot{C}(CH_3)_3$) $-C\equiv N$	Nine lines	—	—	2.2	1.3(N)	[7]

Table 2 (conclusion)

Radical	Number of components of the EPR spectrum	Splitting (in oersteds) on the protons of the substituents and other nuclei				Literature
		Para-	Ortho-	Meta-	Other nuclei	
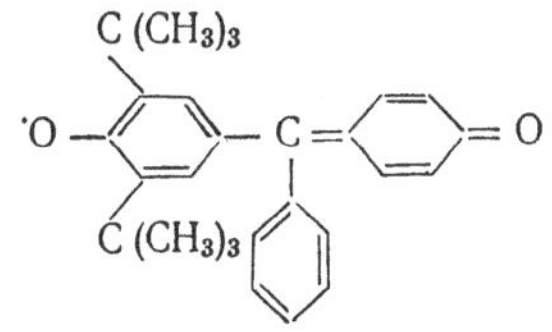 [structure: C(CH₃)₃ substituted phenoxyl radical]	Quintiplet	–	–	1.3	–	[7]
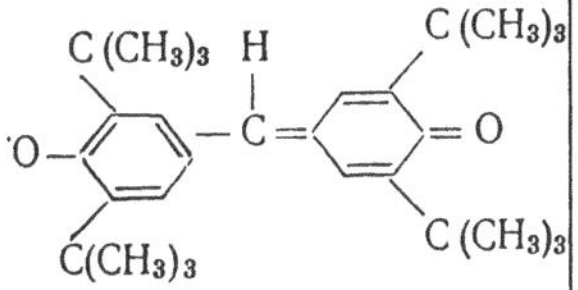 [structure: C(CH₃)₃ substituted radical]	Doublet of quintiplets	5.7	–	1.3	–	[7]
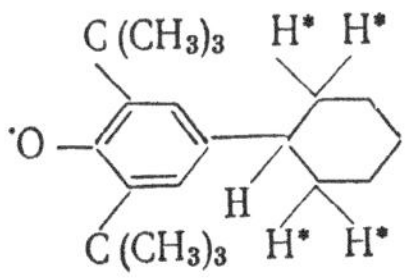 [structure: C(CH₃)₃ substituted cyclohexyl radical with H•]	–	4.6	–	1.8	0.55 (on H•)	[8]
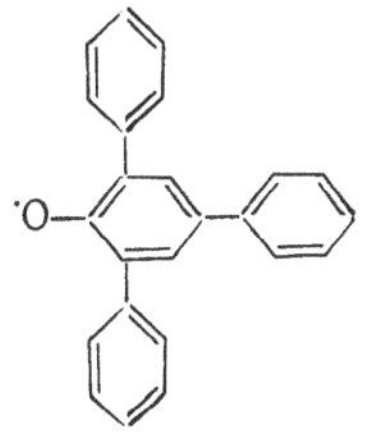 [structure: triphenyl phenoxyl radical]	Singlet	–	–	–	–	[9]
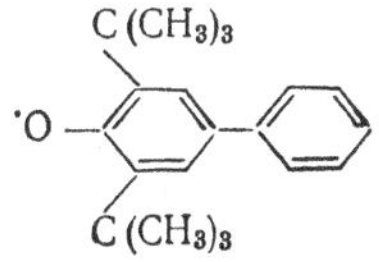 [structure: C(CH₃)₃ substituted biphenyl phenoxyl radical]	13 lines	–	–	–	–	[9]
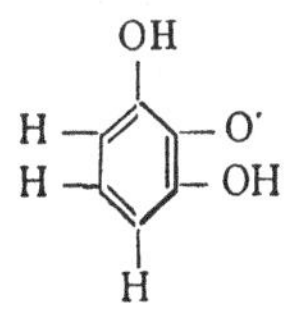 [structure: dihydroxy phenoxyl radical]	Doublet of triplets	5	–	–	–	[10]

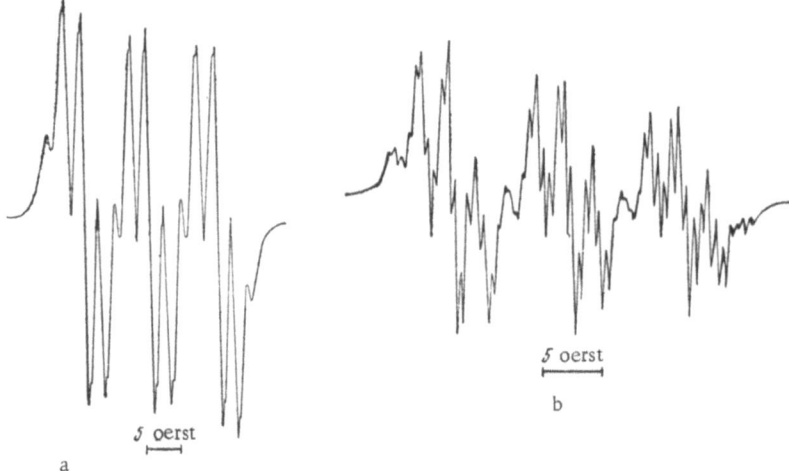

Fig. 40. EPR spectra of radicals [formula (XI)]. a) In concentrated solutions; b) in dilute solutions.

Since the radicals $RO^{\cdot}$ and $RO_2^{\cdot}$, which are carriers of the kinetic oxidation chains, are formed in the catalytic decomposition of hydroperoxides, the appearance of phenoxyl radicals in such model systems of oxidation [14] gives evidence of stripping of the hydroxyl hydrogen of phenol by the active radicals, as the primary elementary event of inhibition, the reactivity of the OH-bond in the phenol being one of the most important characteristics of the effectiveness of the phenol as an inhibitor.

At present the structure of the phenoxyl radicals formed from phenols of various structures has been rather fully investigated. Table 2 presents the characteristics of the EPR spectra of certain phenoxy radicals.

The stability of phenoxy radicals is determined by the effect of conjugation of the unpaired electron with the system of the remaining bonds and by steric effects. The introduction of such voluminous substituents as tertiary butyl and phenyl, which shield the reaction centers of the radicals, sharply increases their stability. Radicals of unsubstituted or incompletely substituted phenols readily recombine or disproportionate, and do not accumulate in significant concentrations [4]. There is no direct and distinct relationship between the stability of the radical and the effectiveness of the corresponding phenol as an inhibitor, since the effectiveness depends not only on the reactivity of the OH-bond, but also on a number of other factors. However, there is a general correspondence between the stability of the radical and the effectiveness of the corresponding phenol.

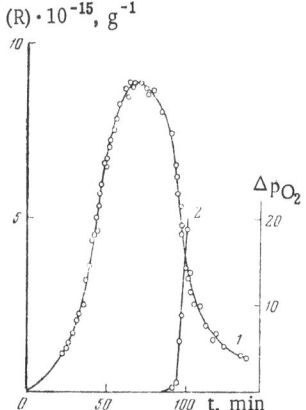

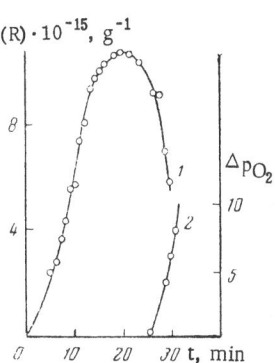

Fig. 41. Oxidation of polypropylene in the presence of α-naphthol. Inhibitor concentration 7×10^{-2} mole/kg, T = 200°C.

Fig. 42. Oxidation of polypropylene in the presence of 7×10^{-2} mole/kg α-naphthol at 220°C.

2. STABLE RADICALS OF NAPHTHOLS

Naphthols are customarily considered typical inhibitors that terminate kinetic oxidations. Actually, in the reaction with peroxide radicals, as well as in oxidation by lead dioxide, α- and β-naphthols form radicals that give singlet EPR spectra (Fig. 36a and 36b); the width of the singlet between the points of maximum slope in the case of the α-naphthyl radical is 3.5 Oe, and about 5 Oe for the β-naphthyl radical. The radicals are probably formed by stripping of the hydroxyl hydrogen and are stabilized by strong conjugation of the unpaired electron

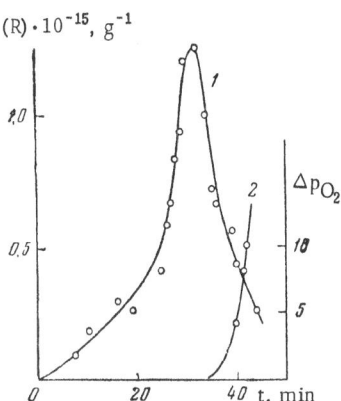

Fig. 43. Oxidation of polypropylene in the presence of 7×10^{-2} mole/kg β-naphthol; T = 200°C.

with the system of π-bonds of the naphthol rings, the magnitude of this conjugation in the α-naphthyl radical being greater than in the β-naphthyl radical. This is also confirmed by calculation and corresponds to the kinetic stabilities of these radicals: α-naphthoxyl is substantially more stable than β-naphthoxyl; the latter is formed only in small concentrations and is rapidly destroyed [11]. Such a series of stability corresponds to the effectiveness of these naphthols as inhibitors, which gives evidence of the predominant role of conjugation effect in such radicals.

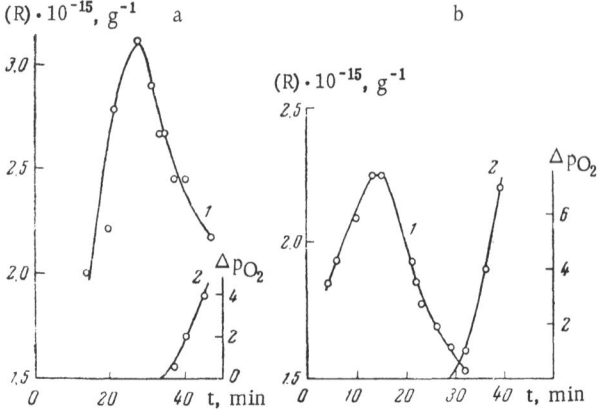

Fig. 44. Oxidation polypropylene (a) in the presence of di-tert-octyl-4-methylphenol in a concentration of 7×10^{-2} mole/kg, T = 180°C; b) in the presence of diphenylamine in a concentration of 9×10^{-2} mole/kg, T = 180°C.

The radicals corresponding to binaphthols possess substantially greater stability. Thus, in the oxidation of binaphthol (VI) and its oxy-derivative (VII), stable radicals are formed, characterized by triplet EPR spectra (Fig. 37). The triplet arises on account of the strong inter-action of the unpaired electron only with the two equivalent protons of the rings (splitting 2.7 Oe); the remaining protons interact weakly with the electron. The radicals are stable and do not react with diphenylpicryl-hydrazyl at room temperature; however, they disproportionate rapidly under the action of light.

VI   VII

In the crystalline state, the radicals mainly recombine. The degree of their dissociation in the crystals is about 0.1% at room temperature. Upon solution, the degree of dissociation increases. The temperature dependence of the degree of reversible dissociation has been investigated for one of these dimers (VIII): according to the scheme $M \overset{K}{\rightleftarrows} 2R$, where M is the molecule of the initial dimer, $R^{\cdot}$ is the radical. The following expression was obtained for the equilibrium constant k:

$$K = 1.8 \cdot 10^{29} \exp\left(- 16500/RT\right) \text{ cm}^{-3}$$

i.e., the energy of the dissociation of (VIII) into radicals is equal to 16.5 kcal/mole [11]. The structure of (VIII) has not been accurately established; hence it is probable that its decomposition takes place not at the O – O bond, but with cleavage of a C – C bond, if the dimer possesses a quinoid structure.

3. STABLE RADICALS OF AROMATIC AMINES

The investigation of the structure of radicals of aromatic amines, produced by their reaction with active radicals of the type of $RO^{\cdot}$ and $RO_2^{\cdot}$, is of considerable interest. It was shown in [12] that the radicals of secondary aromatic amines obtained in this case represent the corresponding azotoxides. The mechanism of their formation can be represented in the following way:

$$RO_2^{\cdot} + Ar_2NH \rightarrow Ar_2N^{\cdot} + RO_2H$$

$$Ar_2N^{\cdot} + RO_2^{\cdot} \rightarrow Ar_2\overset{\cdot}{N}O + RO^{\cdot}$$

The intermediate diarylnitrogen radicals obtained are substantially less stable than the corresponding azotoxides, and readily recombine and disproportionate. The primary diphenylnitrogen radicals cannot be detected by the EPR method even in the thermal dissociation of tetraphenylhydrazine under vacuum. However, the corresponding azotoxides are extremely stable. Figure 38 presents the EPR spectrum of diphenylazotoxide (IX), produced in the catalytic decomposition of hydroperoxides in the presence of diphenylamine.

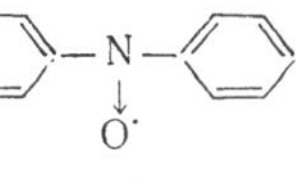

IX

The spectrum is characterized by a triplet structure with splitting of 10 Oe, related to the magnetic moment of the nucleus N^{14}. Each component of the triplet is subdivided into 15 lines. Their intensity ratio gives evidence that the protons of the benzene rings on which this splitting arises are nonequivalent.

An analogous spectrum is given by tetramethyldiphenylazotoxide (X)

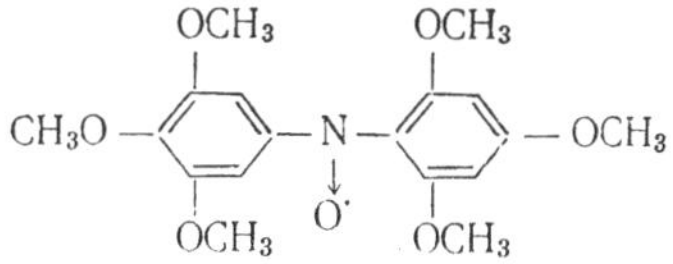

X

(Fig. 39). The splitting between the components of the triplet is 8.2 Oe. The large number of lines of the proton hyperfine structure gives evidence that the unpaired electron easily transmits its influence through the oxygen atom to the protons of the substituents.

When phenothiazine interacts with peroxide radicals, radicals of the corresponding azotoxide (XI), the EPR spectrum of which is shown in Fig. 40a, are also formed.

XI

We can see from the spectrum that the unpaired electron interacts strongly with the four equivalent protons of the benzene rings, which leads to splitting of each component of the triplet into five lines with a distance of 2.3 Oe. These protons are probably ortho- and para-protons, denoted in the formula of the radical by asterisks. However, when the solution of the radicals is further diluted, as the intermolecular interaction is removed, further splitting of the components of the quintiplet on the remaining proton is observed. The value of the splitting is about 0.5 Oe (Fig. 40b).

This means that the probability that the unpaired π-electron will be found in the ortho- and para-positions is almost five times as great as the probability that it will be found in the meta-positions.

A considerable group of radicals – aromatic azotoxides, as well as radicals of aliphatic amines – have now been investigated [13]. The formation of such radicals directly from the corresponding amines under the action of peroxide radicals gives evidence that for amines the primary elementary event of inhibition consists of stripping of the weakly bonded amine hydrogen, forming arylnitrogen radicals, which undergo further transformations.

4. FORMATION OF STABLE RADICALS OF INHIBITIONS DURING OXIDATION PROCESSES

As was indicated in Chapter I, stable radicals of inhibitors, the structure and mechanism of the formation of which were analyzed, are produced under conditions simulating the oxidation process. The question of whether the same radicals are formed in oxidation processes and of the rates of their formation and consumption merits still greater attention, since this will make it possible to characterize the mechanism and rates of the investigated processes.

The first such attempt was made by Harle and Thomas [13], who investigated the oxidation of cetane at 170°C in the presence of phenyl-β-naphthylamine. In this study it was shown that the concentration of the intermediate radical products passes through a maximum with time, followed by the beginning of intensive absorption of oxygen. An unambiguous conclusion on the mechanism of inhibition cannot be drawn from these data, since the structure of the radical product was not identified, while the kinetic analysis of the results satisfied various proposed mechanisms. In addition, attempts to produce radicals using other inhibitors were unsuccessful.

In [14], polypropylene was investigated in the presence of a number of inhibitors – representatives of various classes. The oxidation was conducted directly in ampoules placed in the resonator of an EPR spectrometer and heated with a stream of hot air. In this case, simultaneously with the recording of the EPR signals, the rate of oxygen absorption by the polymer being oxidized could be followed. The oxidation was conducted at an oxygen pressure of 400 mm Hg in the temperature range 180-200°C. The inhibitor concentrations were 7×10^{-2}-9×10^{-2} mole/kg. α- and β-naphthols, binaphthol (VI), 2,6-di-tert-octyl-4-methylphenol (V), and diphenylamine were used as the inhibitors. The EPR spectrum of the inhibitors radicals formed during the oxidation of polypropylene exactly corresponded to the spectra of the same radicals produced under model conditions, considered earlier. Thus, the identity of the structure of the radicals and the identity of the mechanisms of their formation in the two cases was demonstrated.

Figure 41 shows how the concentration of α-naphthol radicals (number of radicals per gram of the polymer) varies during the oxidation of a film of polypropylene (curve 1). Curve 2 illustrates the absorption of oxygen during the oxidation at 200°C. The initial portion of curve 1 is autocatalytic in character, i.e., first the rate of radical formation is small, then it gradually increases and reaches a maximum. Then, as the inhibitor is consumed, the radical concentration drops, since the rate of their destruction exceeds the rate of their formation. Right after the beginning of the drop in the radical concentration, the absorption of oxygen begins, and the reaction emerges from the induction period. The curve of variation of the radical concentration reflects the degenerate-branched chain process that occurs during oxidation. The clearly pronounced autocatalysis gives evidence that most of the radicals are formed as a result of termination of the kinetic oxidation chains by the inhibitors, and not on account of the oxidation of the inhibitor itself.

The same principles are observed in the oxidation of polypropylene film with α-naphthol at 220°C (Fig. 42). The induction period decreases with increasing temperature, while the maximum radical concentration somewhat increases.

Figure 43 also presents the characteristics of the oxidation of poly-propylene in a film in the presence of β-naphthol. The character of the curves remains exactly the same as in the presence of α-naphthol; how-ever, the induction period is almost half the value, while the maximum concentration of β-naphthol radicals is almost an order of magnitude lower than in the case of α-naphthol under the same conditions, which agrees with what was said above on the lower stability of β-naphthol ra-dicals and the greater rate of their destruction in comparison with α-naphthol radicals.

Figure 44 presents the kinetic curves of the accumulation and con-sumption of the inhibitor radicals and curves of the absorption of oxygen in the presence of 2,6-di-tert-octyl-4-methylphenol (a) and diphenyl-amine (b). The character of the curves basically remains the same as for other inhibitors.

The principles described above are common to all the investigated inhibitors. In spite of their qualitative character, they reflect the char-acteristics of inhibited oxidation as a degenerate branched chain pro-cess with linear chain termination on the inhibitor. Undoubtedly a quan-titative investigation of such principles and their comparison with the re-sults of machine calculations will make it possible to solve the problem as a whole.

BIBLIOGRAPHY

1. Ya. S. Lebedev, V. F. Tsepalov, and V. Ya. Shlyapintokh, Doklady Akad. Nauk SSSR, 139(6):58, 1961.
2. L. A. Blyumenfel'd and V. V. Voevodskii, Uspekhi Fiz. Nauk 68(1):31, 1959.
3. J. K. Becconsall, S. Clough, and G. Scott, Trans. Faraday Soc. 56:459, 1960.
4. A. L. Buchachenko, Ya. S. Lebedev, and M. B. Neiman, Zhur. Strukt. Khim. 2:558, 1961.
5. A. Bickel and E. Kooyman, J. Chem. Soc. p. 3211, 1953.
6. E. Müller and K. Ley, Chem. Ber. 87:922, 1954.
7. E. Müller, K. Ley, K. Scheffler, and R. Mayer, Chem. Ber. 91:2682, 1958.
8. K. Dimroth, Angew. Chem. 72:102, 1960.
9. K. Dimroth, K. Kalk, R. Self, and K. Schlomer, Ann. Chem. 624:51, 1959.
10. R. M. Hoskins and B. R. Loy, J. Chem. Phys. 23:2461, 1955.
11. M. L. Khidekel', A. L. Buchachenko, L. V. Gorbunova, G. A. Razuvaev, and M. B. Neiman, Doklady Akad. Nauk SSSR 140:5, 1961.

12. A. L. Buchachenko, Optika i Spektroskopiya 13:6, 795, 1962.

13. O. L. Harle and J. R. Thomas, J. Am. Chem. Soc. 79:2973, 1957.

14. A. L. Buchachenko, M. S. Khloplyankina, and M. B. Neiman, Doklady Akad. Nauk SSSR 143:146, 1962.

15. A. L. Buchachenko, Stable Radicals, Moscow, Academy of Sciences USSR Press, 1963.

Chapter III

SYNTHESIS OF STABILIZERS FOR POLYMER MATERIALS

Depending on the purpose of stabilizers, it is now customary to divide them basically into photostabilizers, antioxidants, and thermal stabilizers. However, such a division is rather arbitrary in character, since many of them can simultaneously perform various functions, depending on the type of material to be stabilized, the conditions of its reprocessing, use, etc. For example, the so-called antioxidants and thermal stabilizers are equally designed to inhibit oxidation processes and differ only in that they operate under different conditions.

I. PHOTOSTABILIZERS

Among the many organic compounds recently proposed for the photostabilization of polymers, derivatives of benzophenone, complex esters of salicyclic acid, derivatives of benzotriazole, various organic tin compounds, thiazolidones, etc. are finding use. The last two classes of compounds are not considered in this chapter, since the methods of synthesis, properties, and use of organic tin compounds [1, 2] and thiazolidones [3, 4] have been rather fully treated in the literature.

1. Benzophenone Derivatives

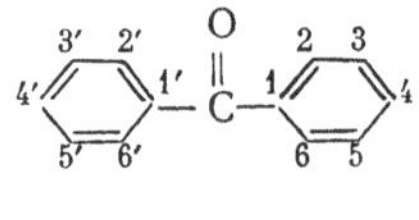

This class of photostabilizers is characterized by the presence in the benzophenone molecule of no less than one hydroxy group in the ortho-position to the carbonyl.

Depending on the number of OH groups, hydroxy derivatives of benzophenone used as photostabilizers can be subdivided into four groups.

1) monohydroxy derivatives (for example, 2-hydroxybenzophenone);

2) dihydroxy derivatives (2,4-dihydroxybenzophenone);

3) trihydroxy derivatives (2,2',4-trihydroxybenzophenone and 2,4,4'-trihydroxybenzophenone);

4) tetrahydroxy derivatives (2,2',4,4'-tetrahydroxybenzophenone).

1. Monohydroxy Derivatives of Benzophenone

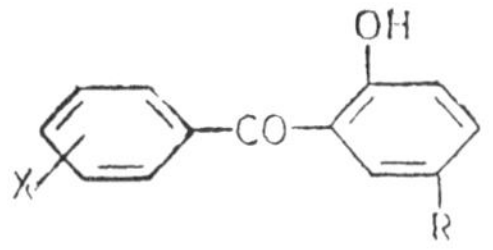

where R = H, Cl, CH_3, C_8H_{17}, $C(CH_3)_2 - CH(CH_3)_2 - CH_3$, etc.; X = H, Cl, CH_3.

The synthesis of monohydroxy derivatives of benzophenone is accomplished by condensing the corresponding p-alkyl- or p-halophenol with a substituted or unsubstituted benzoic acid in an inert solvent medium during the passage of BF_3 [5]:

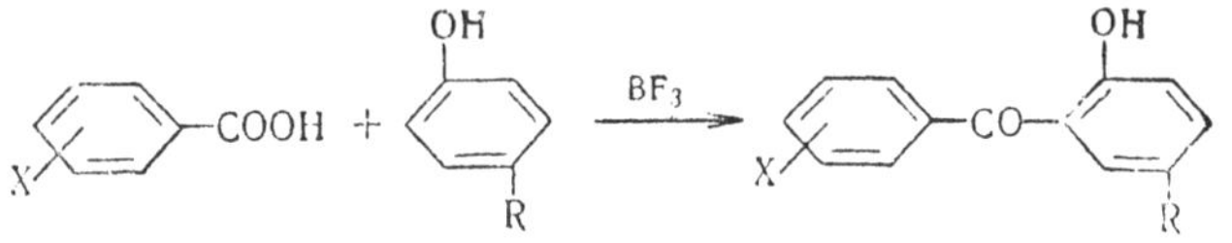

2-Hydroxybenzophenone and its derivatives are also produced by condensing phenols with halo- and alkyl-substituted derivatives of benzoic acid in the presence of $AlCl_3$ at a temperature of 90-190°C [6] or by the reaction of Hoesch [8] (see below).

Derivatives of monohydroxybenzophenone which possess R = H, lower alkyl, or halogen, and finding use for the stabilization of vinyl halide polymers [9-13]; derivatives containing normal or branched alkyl radicals with six carbon atoms or more are finding use for the stabilization of polyethylene [5].

2. Dihydroxy Derivatives of Benzophenone

One of the early methods of producing 2,4-dihydroxybenzophenone was the melting of benzoic acid or benzoyl chloride with resorcinol at a temperature of 150-160°C in the presence of anhydrous $ZnCl_2$ [14-17], a mixture of $ZnCl_2$ and $POCl_3$ [25], or in the presence of BF_3 [26], which serve as catalysts of the reaction:

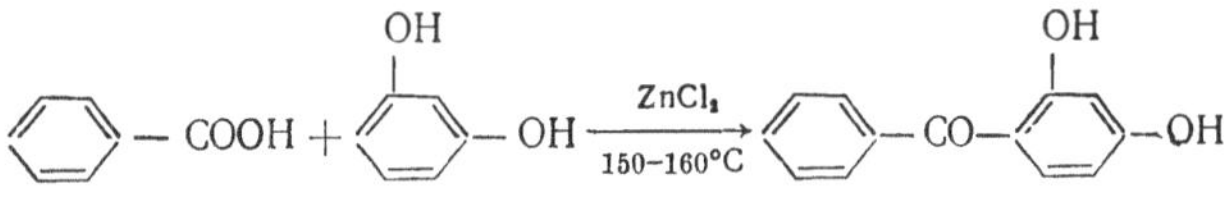

Later K. Hoesch [8, 18, 19] proposed the condensation of benzo-
nitrile with resorcinol in ether solution, heating the mixture with $ZnCl_2$
and with simultaneous saturation of the reaction mass by dry hydrogen
chloride. After hydrolysis of the intermediate ketoimine compound, 2,4-
dihydroxybenzophenone is formed:

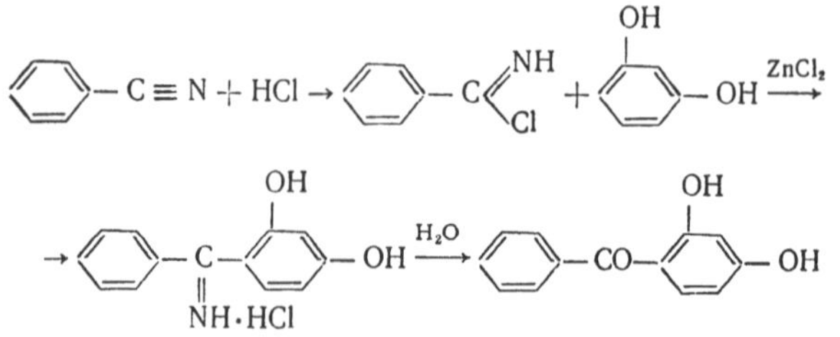

The correspondingly substituted benzophenones are obtained in a yield
not exceeding 35%, using substituted benzonitriles [20]. E. N. Zil'-
berman and N. A. Rybakova [21-23] modified the procedure for conducting the Hoesch reaction and synthesized a number of aromatic hydroxyketones in 60-80% yield. According to the procedure they developed,
hydrogen chloride is passed into an ether solution of the nitrile and zinc
chloride for 2-3 hr at a temperature of 2-5°C, and then resorcinol is
added.

R. Shah and P. Mehta [24] synthesized 2,4-dihydroxybenzophenone
by condensing benzamide or benzanilide with resorcinol in the presence
of $ZnCl_2$ and $POCl_3$ or $SnCl_4$ and P_2O_5. The yield of the benzophenone according to this method is 30-40%.

Of the derivatives of 2,4-dihydroxybenzophenone, 2-hydroxy-4-
alkoxybenzophenones, in which the alkyl radical contains from one to 20
carbon atoms, are finding use. The synthesis of alkoxybenzophenones
is accomplished by alkylating dihydroxybenzophenone using alkyl halides
(better the bromides or iodides) [27-29] or dialkyl sulfates [14, 30]:

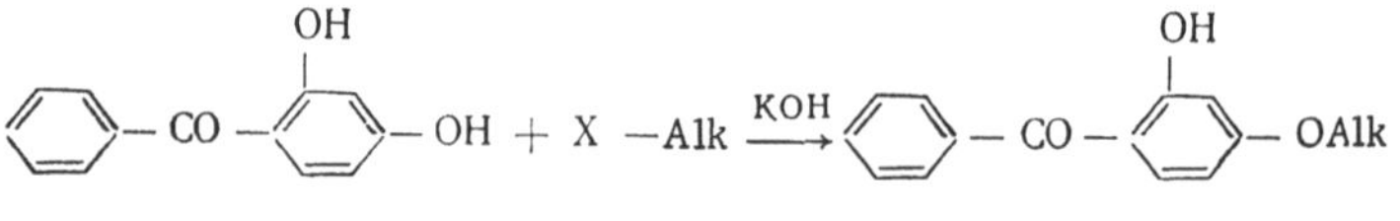

as well as by condensation of benzoic acid or, more often, benzoyl chloride with 1,3-dialkoxybenzene according to Friedel-Krafts [29, 31-34],
followed by dealkylation of the ortho-alkoxy group at increased temperature:

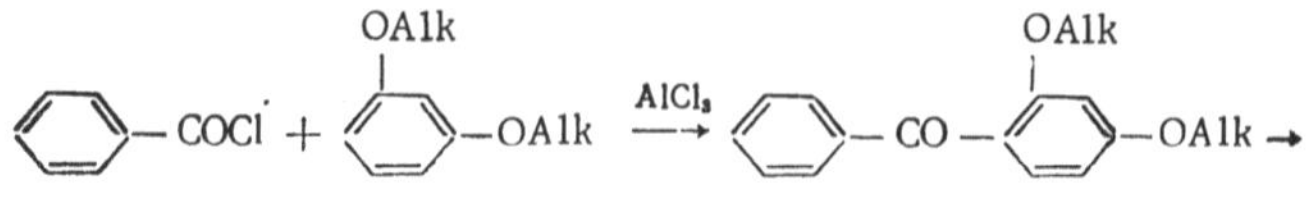

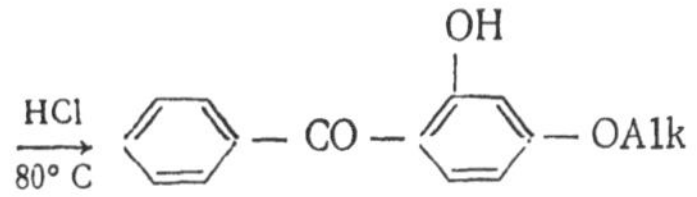

$$\xrightarrow[80° C]{HCl} \quad \bigcirc - CO - \bigcirc(OH) - OAlk$$

The second method is suitable only for producing alkoxy derivatives containing lower alkyls, since the dealkylation proceeds with great difficulty when the aliphatic chain in the alkoxy group is lengthened [35].

Some authors propose the use of mixtures of NaCl, KCl, NaBr with AlCl$_3$ in various ratios as condensation catalysts [40].

One of the possible methods of synthesizing benzophenones can be the oxidation of diphenylmethane and its derivatives [41].

Of the alkoxy derivatives, 2-hydroxy-4-methoxybenzophenone (ciasorb UV-9), the 4-octoxy- (ciasorb UV-531) and 4-dodecyloxy-derivatives are finding industrial use for the stabilization of polymers. The latter are especially recommended for the photostabilization of polyolefins, since, thanks to the presence of a long hydrocarbon chain, they possess good compatibility with the polymers and do not migrate to the surface.

Of the other derivatives of 2,4-dihydroxybenzophenone used as stabilizers, the condensation products of one mole of 1,3-dialkoxybenzene with two moles of benzoyl chloride are indicated [37-39]:

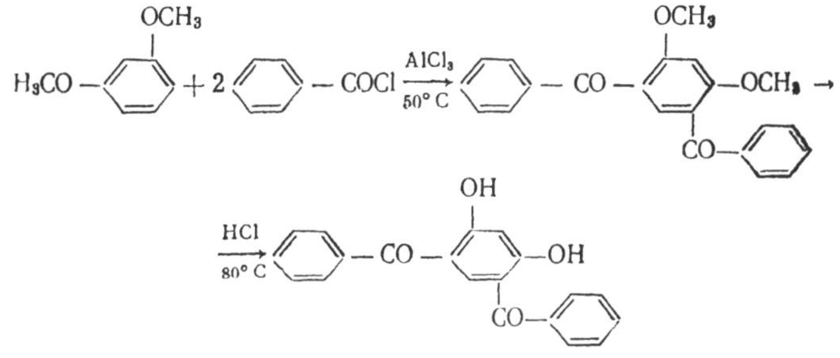

2,4-Dihydroxy-5-benzoylbenzophenone
or 4,6-dibenzoylresorcinol

as well as the products obtained by the reaction of 2,4-dihydroxybenzophenones with phenylene diisocyanates [43] or with aliphatic dicarboxylic acid chlorides [42]:

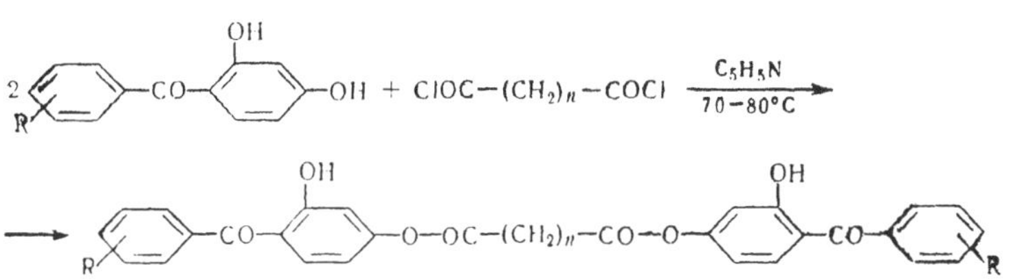

where R = H, halogen, alkyl, n = from 1 to 16;

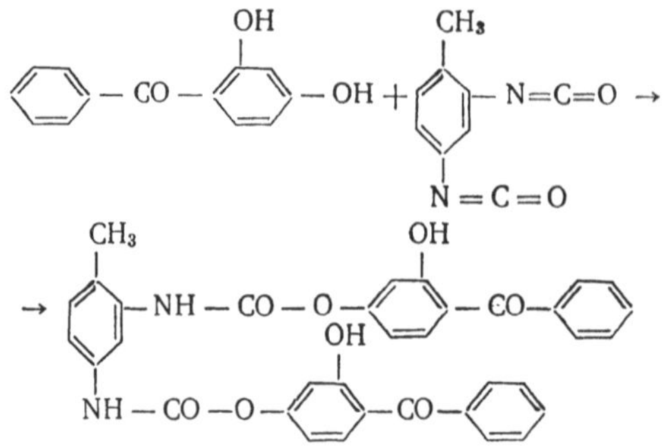

Bis-(p-salicyloylphenyl)-2-methyl-m-phenylene dicarbamate

3. Trihydroxy Derivatives of Benzophenones

Derivatives of trihydroxybenzophenone, containing alkoxy groups with a long carbon chain in the para-position, are the most effective; they are considerably more effective than the mono-, di-, and tetra-hydroxy derivatives of benzophenone [44].

P. K. Grover et al., [45] describes several possible methods of synthesizing 2, 2', 4-trihydroxybenzophenone and its 2, 2'-dihydroxy-4-alkyloxy derivatives:

a) by condensation of β-resorcylic acid and phenol in the presence of a catalyst consisting of a mixture of $POCl_3$ and $ZnCl_2$, at a temperature of 70-80°C;

b) by condensation of the methyl ester of salicylyl chloride with the dimethyl ester of resorcinol according to Friedel-Crafts, followed by partial dealkylation of the condensation product:

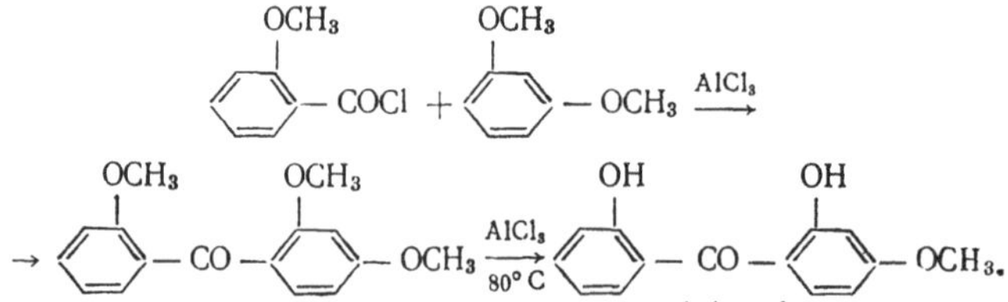

(ciasorb UV-24)

Using the dioctyl ester of resorcinol, one obtains 2, 2'-dihydroxy-4-octoxybenzophenone [31, 45] – a photostabilizer known under the trade name of ciasorb UV-314;

c) by phosgenization of a mixture of methyl esters of resorcinol and phenol:

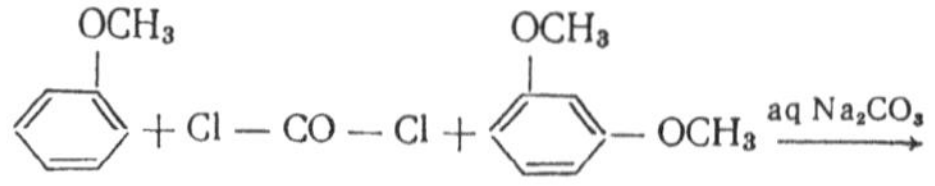

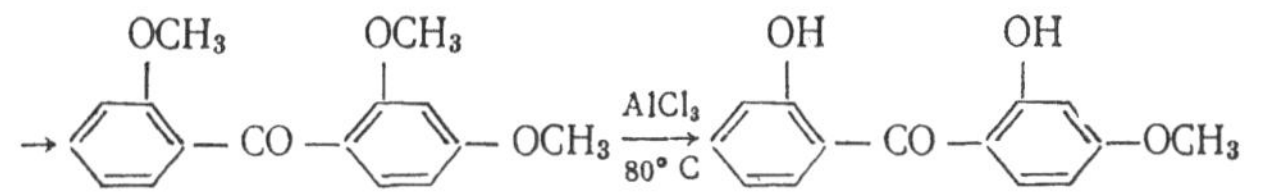

2,2'-Dihydroxy- and 2,2',4,4'-tetrahydroxybenzophenones are formed as side products.

2,2'-Dihydroxy-4-alkoxy derivatives of benzophenone are also produced by alkylation of 2,2',4-trihydroxybenzophenone with the corresponding alkyl halide [29, 31]; in this case the alkyl group is inserted primarily in the para-position. When reliable methods of synthesizing the initial 2,2',4-trihydroxybenzophenone are available, this method is the most expedient for the industrial production of 2,2'-dihydroxy-4-alkoxy-substituted benzophenones.

4. Tetrahydroxy Derivatives of Benzophenone

The synthesis of 2,2',4,4'-tetrahydroxybenzophenone and its alkyl derivatives is accomplished basically by three methods:

a) by condensation of dialkoxyresorcinol with phosgene in the presence of $AlCl_3$ [36, 47, 49] or a mixture of $AlCl_3$ and NaCl [48] or $AlCl_3$ and $ZnCl_2$ [49]; depending on the temperature conditions of the reaction, tetraalkoxy-substituted or 2,2'-dihydroxy-4,4'-dialkoxy-substituted benzophenones are obtained; this method may be of industrial significance;

b) by condensation of the chloride or unsubstituted or dihydroxy-alkylated β-resorcylic acid with dialkoxyresorcinol in the presence of $AlCl_3$ [34, 50] or $ZnCl_2$ [45, 51]. The formation of β-resorcylyl chloride occurs directly during the reaction under the action of $POCl_3$ [45, 51] or $SOCl_2$ [50] on the acid:

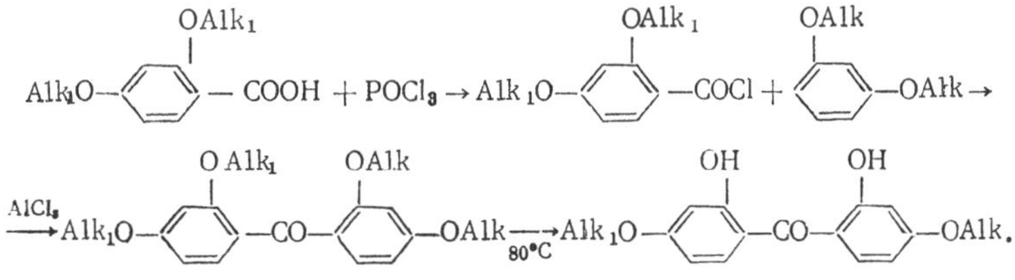

This method is recommended for producing benzophenones with the same or different alkyl radicals;

c) by alkylation of the 2,2'-dihydroxy-4,4'-disodium or dipotassium salt of benzophenone with the corresponding alkyl halide, analogously to dihydroxybenzophenones [14, 28, 33].

Of the other derivatives of aromatic hydroxyketones used as photostabilizers, the reaction products of aliphatic or arylaliphatic mono- and dinitriles with phenols are indicated [23, 52-54]:

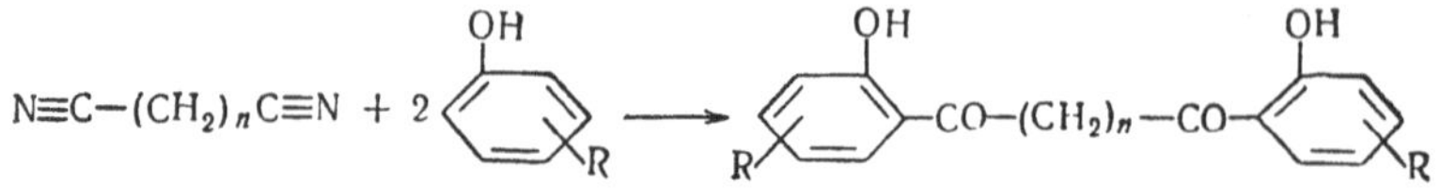

where R = H, halogen, OH; n = from 1 to 20.

It is also proposed that analogous compounds be produced by acylation of aluminum phenolates with acid chlorides in the presence of $AlCl_3$ [55, 56]:

$$2 (C_6H_5O)_3 Al + 3ClOC - (CH_2)_n - COCl \xrightarrow{AlCl_3} HO - C_6H_4 - OC - (CH_2)_n - CO - C_6H_4 - OH$$

The hydroxybenzophenones used as photostabilizers should possess a high degree of purity. However, in the case of production by any of the methods described, they are highly contaminated by resinous products of side reactions, which are difficult to remove by the known methods. It is proposed that they be purified by isolating the hydroxybenzophenones from their solutions in organic solvents, containing not only the benzophenone, but also an aromatic or aliphatic monobasic acid. When such a solution is poured into water acidified by hydrochloric acid, containing small amounts of the same organic acid, extremely pure crystalline products are obtained [57].

2. Complex Esters of Salicyclic Acid

Phenyl esters of salicylic acid, with the general formula

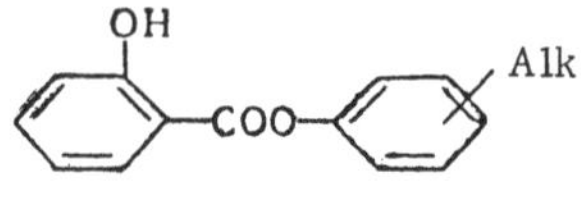

where Alk is a normal or branched hydrocarbon chain containing from 1 to 10 carbon atoms, are finding use as photostabilizers of polyolefins and halogen-containing polymers.

The general method for producing the phenyl esters is heating the salicylic acid with the corresponding phenol derivatives in the presence of phosphorus oxychloride [58-66], phosphorus pentachloride [61, 67], phosphorus trichloride [61, 64, 66], phosgene, or thionyl chloride [61].

When the reaction mixture is heated, first salicylyl chloride is formed, which reacts with phenol at the moment of formation:

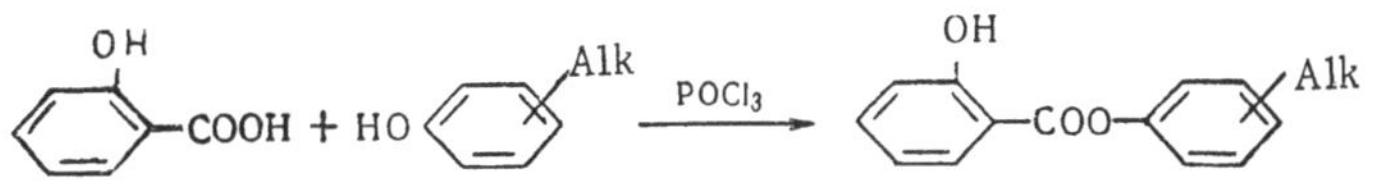

Methods are indicated for producing esters directly from the acid chlorides; in this case the reaction is conducted in a solvent (for example, in nitrobenzene) or without it [68, 69]. However, the synthesis of acid chlorides containing substituents with labile hydrogen atoms in the ortho-position to the carboxyl group involve definite difficulties. Protection of the OH group in salicylic acid, for example, by methoxylation, followed by its saponification, complicates the technology and leads to considerable losses of the finished product.

Benzoyl derivatives of hydroxyphenylsalicylic acid of the general formula

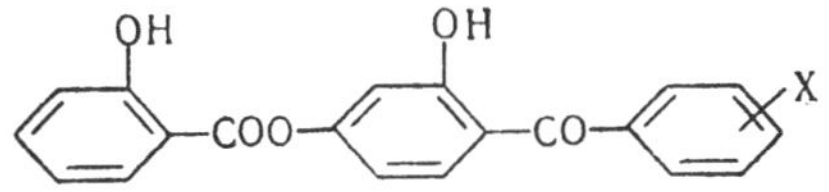

(where X = H, OH, halogen) are produced by condensation of monobenzoylresorcinol or its derivatives with salol at 180-200°C [69]:

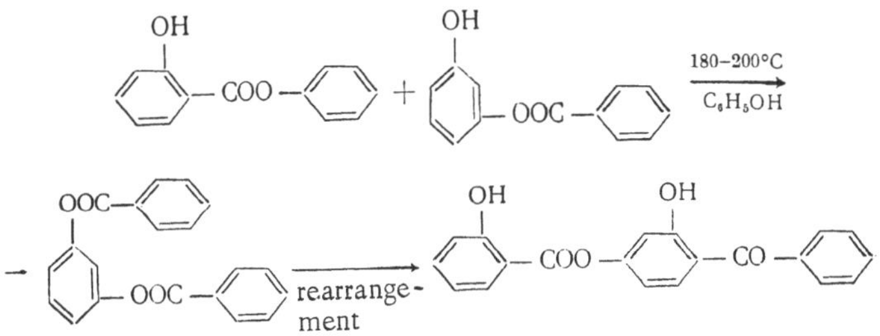

The process is conducted under vacuum to remove the phenol formed. The reaction evidently proceeds through the formation of a diester of resorcinol, which is rearranged to the hydroxyketone at high temperatures. The shortcoming of the indicated method is the absence of reliable methods for producing the initial monobenzoylresorcinol.

L. Ya. Kotikovskaya and V. V. Mikhailov [70] have proposed a method for producing hydroxybenzoylphenyl salicylates by the condensation of 2,4-dihydroxybenzophenone with salicylic acid in the presence of phosphorus oxychloride at a temperature of 135-140°C; yield 45-50% of the theoretical:

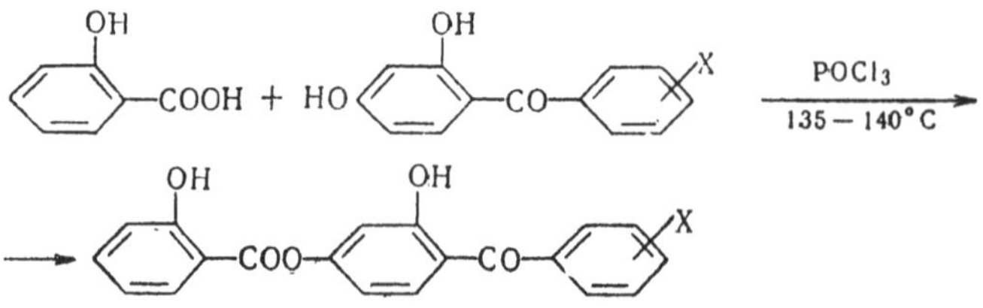

3. Derivatives of Benzo-1,2,3-Triazole

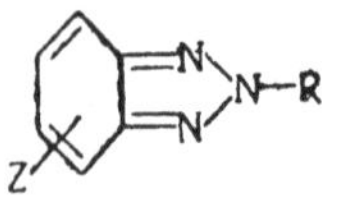

where R = C_6H_5, C_6H_4X, C_6H_4Y, C_6H_3XY, $C_{10}H_8$, etc.; X = H, OH, OCH_3, NH_2, halogen, $N(CH_3)_2$, etc.; Y = H, alkyl, halogen, OH, COOH, etc.; Z = H, CH_3, NH_2, halogen.

The general method of synthesizing derivatives of benzo-1,2,3-triazole consists of reducing the corresponding o-nitroazo dyes. The reduction process most often proceeds in two steps: first the benzotriazole oxide is formed, and then is reduced to the benzotriazole; in certain cases the reduction to the benzotriazole proceeds in one step. Aqueous or aqueous alcohol solutions of Na_2S or $(NH_4)_2S$ [71-73], $NaHSO_3$ in alkaline medium [74, 75], or Zn in ammonia medium are used as reducing agents in the first step.

The reduction of the benzotriazole oxide obtained is conducted with $ZnCl_2$ in hydrochloric acid solution [72], $SnCl_2$ [75], or Zn in alkaline medium [70, 71].

Derivatives of benzo-1,2,3-triazole are also produced by oxidizing ortho-aminoazo dyes through the action of sodium hypochlorite, chromic anhydride, or the copper ammonia complex [76, 78]:

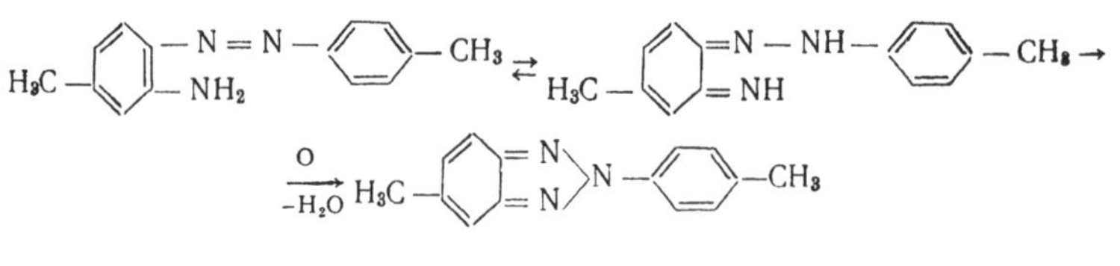

Of the derivatives of benzo-1,2,3-triazole, derivatives in which R is phenyl [79], containing an unsubstituted OH group in the ortho-position to the nitrogen, are used as photostabilizers.

As an example, let us cite the recently widely used preparation SN-3457, known under the trade name of tinuvin P. This preparation is synthesized according to the scheme:

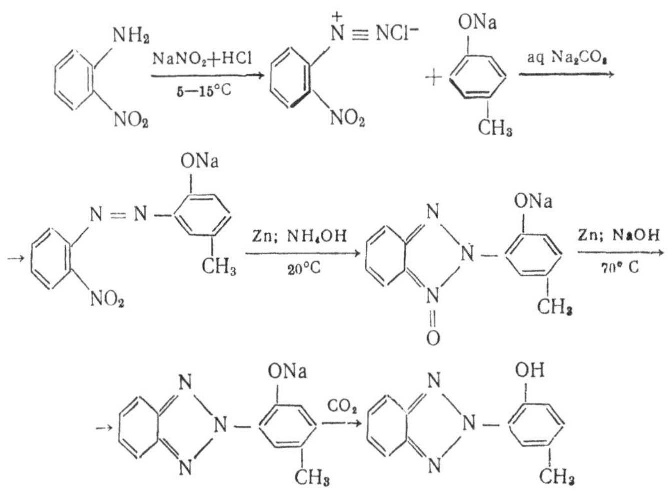

The dye obtained after the combination of o-nitrodiazobenzene with p-cresol is reduced with zinc dust in ammonia-alcohol medium, first at a temperature of 20°C (about 2 hr), and then at increased temperature (70-75°C) for approximately 2-2.5 hr. After filtration of the excess zinc dust, the finished product is isolated from solution by the passage of CO_2.

II. ANTIOXIDANTS

The antioxidants presently used in the practice of reprocessing polymer materials mainly represent derivatives of the following classes of compounds: phenols, sulfide phenols, and arylamines, as well as phosphorous acid esters (phosphites).

1.   Phenol Derivatives

Alkylphenols can be subdivided into three basic groups according to the number of phenol rings:

1) alkylphenols;

2) monoalkylenedialkylphenols;

3) dialkylenetrialkylphenols.

1. Alkylphenols

The presence of from two to three alkyl radicals in the benzene ring of the phenol, situated in various positions with respect to the OH group, is characteristic of stabilizers of this class of compounds.

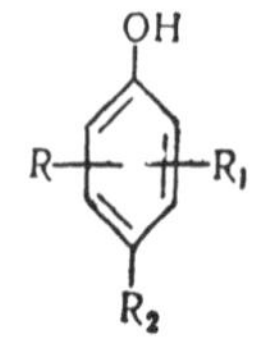

where R represents alkyl radicals with a normal or branched hydro-carbon chain, containing from 1 to 10 carbon atoms; $R_1 = R_2$, arylalkyl, cycloalkyl; $R_2 = R$, H, OH, OCH_3, halogen.

It has been established that the effectiveness of these compounds is increased when R and R_1 are in the ortho-position to the OH-group and represent alkyl radicals with a branched hydrocarbon chain (for example, tert-butyl), while R_2 are lower alkyls of normal structure (CH_3, C_2H_5) [80, 81].

Such compounds are produced by alkylating phenol or its mono- or dialkyl derivatives with various alkylating agents. Alkyl halides [82–87, 116] or arylalkyl halides [89, 90], alcohols [85, 91–105, 116], aliphatic [87, 90, 95, 104, 106–127], aromatic [88, 90, 127–135], or cyclic [136–138] olefins are used as such alkylating agents.

Alkylation is usually conducted in the presence of catalysts, of which H_2SO_4, H_3PO_4, $AlCl_3$, BF_3, etc., are most frequently used. Cation exchange resins (cation exchange resins of a strongly acid type with active group SO_3H) [105, 106, 139, 140], phenolates of various metals [108, 116], iodine [126], sulfonated coal [107], etc., are also proposed as catalysts.

In their extensive survey [141], devoted to methods of catalytic synthesis of alkylphenols, N. I. Shuikin and E. A. Viktorova cite numerous literature data on the reactions of alkylation of phenol by olefins, alcohols, and alkyl halides. In practice, methods of alkylation by olefins and alcohols have received wide circulation.

In the case of alkylation with gaseous olefins, for example, iso-butylene, the alkylation process is conducted by passing the olefin into the heated reaction mixture, containing the phenol to be alkylated and a catalyst [117, 118]:

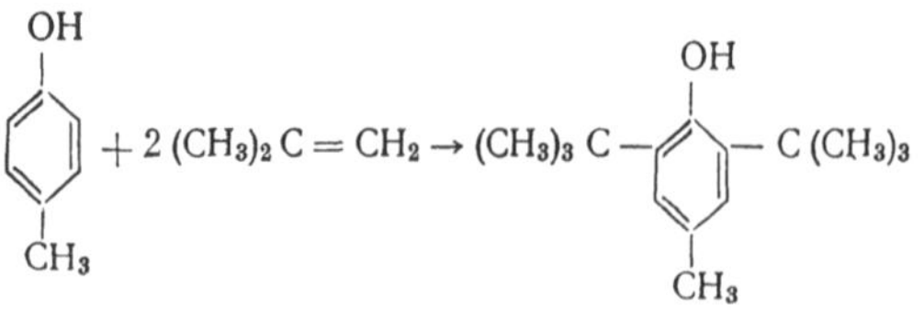

2,6-Di-tert-butyl-4-methylphenol (ionol)

The synthesis of alkylphenols in the liquid phase is more convenient from the technological standpoint and is accomplished by mixing the initial reagents with heating [142, 143].

Mono-, di-, or trialkylphenols are produced by condensing phenols with alcohols, depending on the ratios of the reagents used and the reaction conditions. Thus, 2,6-di-tert-butyl-4-methylphenol, indicated above, is produced in about 56% yield by heating p-cresol with isobutanol at 200-220°C in the presence of a catalyst (Al-shavings +NaCl) [96].

Identical products are formed by heating substituted or unsubstituted phenol with an alcohol in the presence of H_3PO_4 or $AlCl_3$ [91-94].

Methods have been described for producing triarylphenols from 2,6-dialkylphenols by heating the latter in an autoclave at 200°C with an aldehyde in a large excess of the corresponding alcohol in the presence of sodium hydroxide [144]:

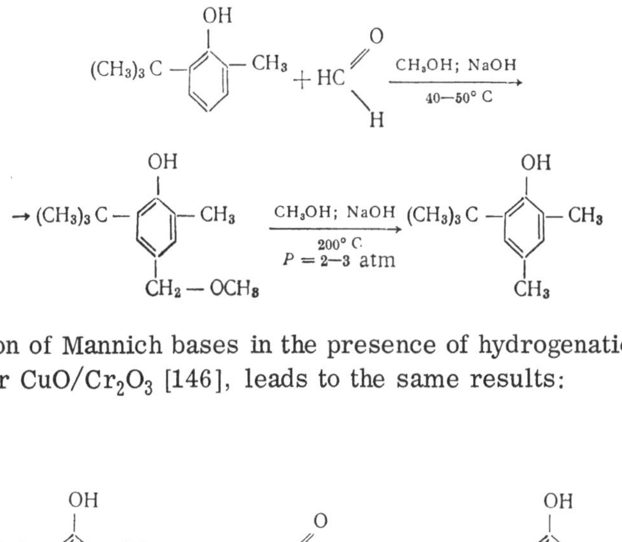

Hydrogenation of Mannich bases in the presence of hydrogenation catalysts, MoS_2 [145] or CuO/Cr_2O_3 [146], leads to the same results:

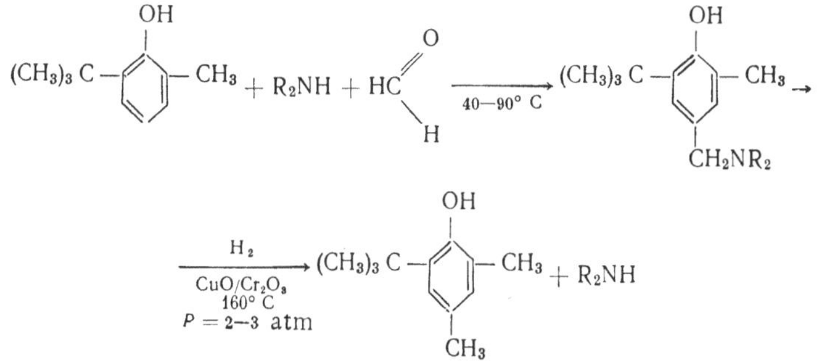

Reports on methods of producing alkylphenols from halophenols [147], as well as from boric esters of phenols in the presence of Friedel-Crafts catalysts have recently appeared in the literature [148].

W. Baker [149], E. Armstrong [150], and other authors propose that alkylphenols and alkylhydroquinones be produced by alkylating phenols or hydroquinones with aliphatic acids in the presence of BF_3, followed by reduction of the acyl derivatives in the presence of hydrogenation catalysts (Pd/C; $CuCrO_2$):

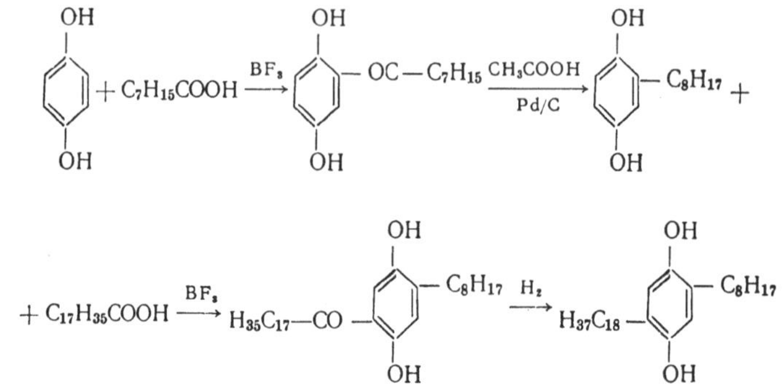

This method makes it possible to introduce several alkyl radicals with different lengths of the hydrocarbon chain into the benzene ring.

The synthesis of p-alkylphenols or p-alkylphenylamines is accomplished by a method developed by L. N. Nikolenko [151]. According to this method, an alkyl radical with practically any length of the hydrocarbon chain is introduced into the phenol or aniline nucleus:

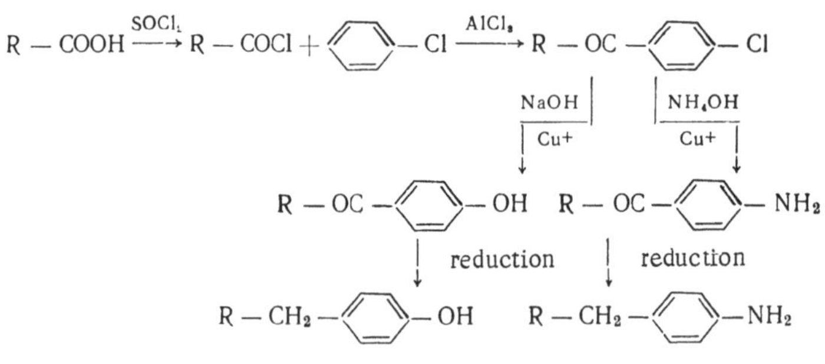

When phenol or its alkyl derivatives are reacted with aliphatic carboxylic acids in the presence of $POCl_3$ or with acid chlorides of these acids, phenyl esters are formed which are readily converted to o- or p-hydroxyketones when heated with anhydrous $AlCl_3$ (the Fries rearrangement) [152]. It is indicated that predominant formation of the ortho- or para-isomers depends on the reaction temperature [153]:

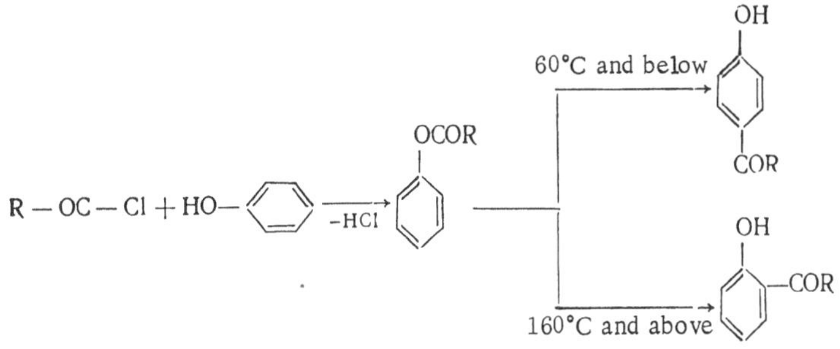

Further reduction of the hydroxyketones leads to the formation of o- or p-alkylphenols, respectively.

One of the methods of producing alkylphenols can be the method used by R. Stroh, J. Ebersberger, et al. [154] to identify alkylphenylamines. It consists of diazotization of the alkylphenylamine, followed by decomposition of the diazo compound by boiling in an aqueous solution of alcohol:

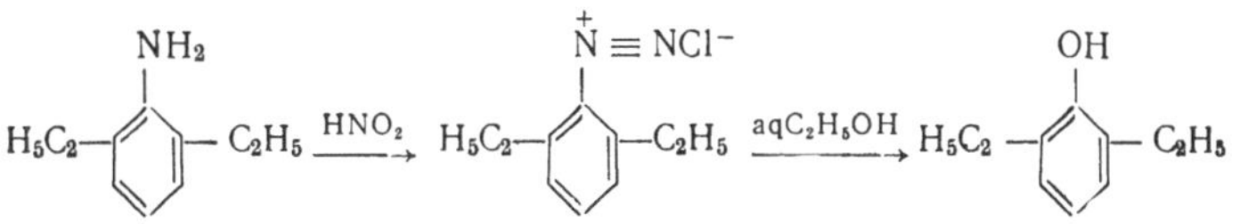

Methods have been described for producing the initial alkylphenylamines [154-157].

The condensation products of phenol and its alkyl derivatives with styrene or α-methylstyrene are receiving especially widespread use as antioxidants and thermal stabilizers of polymer materials. Under the action of styrene on phenol in the presence of an acid catalyst, a mixture of mono-, di-, and trisubstituted derivatives is formed [127, 128, 131, 132, 158]:

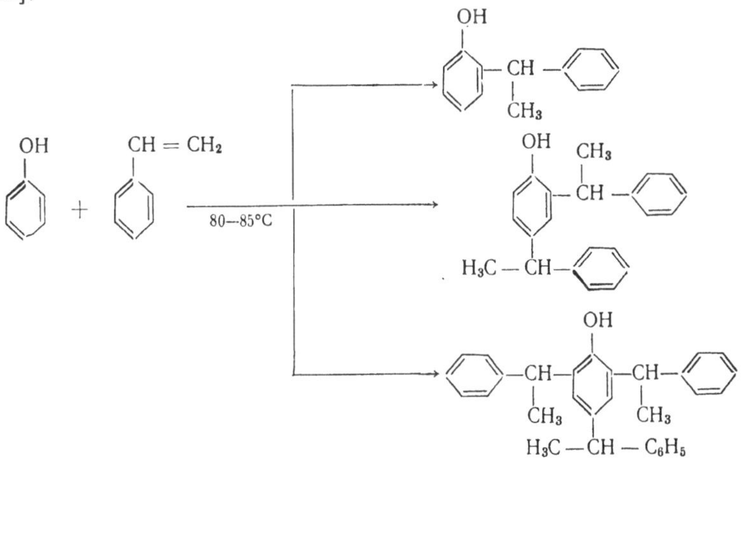

The reaction product represents a viscous yellowish liquid, which distills at reduced pressure within a broad temperature interval.

It is proposed that dialkyl-substituted arylphenols be produced by two methods:

a) by successive reaction of phenol or monoalkylphenol first with aliphatic, and then with aromatic olefins [88, 90, 129, 130, 134]:

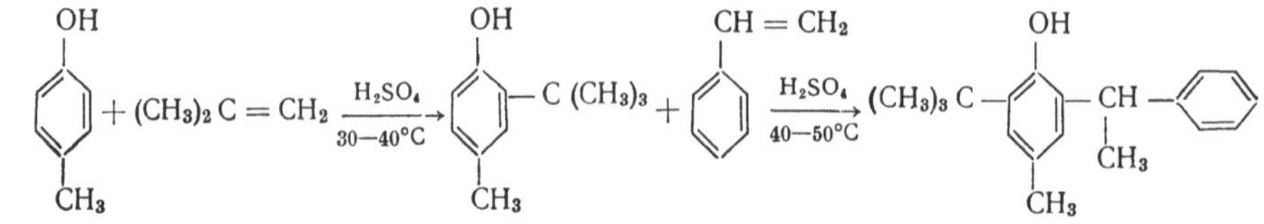

In this case 2,6-di-tert-butyl-4-methylphenol, formed in the first step of the process, may be present in the final product;

b) by the reaction of phenol first with aromatic, and then with aliphatic olefins [128, 131-133]:

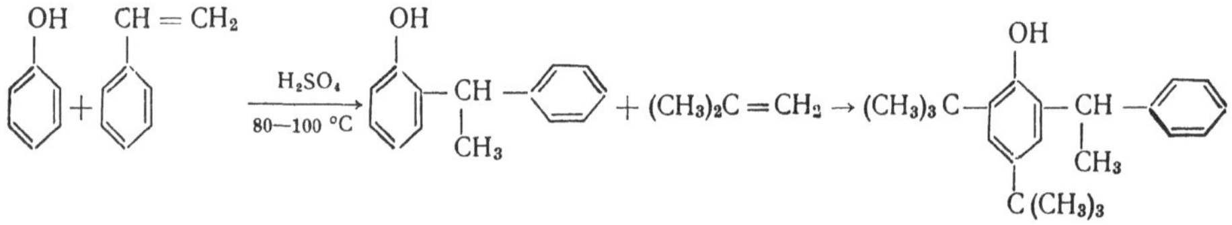

In the second case the intermediate monoarylphenol probably consists of a mixture of ortho- and para-isomers with an admixture of small amounts of diaryl-substituted phenol.

Alkylarylphenols with a definite structure are produced by condensing dialkylphenols with aromatic olefins:

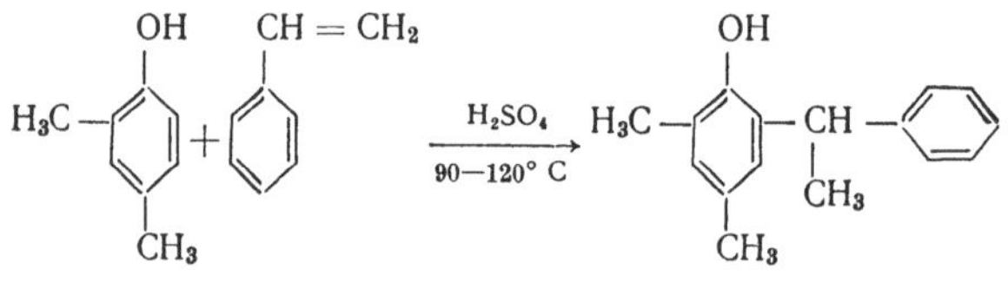

2. Monoalkylenedialkylphenols

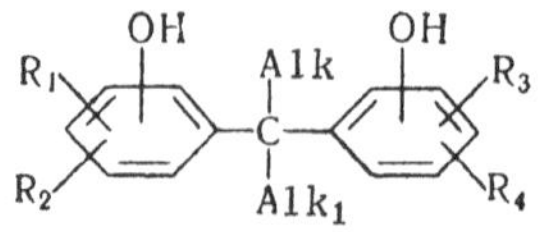

where Alk and Alk_1 = H, normal or isoaliphatic radicals of saturated or unsaturated structure, aromatic, arylaliphatic, or cyclic radicals; R_1, R_2, R_3, and R_4 represent the same or different alkyls, alkylaryls, or cycloalkyls.

In practice, compounds of symmetrical structure, in which $R_1 = R_3$ and $R_2 = R_4$, have received the most widespread use for technological reasons. These radicals can be present in the initial phenol or can be introduced into the condensation products.

The general method of producing monoalkylenedialkylphenols of sym-metrical structure consists of condensing the phenol or its alkyl deriv-atives with saturated [84, 159-174, 178, 185-188, 193] and unsaturated [175, 176] aliphatic, arylaliphatic [177], aromatic [177], and cyclic [179, 180, 184, 185, 189] aldehydes and ketones, as well as with various ter-penes [190-192]. The reaction is conducted in the presence of catalysts, of which HCl, H_2SO_4, BF_3, cation exchange resins, and others are used:

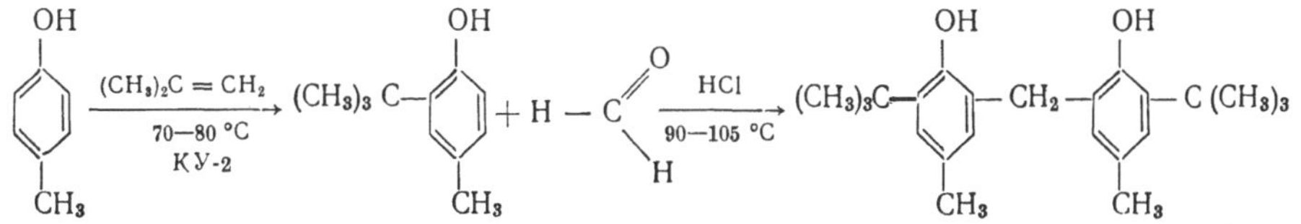

2,2-Methylene-bis-(4-methyl-6-tert-butylphenol) (antioxidant 2246)

In the condensation of naphthols with aldehydes in the presence of H_2SO_4, monoalkylene-dialkylnaphthols are formed [181].

Under the action of aldehydes, ketones, or olefins on unsubstituted or partially substituted monoalkylenedialkylphenols, the hydrogen atoms of the benzene rings in the ortho-position to the hydroxy group are re-placed, forming symmetrical compounds [182-184, 194]:

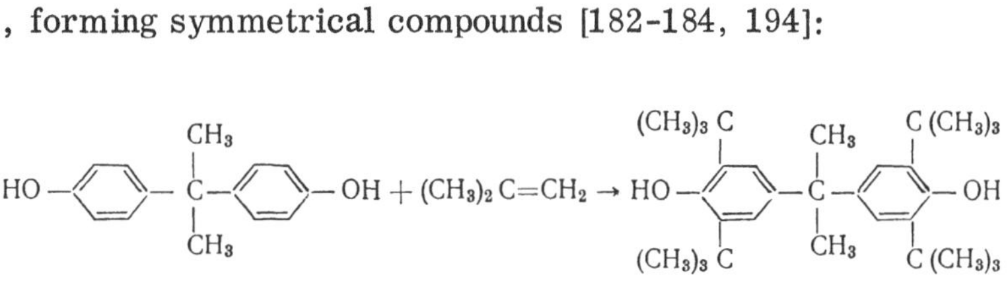

The reaction is conducted in indifferent solvents (hexane, heptane, tolu-ene, benzene) with an acid catalyst (H_2SO_4, HCl, BF_3) at a temperature from 50°C to the boiling point of the solvent.

It is proposed that monoalkylenedialkylphenols, containing different alkyl radicals in the benzene nuclei, be synthesized in two steps [195]. First the phenol or alkylphenol is treated with formaldehyde with heating in the presence of alkali [195-198]:

and then the hydroxyalkylbenzyl alcohol obtained is condensed with one mole of an alkylphenol containing other alkyl substituents:

2,2'-Dihydroxy-3-methyl-3'-isobornyl-5,5'-dimethyldiphenylmethane

The condensation is conducted in acid medium at the melting point of the reaction mass. The condensation occurs under considerably more rigorous conditions (at 20°C) if the corresponding benzyl chloride is introduced in the reaction in place of the substituted benzyl alcohol:

2,2'-Dihydroxy-3-tert-butyl-3'-cyclohexyl-5,5'-dimethyldiphenylmethane

Of the other stabilizers containing alkylene bridges bonding alkylphenols through sulfur or oxygen atoms, the condensation products of alkylbenzyl alcohols with alkyl thioalcohols are indicated [199]:

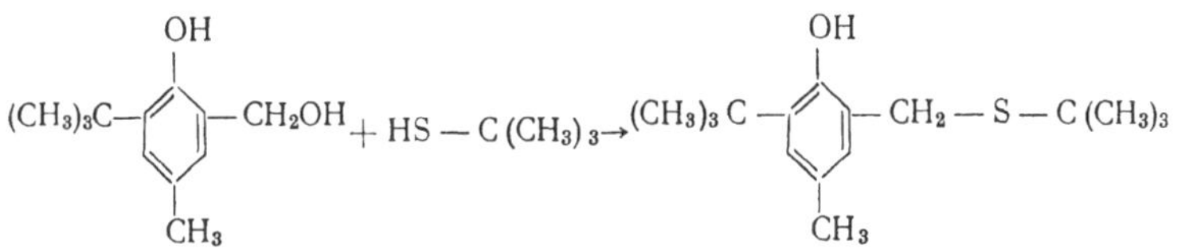

$$(CH_3)_3C-\underset{\underset{CH_3}{|}}{\overset{\overset{OH}{|}}{\bigcirc}}-CH_2OH + HS-C(CH_3)_3 \rightarrow (CH_3)_3C-\underset{\underset{CH_3}{|}}{\overset{\overset{OH}{|}}{\bigcirc}}-CH_2-S-C(CH_3)_3$$

Analogous compounds are also produced in the reaction of phenols with halogen derivatives of dialkyl sulfides in the presence of $AlCl_3$ [200]:

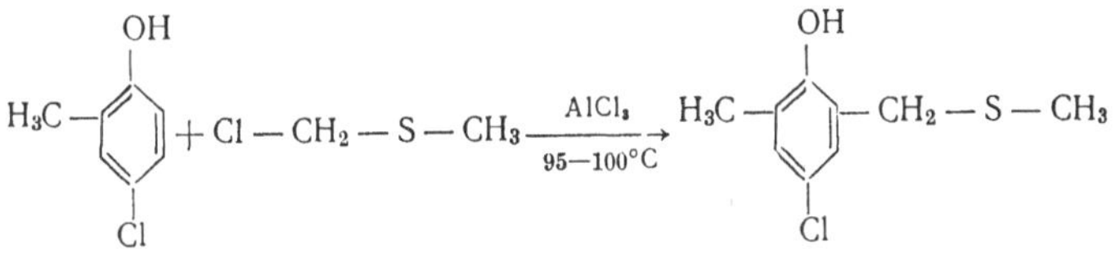

$$H_3C-\underset{\underset{Cl}{|}}{\overset{\overset{OH}{|}}{\bigcirc}}+Cl-CH_2-S-CH_3 \xrightarrow[95-100°C]{AlCl_3} H_3C-\underset{\underset{Cl}{|}}{\overset{\overset{OH}{|}}{\bigcirc}}-CH_2-S-CH_3$$

In the condensation of alkyl derivatives of hydroquinone with di-haloalkanes, compounds recommended for the inhibition of oxidation processes in polyolefins are formed [201]:

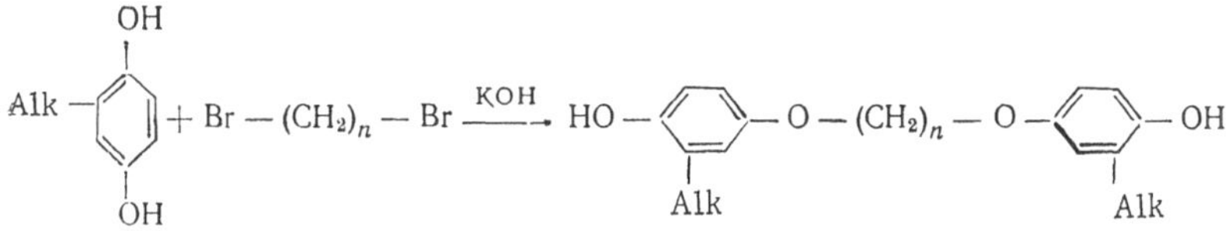

$$Alk-\underset{\underset{OH}{|}}{\overset{\overset{OH}{|}}{\bigcirc}}+Br-(CH_2)_n-Br \xrightarrow{KOH} HO-\underset{\underset{Alk}{|}}{\bigcirc}-O-(CH_2)_n-O-\underset{\underset{Alk}{|}}{\bigcirc}-OH$$

3. Dialkylenetrialkylphenols

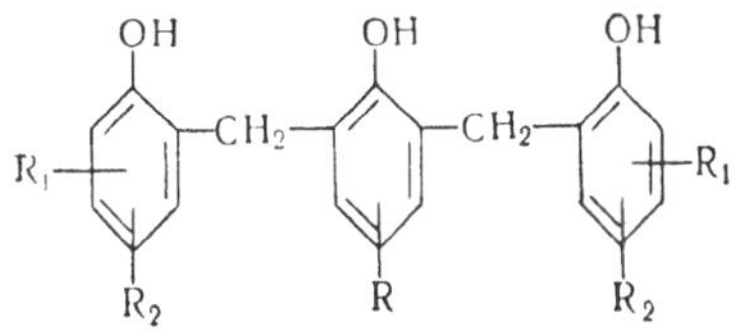

$$R_1-\underset{\underset{R_2}{|}}{\overset{\overset{OH}{|}}{\bigcirc}}-CH_2-\underset{\underset{R}{|}}{\overset{\overset{OH}{|}}{\bigcirc}}-CH_2-\underset{\underset{R_2}{|}}{\overset{\overset{OH}{|}}{\bigcirc}}-R_1$$

where R represents alkyl or normal or branched structure, aryl; R_1 represents H, alkyl, cycloalkyl; R_2 represents alkyl.

Substances of the indicated structure are finding use for the thermal stabilization of polyethylene [203], various rubber mixtures [202], etc.

The synthesis of such stabilizers can be accomplished by several methods. The most widespread method is the condensation of 2,6-di-hydroxymethylene-4-alkylphenols (mesitylene diols) with alkylated phenols [204-210]:

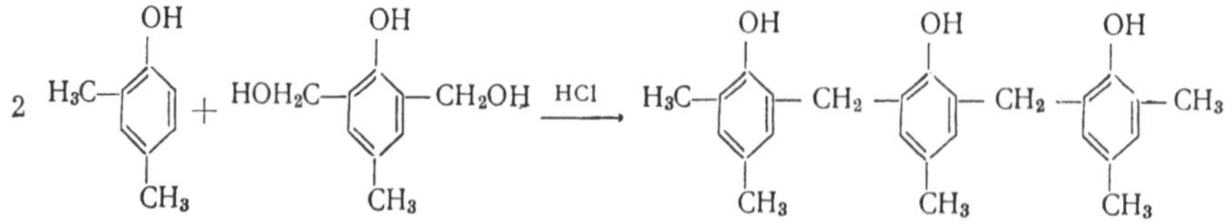

The reaction is conducted in a solvent (glacial acetic acid, heptane, etc.), saturated with dry hydrogen chloride. The initial mesitylene diols are produced by the reaction of p-alkylphenols with two moles of formaldehyde in the presence of alkaline agents [NaOH, Na_2CO_3, $Ca(OH)_2$, etc.], which serve as catalysts of the reaction [205-212].

The resinous condensation products produced according to this method represent a mixture of polyalkylenealkylphenols, from which it is very difficult to isolate the dialkylenetrialkylphenols.

J. Mleziva and M. Lidarik [185] propose that dialkylenetrialkylphenols be produced by melting a mixture of 2,6-dimethyl-4-alkylphenol with an excess of a parasubstituted alkylphenol in the presence of concentrated hydrochloric acid:

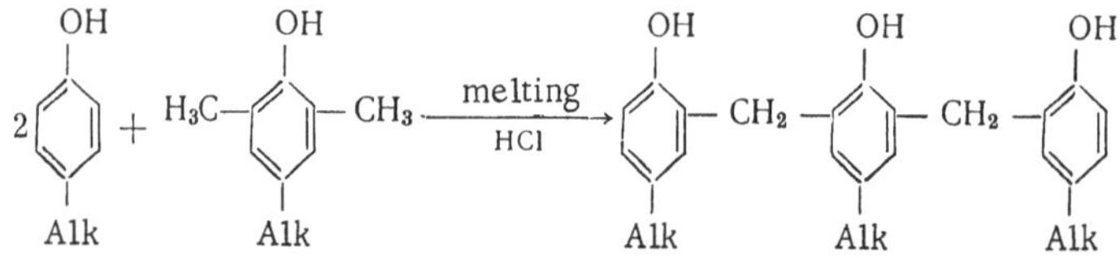

An interesting method, used by A. Davis [213] to synthesize fragments of phenolformaldehyde resins, can be used to produce the products under consideration. By heating an alcoholic solution of a mixture of cyclohexanone with p-hydroxybenzaldehyde in the presence of sodium hydroxide, the authors produced the intermediate 2,6-bis-hydroxy-benzylidenecyclohexanone, which was converted to a dialkylenetriphenol after hydrogenation with heating with Pd/C:

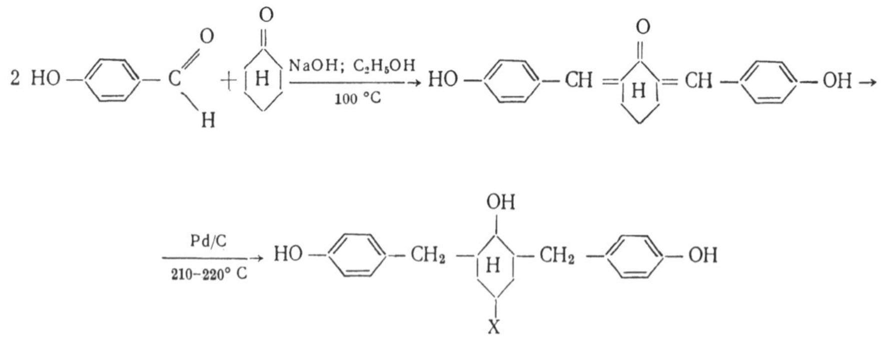

In their work the authors describe a method for introducing alkyl radicals into the position X. Using alkyl-substituted o- and p-hydroxybenzaldehyde in this synthesis, rather pure final products can apparently be produced.

Finally, the synthesis of the compounds under consideration can be effected by the condensation of two moles of a hydroxylalkylbenzyl alcohol or the corresponding benzyl halide with one mole of a p-alkylphenol in the presence of catalysts, as which HCl, H_2SO_4, $ZnCl_2$, aromatic sulfonic acids, etc., are recommended [214]:

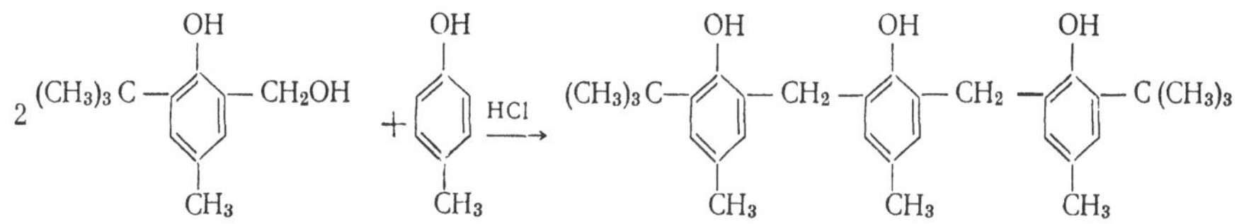

It is recommended that tri- and polyalkylphenols, containing the same or different alkyls in each phenol ring, be produced by successive condensation of phenols substituted by different alkyl radicals [214, 215]:

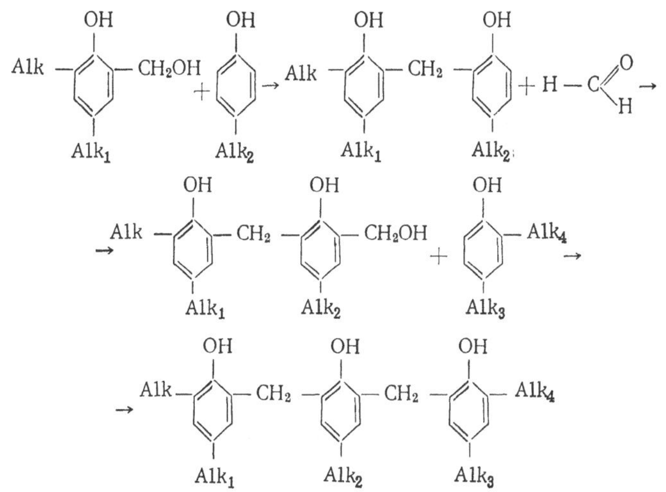

2. Sulfides of Phenol

A common feature of the stabilizers of this class is the presence of a sulfur atom, bonding two molecules of phenol or its alkyl derivatives in the ortho- or para-positions to the hydroxy group. Depending on the number of sulfur atoms contained in the stabilizer molecule, they can be separated into bis-phenol monosulfides and bis-phenol disulfides.

1. Bis-Phenol Monosulfides

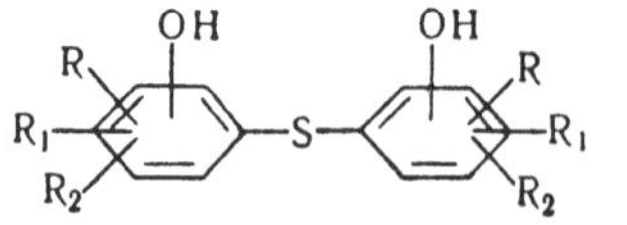

where R, R_1, and R_2 are the same or different substituents [216], halogen [217, 218, 231], hydroxylalkyl [7], alkyl [82, 200, 219-230] with normal or branched hydrocarbon chain, containing from 1 to 20 carbon atoms, aryl alkyl [134, 200], etc.

Bis-phenol monosulfides are formed in the condensation of two moles of mono-, di-, or tri-substituted phenols with one mole of sulfur dichloride SCl_2. The reaction is conducted in inert solvents, more often in carbon tetrachloride medium, at temperatures from 5 to 25°C. Methods of condensation without a solvent are also proposed [219].

The position of the sulfur atom bonding the benzene rings with respect to the hydroxy group depends on the position of the substituents present in the nucleus of the initial phenol.

A mixture of isomeric bis-phenol monosulfides is formed in the condensation of phenol with SCl_2 [216]:

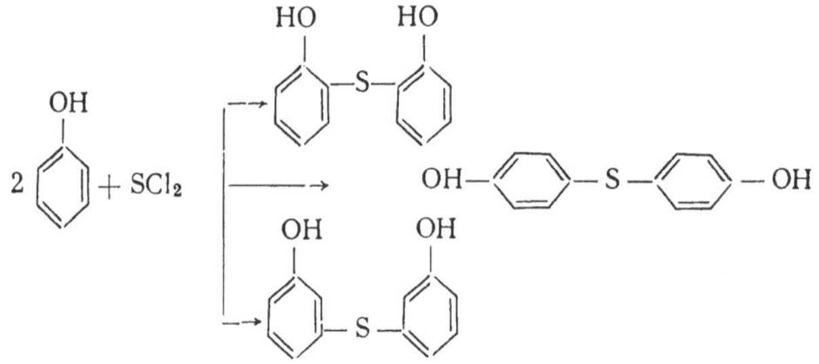

If the substituent stands in the para-position to the hydroxy group in the substituted phenol, or if two substituents occupy the ortho- or para-position, the sulfur atom bonds two phenol molecules only in the ortho-position to the OH group:

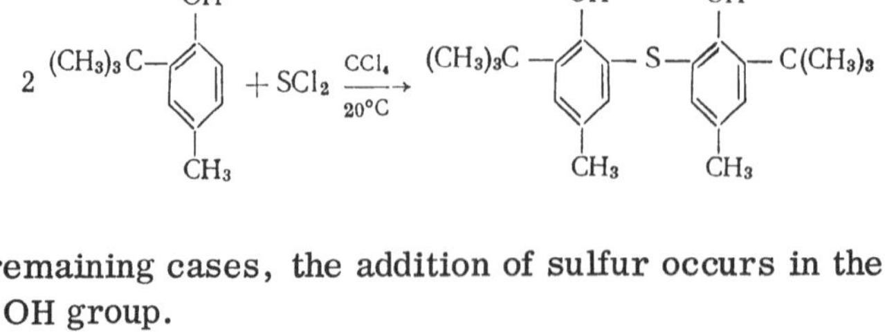

In all the remaining cases, the addition of sulfur occurs in the para-position to the OH group.

It is indicated [95] that phenolates produced by the treatment of phenol sulfides with metal hydroxides are more effective thermal stabilizers for motor fuels and oils.

Of the other bis-phenol monosulfides, compounds are cited in which sulfur bonds two phenol nuclei through methylene bridges [210]:

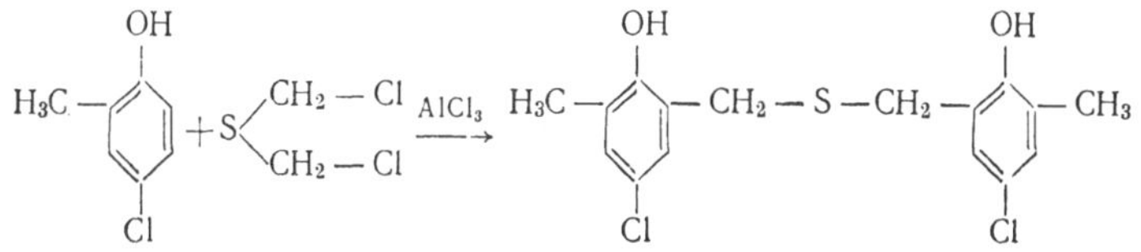

2. Bis-Phenol Disulfides

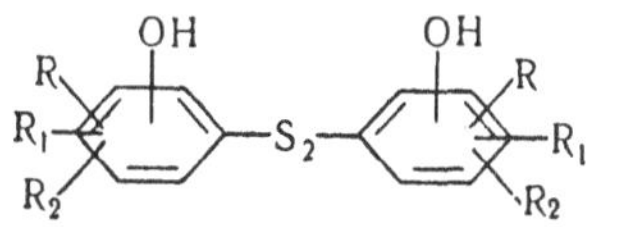

where R, R_1, and R_2 have the same significance as above.

Stabilizers of the indicated structure are used to protect polyolefins [223, 232, 233], polyvinyl chloride, and its copolymers [234] from thermooxidative destruction, and as antioxidizing additives for fuels and oils [95, 140].

The synthesis of such compounds is accomplished by condensing di- or tri-alkylphenols possessing one unsubstituted ortho-position with sulfur monochloride, S_2Cl_2, in an inert solvent at normal or increased temperatures [95, 140, 222]:

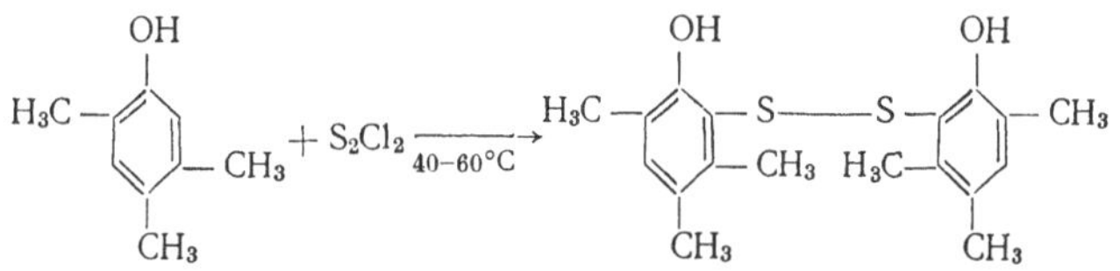

3. Derivatives of Arylamines and Other Nitrogen-Containing Stabilizers

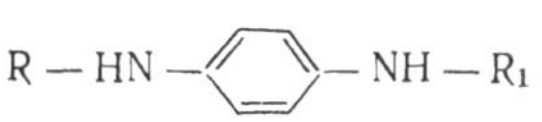

where R and R_1 are the same or different alkyls (C_3-C_{10}), cycloalkyls, aryls, arylalkyls.

Arylamines used as stabilizers of polymer materials can be assigned to the following basic groups – derivatives of: p-phenylenediamine, diphenylamine, and quinoline.

1. Derivatives of p-Phenylenediamine

Of the derivatives of p-phenylenediamine, the most effective are the N, N'-di-alkyl derivatives, in which the alkyl possesses a branched hydrocarbon chain on the carbon closest to the nitrogen atom [235]. They are usually produced by reductive alkylation of p-nitroaniline, p-phenylenediamine, p-nitrodiphenylamine, p-nitrosodiphenylamine, or p-aminodiphenylamine with aldehydes and ketones [236-252] with hydrogen at a temperature of 100-250°C and a pressure from 5 to 200 atm in the presence of catalysts. Copper-chromium catalysts (mixtures of oxides of the metals chromium, copper, barium, etc.) [241, 242, 245-247, 249, 250], iodine in the presence of HCl, HBr, or HI [253, 254], Pt/C [245, 252], and Raney nickel [244] are most often used as the catalysts.

As an example let us cite the method for producing N-isopropyl-N'-phenyl-p-phenylenediamine, known under the trade name of "Antioxidant 4010NA." p-Aminodiphenylamine, acetone, and a copper-chromium catalyst are loaded into an autoclave. The reaction mass is heated with mixing to 160°C, and the hydrogen pressure is raised to 60 atm. The temperature is reduced to 30°C after a three-hour exposure, the mass is filtered off from the catalyst, and the excess acetone is distilled off together with the isopropanol and water formed during the reaction. The yield of the product after purification by vacuum distillation is 95%, converted to the initial amine:

N-Isopropyl-N'-phenyl-p-phenyl-
enediamine (Antioxidant 4010NA)

A second method of producing p-phenylenediamine derivatives consists of treating p-phenylenediamine or p-aminodiphenylamine with primary or secondary alcohols with a Raney nickel catalyst. Diphenyl- or dinaphthyl-N, N'-derivatives of p-phenylenediamine are produced by this same method at atmospheric pressure.

Continuous methods of producing derivatives of p-phenylenediamine by passing a reaction mixture consisting of the amine to be alkylated, an aldehyde or ketone, and a catalyst through a series of successively connected cylindrical vessels, with simultaneous heating of the mass to 150-

210°C, are also described [237, 243, 249]. The reaction mixture is mixed with hydrogen, delivered under a pressure of 180-200 atm. The outgoing hydrogen and the catalyst liberated are returned to the process.

In the treatment of alkyl derivatives of p-phenylenediamine with sodium nitrite and hydrochloric acid at 0-20°C, nitrosoalkylphenylenediamines are formed in a yield almost equal to the theoretical yield [254, 259]:

$$R-HN-\langle\ \rangle-NH-R \xrightarrow[0-20°C]{HNO_2} R-\underset{\underset{NO}{|}}{N}-\langle\ \rangle-\underset{\underset{NO}{|}}{N}-R$$

The nitrosoalkylphenylenediamines, exhibiting a stabilizing action in diene copolymers, simultaneously improve the technology of reprocessing of rubber mixtures in the presence of sulfenamide accelerators.

The reaction products of N-monoalkyl-substituted p-phenylenediamines with alkyl isothiocyanates are proposed as effective antiozonants for rubber [260]:

$$Alk-HN-\langle\ \rangle-NH_2 + \underset{\underset{S}{\|}}{C}=N-(CH_2)_3-CH_3 \xrightarrow[boiling]{C_6H_6}$$

$$\rightarrow Alk-HN-\langle\ \rangle-NH-\underset{\underset{S}{\|}}{C}-NH-(CH_2)_3-CH_3$$

2. Derivatives of Diphenylamine

$$R-\langle\ \rangle-\underset{\underset{N}{|}}{\overset{H}{\ }}-\langle\ \rangle-R_1$$

where R = alkyl, arylalkyl, aryl; R_1 = H or R.

The known methods for producing alkylated diphenylamines reduce to heating in an autoclave at a temperature of 50-300°C and a small pressure of the mixture of diphenylamine with the alkylating agent in the presence of catalysts: $AlCl_3$ or $ZnCl_2$ [261-264], aluminum or magnesium silicates [265], HCl, H_3PO_4, H_2SO_4, $NaHSO_4$, BF_3 [262], etc. In practice Friedel-Crafts catalysts are more often used. The amount of the catalyst is usually 0.1-10% of the amount of the reaction mixture.

The alkylation of diphenylamine is accomplished with aliphatic olefins (for example, diisobutylene) or alkyl halides:

$$\langle\ \rangle-NH-\langle\ \rangle + (CH_3)_3-CH_2-C\underset{\diagdown CH_2}{\overset{\diagup CH_3}{\ }} \xrightarrow[150-185°C]{AlCl_3}$$

$$\rightarrow \langle\ \rangle-NH-\langle\ \rangle-C_8H_{17} + H_{17}C_8-\langle\ \rangle-NH-\langle\ \rangle-C_8H_{17}$$

Alkylation results in the production of a mixture of mono- and di-substituted diphenylamines, the separation of which is technologically difficult. Depending on the conditions of conducting the process, the products obtained can be solid (resinous) [261] or liquid (oily) [263].

3. Derivatives of Hydrogenated Quinoline

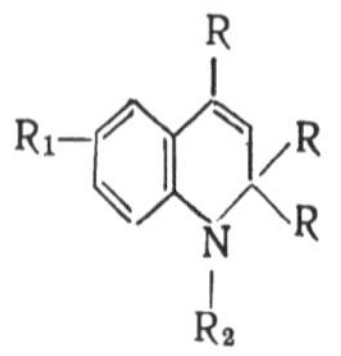

where R = alkyl; R_1 = H, alkyl, alkoxy, dialkylamine; R_2 = H, aryl, alkyl.

Alkyl derivatives of dihydroquinoline are the earliest and best studied anti-aging agents, which have found wide use for the stabilization of polymers.

The basic method of synthesizing derivatives of dihydroquinoline is condensation of aniline [70, 266-268] and its p-alkyl- [269], alkoxy- [270, 271], mono- and di-alkylamines [269] with the corresponding aliphatic aldehydes and ketones in the presence of catalysts (HCl, H_2SO_4, iodine, bromine, aromatic sulfonic acids, etc.):

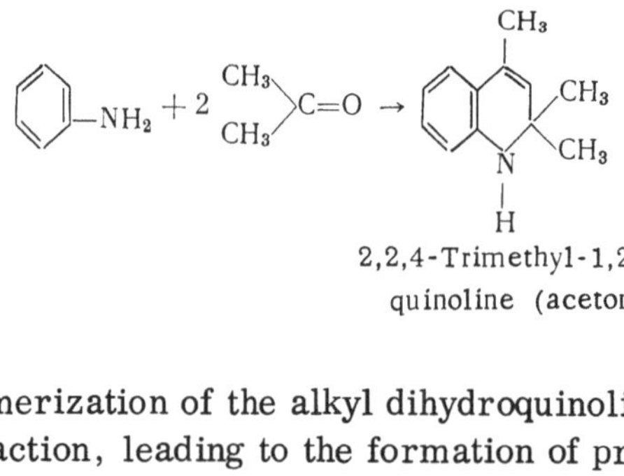

2,2,4-Trimethyl-1,2-dihydro-
quinoline (acetone anil)

However, isomerization of the alkyl dihydroquinolines formed occurs during the reaction, leading to the formation of products of various degrees of polymerization. It is believed [272, 273] that the polymerization process is initiated by inorganic acids.

It is indicated [274] that the introduction of a substituent into the 6-position increases the stabilizing action. The greater the electropositive properties of the substituents, the higher the effectiveness of the compound. For example, compounds with alkylamine or dialkylamine groups are more effective than compounds containing alkoxy groups.

It is proposed that tetrahydroquinolines be produced by catalytic hydrogenation of the corresponding alkyl derivatives with hydrogen on a nickel catalyst at a temperature of 20-150°C [266, 271] or on a Ni/Cr_2O_3 catalyst at a temperature of 130-140°C and a pressure of 140-150 atm [275].

Monomer dihydroquinolines, the production of which is difficult, are required for the synthesis of tetrahydroquinolines; moreover, an expensive catalyst is needed for this process – crystalline iodine. L. P. Zalukaev et al. [70] proposed that sulfanilic acid be used as a catalyst for this purpose in the condensation of the amine with acetone. According to the data of the authors, the addition of 1-2% sulfanilic acid to the reaction mass makes it possible to direct the reaction toward the formation of a large amount of the monomer, readily isolated by vacuum distillation.

Methods are described [267, 275] for producing derivatives of dihydroquinoline possessing substituents on the nitrogen. The synthesis of such products is accomplished by heating 1,2-dihydro-2,2,4-trimethylquinoline with substituted or unsubstituted aromatic olefins, using Friedel-Crafts catalysts.

When alkyl derivatives of dihydroquinoline are heated to 50-70°C with SCl_2 in an inert solvent, compounds are obtained in which two quinoline molecules are connected by sulfur at the nitrogen [276]:

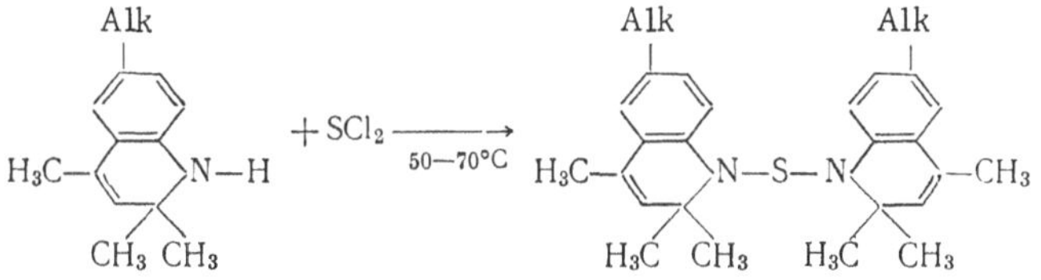

Of the other nitrogen-containing stabilizers, F. Yu. Rachinskii et al., have proposed new compounds for stabilizing polyolefins and polyvinyl acetals [158, 277, 278] from the class of azomethynes, of the general formula

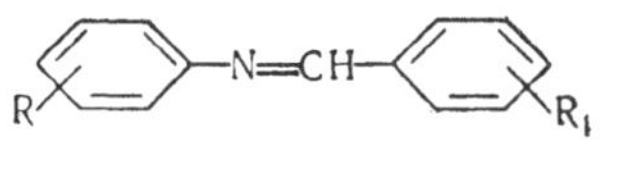

where R = OH, NH_2; R_1 = OH, $N(CH_3)_2$.

The indicated compounds are produced by condensing o- and p-aminophenols or o- and p-phenylenediamines with aromatic substituted and unsubstituted aldehydes. It is noted [278] that azomethynes, produced by condensation of unsubstituted aniline with benzaldehyde and p-dimethylaminobenzaldehyde, are not stabilizers of polyolefins.

4. Esters of Phosphoric Acid
and Other Phosphorus-Containing Stabilizers

In recent years reports have appeared on the use of organophosphorus compounds as noncoloring antioxidants for the stabilization of
many polymers (especially at high temperatures). Of them, complete
esters of phosphorous acid are finding the most widespread use.

It is indicated that they can be used to protect natural and synthetic
rubber [279-285], polyvinyl resins [286-290], ethylcelluloses [291],
phenol-formaldehyde resins [292], polyorganosiloxane polymers [293],
polyolefins [158, 294], polyethylene terephthalate [295], polyurethans
[296], etc.

Complete Esters of Phosphorous Acid

$$\begin{array}{c} R_1O \\ \diagdown \\ R_2O \diagup \end{array} P—OR_3$$

where R_1, R_2, and R_3 are the same or different aliphatic, aromatic,
arylaliphatic, or cyclic radicals.

The most widespread method of producing complete esters of phosphorous acid, in which $R_1 = R_2 = R_3$, is the treatment of alcohols or alkylated phenols, taken in small excesses (0.1-0.5 mole) with phosphorus trihalide (most often trichloride) at a temperature of 150-200°C
[70, 158, 285, 294]:

$$3R — OH + PCl_3 \rightarrow (R — O)_3P + 3HCl$$

It is recommended that the hydrogen chloride formed during the reaction
be removed from the reaction zone by passing through inert gases (CO_2,
nitrogen, etc.) under a small pressure [158].

Mixed esters containing radicals ($R_1 \neq R_2 \neq R_3$ or $R_1 = R_2 \neq R_3$) are
produced either by successive interaction of various alcohols or phenols
with the corresponding acid chlorides:

$$R_1—OH + PCl_3 \rightarrow R_1—O — PCl_2 + R_2 — OH$$

$$\begin{array}{c} R_1O \\ \diagdown \\ R_2O \diagup \end{array} PCl + R_3 — OH \rightarrow \begin{array}{c} R_1O \\ \diagdown \\ R_2O \diagup \end{array} P — OR_3$$

or by treating a mixture of various alcohols or alkylphenols with phosphorus trichloride [297-301].

It has been established [70] that alkylaryl phosphites are more effective stabilizers for many polymers and better inhibitors of radical chain processes in comparison with alkyl phosphites. Especially effective stabilizers are mixed esters of pyrocatecholphosphorous acid [70]:

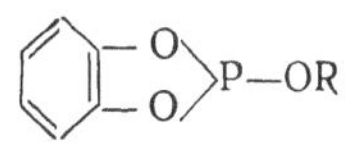

where R represent alkyl (normal or branched structure, containing from 1 to 18 carbon atoms), phenyl, naphthyl, diphenyl, benzyl, etc., containing a cycloalkyl in the nucleus of the substituent (alkyl, halogen, sulfur, etc.).

Their synthesis is accomplished by treating pyrocatechol with phosphorus trichloride, followed by reaction of the chloride of pyrocatecholphosphorous acid obtained with alcohols, alkylphenols, naphthols, etc.

P. A. Kirpichnikov [302] has synthesized a number of alkyl-, aryl-, and aryl-alkylated esters of pyrocatecholphosphorous acid, of which the most effective and universal in the sense of protecting various polymers from aging proved to be the 2,6-di-tert-butyl-4-methylphenyl ester of pyrocatecholic acid.

Polyphosphites

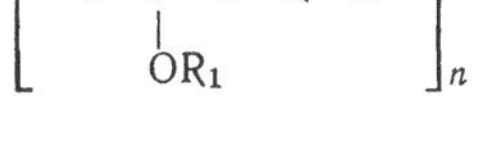

$$\left[-O-P-O-R-O-\atop\qquad\;\; |\atop\qquad\; OR_1\right]_n$$

where

$$R = -CH_2-CH_2-, \quad -CH_2-CH_2-O-CH_2-CH_2-, \quad C_6H_4,$$

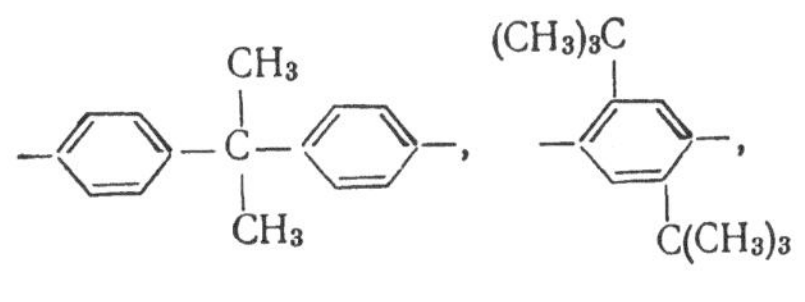

etc.; $R_1 = C_2H_5$, C_4H_9, C_6H_5, ClC_6H_4, $C_{10}H_7$; and n = from 2 to 10 are of great interest as stabilizers [70].

These substances are readily miscible with many polymers, and in view of their large molecular weights, they are stably retained in them. This imparts valuable properties to polyphosphites, which are not possessed by the usual arylalkyl esters of phosphorous acid. Polyphosphites are produced from the corresponding dichlorides of phosphorous acid by their reaction with glycols or aromatic dihydroxy compounds:

$$n\left(HO-\!\!\langle\ \rangle\!\!-\overset{\overset{\displaystyle CH_3}{|}}{\underset{\underset{\displaystyle CH_3}{|}}{C}}-\!\!\langle\ \rangle\!\!-OH\right) + n\left(\overset{Cl-P-Cl}{\underset{OR}{|}}\right) \rightarrow \left[-O-\overset{}{\underset{OR}{P}}-O-\!\!\langle\ \rangle\!\!-\overset{\overset{\displaystyle CH_3}{|}}{\underset{\underset{\displaystyle CH_3}{|}}{C}}-\!\!\langle\ \rangle\!\!-O-\right]_n$$

Of no less interest are organophosphorus stabilizers containing the epoxide group and a phosphorus-carbon bond in the molecule simultaneously. As investigations have shown [70], esters of 1,2-epoxy-2-propylphosphinic acid:

$$(RO)_2-\overset{}{\underset{\displaystyle O}{\overset{\displaystyle \|}{P}}}-\overset{\overset{\displaystyle CH_3}{|}}{C}-CH_2$$

where R = alkyl, aryl, or cycloalkyl, are such stabilizers.

It is indicated that these esters are especially suitable for the stabilization of polyvinyl chloride.

We should mention that the overwhelming majority of phosphorus-containing stabilizers are produced on the basis of readily available starting materials in a high yield, and their synthesis is simply accomplished under the conditions of industrial production.

BIBLIOGRAPHY

1. G. Kerk, J. Luijten, and J. Noltes, Angew. Chem. 70:298, 1958.
2. R. K. Ingham, Chem. Revs. 60:459, 1960.
3. F. C. Brown, Chem. Rev. 61:463, 1961.
4. J. W. Tamblyn, G. C. Newland, and M. T. Watson, Plastics Technol. 4:427, 1958; C. A. 52:11463, 1958.
5. U. S. Patent 2877466, 1959; C. A. 53:15645, 1959; Referat. Zhur. Khim. 1961, 8P 149.
6. Canadian Patent 494220, 1954; Referat. Zhur. Khim. 1956, 48292.
7. C. Thomas, Anhydrous Aluminum Chloride in Organic Chemistry [Russian translation], Moscow, Foreign Literature Press, 1949, p. 169.
8. K. Hoesch, Ber. 48:1122, 1915.
9. U. S. Patent 2787607, 1957; Referat. Zhur. Khim. 1959, 80388.
10. Canadian Patent 493516, 1953; Referat. Zhur. Khim. 1955, 47629.
11. U. S. Patent 2904529, 1959; C. A. 48:1931, 1954.
12. M. Giesen, Farbenchemiker 61:13, 1959; C. A. 54:21836, 1960.
13. Great Britain Patent 706151, 1954.
14. O. Döbner, Ann. 210:256, 1881.
15. A. Baeyer, Ann. 372:86, 1910.
16. A. Komarowsky and S. Konstaneski, Ber. 27:1997, 1894.

17. E. H. Cox, Rec. Trav. chim. 50:848, 1931; C. A. 25:48658, 1931.

18. Organic Reactions, Vol. 5, J. Wiley and Sons, Inc., 1949, p. 387.

19. E. Klarmann, J. Am. Chem. Soc. 48:793, 1926.

20. S. Gabriel and M. Herzberg, Ber. 16:2000, 1883.

21. E. H. Zil'berman and N. A. Rybakova, Zhur. Obshchei Khim. 30:1992, 1960.

22. N. A. Rybakova and E. N. Zil'berman, Zhur. Obshchei Khim. 31:1272, 1961; Referat. Zhur. Khim. 1961, 22J 100.

23. E. N. Zil'berman and A. M. Sladkov, Zhur. Obshchei Khim. 31:245, 1961.

24. R. Shah and P. Mehta, J. Indian. Chem. Soc. 13:368, 1936; C. A. 30:8192, 1936.

25. P. K. Grover, G. D. Shah, and R. C. Shah, J. Chem. Soc. 1955, 3982.

26. H. Oelschläger, Arch. Pharm. 288:102, 1955.

27. U. S. Patent 2861053, 1958.

28. N. A. Larin, E. N. Matveeva, and V. S. Smirnova, Zhur. Obshchei Khim. 30:2377, 1960.

29. U. S. Patent 2853521, 1958; C. A. 53:5206, 1959.

30. F. Ullman, Ann. 327:104, 1903.

31. U. S. Patent 2892872, 1959.

32. U. S. Patent 2861104, 1958; Referat. Zhur. Khim. 1960, 58137.

33. B. König and S. Konstaneski, Ber. 39:4027, 1906.

34. Indian Patent 51712, 1956; Zbl. 1961, 6327.

35. U. S. Patent 2892878, 1959.

36. U. S. Patent 2694729, 1954.

37. U. S. Patent 2794052, 1957.

38. U. S. Patent 2852488, 1958.

39. U. S. Patent 2917402, 1959.

40. U. S. Patent 2861105, 1958.

41. German Federated Republic Patent 934525, 1955.

42. U. S. Patent 2894022, 1959.

43. U. S. Patent 2853466, 1958. Referat. Zhur. Khim. 1961, 5P 200.

44. R. A. Coleman, Cyanamid Bulletin, June 1959.

45. P. K. Grover, G. D. Shah, and R. C. Shah, Chem. Ind. 15:62, 1955.

46. Novosti Khim. Prom. No. 2:26, 1961.

47. U. S. Patent 2692492, 1954.

48. U. S. Patent 2853522, 1958.

49. U. S. Patent 2853523, 1958.

50. U. S. Patent 2773903, 1956.

51. U. S. Patent 2854485, 1958.

52. U. S. Patent 2807604, 1957.

53. U. S. Patent 2807605, 1957.

54. E. N. Zil'berman, Z. V. Popova, N. A. Rybakova, and D. M. Yakovskii, Zhur. Obshchei Khim. 31:1272, 1961.

55. V. K. Kuskov and Yu. A. Naumov, Zhur. Obshchei Khim. 31:54, 1961.

56. V. K. Kuskov and L. P. Yur'eva, Doklady Akad. Nauk 109:319, 1956.

57. U. S. Patent 2682559, 1954.

58. R. Seifert, J. prakt. Chem. 31:472, 1885.

59. M. Nencki and F. Heyden, Ber. 20R:351, 1887.

60. M. Nencki and F. Heyden, Ber. 21R:554, 1888.

61. F. Heyden, Ber. 26R:967, 1893.

62. C. K. Krauz and A. J. Remenek, Collection Czeck. Chem. Commun. 1:610, 1929; C. A. 24:1365, 1930.

63. S. E. Harris and W. G. Christiansen, J. Am. Pharm. Assoc. 24:553, 1935.

64. H. G. Kolloff and J. O. Page, J. Am. Chem. Soc. 60:948, 1938.

65. U. S. Patent 2141072, 1938; C. A. 33:2655, 1939.

66. B. T. Tozer and S. Smiles, J. Chem. Soc. 1938, 1897.

67. Organic Reactions [Russian Translation], Collection 1, Moscow, Foreign Literature Press, 1948, p. 470.

68. K. W. Rosemund and W. Schnurr, Ann. 460:56, 1928.

69. U. S. Patent 2898323, 1959.

70. Summaries of Reports at the Conference on the Aging and Stabilization of Polymers, Moscow, Academy of Sciences USSR Press, 1961, pp. 15, 17, 19, 25.

71. A. Rosenstiehl and E. Suais, Compt. rend. 134:606, 1902.

72. E. Bamberger and R. Hübner, Ber. 36:3822, 1903.

73. U. S. Patent 24122767, 1946.

74. E. Grandmougin and J. Guisan, Ber. 40:4205, 1907.

75. E. Grandmougin, J. Prakt. Chem. 76:134, 1907.

76. U. S. Patent 2704228, 1955.

77. U. S. Patents 2713054, 2713055, 1955.

78. Swiss Patent 323318, 1957.

79. F. Kirchhof, Gummi u. Asbest 10:614, 1958.

80. V. I. Isagulyants, V. N. Tishkova, and N. A. Favorskaya, Trudy Groznensk. Neft. Inst. Sb. No. 23:123, 1960.

81. R. Rosenwald, J. Hostsun, and J. Cheniok, Ind. Eng. Chem. 42:162, 1950; C. A. 43:3704, 1950.

82. U. S. Patent 2841619, 1958; Referat. Zhur. Khim. 1960, 31814.

83. U. S. Patent 2947724, 1960; C. A. 55:2165, 1961.

84. U. S. Patent 2947789, 1960; C. A. 55:4443, 1961.

85. A. Baur, Ber. 27:1615, 1894.

86. A. Gurewitsch, Ber. 32:2428, 1899.

87. S. P. Malenik and R. B. Hannau, J. Am. Chem. Soc. 81:2119, 1959.

88. Japanese Patent 9785/1958, 1955; Zbl. 1961, 6706.

89. Great Britain Patent 729780, 1955; Referat. Zhur. Khim. 1957, 39357.

90. Great Britain Patent 818035, 1959; C. A. 54:15992.

91. I. P. Tsukervanik and Z. N. Nazarova, Zhur. Obshchei Khim. 5:767, 1935.

92. I. P. Tsukervanik and N. G. Sidorova, Zhur. Obshchei Khim. 7:623, 641, 1937.

93. N. G. Sidorova, Zhur. Obshchei Khim. 22:962, 1952.

94. I. P. Tsukervanik, Investigations in the Field of Alkylation in the Nucleus of Aromatic Compounds, Erevan, Academy of Sciences Armenian SSR Press, 1955.

95. V. I. Isagulyants, Khim. Prom. No. 2:20, 1958.

96. German Federated Republic Patent 944014, 1955; Referat. Zhur. Khim. 1957, 67199; German Federated Republic Patent 1044825, 1959; Referat. Zhur. Khim. 1961, 9L 141.

97. U. S. Patent 2822415, 1958; Referat. Zhur. Khim. 1960, 67739.

98. U. S. Patent 2744938, 1956; C. A. 51:1268, 1957.

99. A. Liebmann, Ber. 14:1842, 1881; 15:150, 1882.

100. C. Hartmann and L. Gatermann, Ber. 25:3532, 1892.

101. H. Meyer and K. Bernhauer, Monatsch, 53/54:721, 1929.

102. U. S. Patent 2802884, 1957; Zbl. 1959, 297.

103. V. I. Isagulyants and E. V. Panidi, Zhur. Priklad. Khim. 1961, 1849.

104. U. S. Patent 2722556, 1955; C. A. 50:4214, 1956.

105. R. C. Huston and T. J. Hsieh, J. Am. Chem. Soc. 58:439, 1936.

106. USSR Patent No. 125802, 1960.

107. USSR Patent No. 121133, 1959.

108. U. S. Patent 2831898, 1958; C. A. 52:16293, 1958; Referat. Zhur. Khim. 1960, 70513.

109. U. S. Patent 2837752, 1958.

110. U. S. Patent 2581906, 1952.

111. U. S. Patent 2605252, 1952.

112. Australian Patent 166019, 1955; Referat. Zhur. Khim. 1956, 79263.

113. Great Britain Patent 717863, 1954; Rubb. Abs. 33:75, 1955; Referat. Zhur. Khim. 1955, 56760.

114. Australian Patent 166544, 1956; Referat. Zhur. Khim. 1957, 13435.

115. U. S. Patent 2942033, 1960; Offic. Gaz., U. S. Pat. Office, 755, 851, 1960.

116. R. Stroh, R. Seydel, and W. Hahn, Angew. Chem. 69:699, 1957.

117. O. F. Joklik, J. Prakt. Chem. 10:499, 1959; Referat. Zhur. Khim. 1960, 93958.

118. O. F. Joklik, Metano 14:269, 1960; Referat. Zhur. Khim. 1961, 5M 219.
119. W. Weinrich, Ind. Eng. Chem. 35:264, 1943.
120. F. Uénaka and T. Sëna, Koru Taru 12:543, 1960; Referat. Zhur. Khim. 1961, 19L 22.
121. U. S. Patent 2051473, 1933.
122. U. S. Patent 2248828, 1937.
123. H. M. Crawford, J. Am. Chem. Soc. 74:4087, 1952.
124. D. R. Stevens, J. Org. Chem. 20:1233, 1955.
125. A. Helmel, Kunststoffe p. 514, 1960.
126. U. S. Patent 2900418, 1959.
127. German Federated Republic Patent 1086887, 1960; C. A. 55:17064, 1961.
128. U. S. Patent 2714120, 1955.
129. U. S. Patent 2900362, 1959; C. A. 53:23048, 1959; Referat. Zhur. Khim. 1961, 14P 380.
130. M. E. Mc Greal and J. B. Niederl, J. Am. Chem. Soc. 57:2625, 1935.
131. S. V. Zavgorodnii, B. A. Zaitsev, and D. P. El'chikov, Zhur. Obshchei Khim. 30:2196, 1960.
132. Great Britain Patent 820486, 1959; C. A. 54:4026, 1960.
133. U. S. Patent 2909504, 1959; C. A. 54:1916, 1960.
134. U. S. Patent 2849517, 1958; Referat. Zhur. Khim. 1960, 7310.
135. Ng. Ph. Buu-Hoi, H. LeBihan, and F. Binon, J. Org. Chem. 17:243, 1952.
136. Great Britain Patent 923838, 1955; Referat. Zhur. Khim. 1956, 37559, Rubb. Abs. 33:212, 1955.
137. French Patent 1128968; Referat. Zhur. Khim. 1958, 68375.
138. Great Britain Patent 788168, 1957; Zbl, 1960, 12522.
139. Ya. E. Vertlib, V. I. Grushevenko, and I. P. Pavlova, Khim. i Tekhnol. Topliva i Massel, No. 5:12, 1960.
140. V. I. Isagulyants and N. A. Favorskaya, Trudy Mosk. Inst. Neftekhim. Gaz. Prom. No. 28:56, 1960; Referat. Zhur. Khim. 1961, 2M 239.
141. N. I. Shuikin and E. A. Viktorova, Uspekhi Khim. 29:1229, 1960.
142. U. S. Patent 2930820, 1960; Referat. Zhur. Khim. 1961, 19L 127.
143. Great Britain Patent 846516, 1960; Referat. Zhur. Khim. 1961, 14L 157.
144. U. S. Patent 2841624, 1958; Referat. Zhur. Khim. 1960, 70514.
145. U. S. Patent 2882319, 1958.
146. E. P. Previc, E. B. Hotteling, and M. B. Neuworth, Ind. Eng. Chem. 53:137, 1961.
147. USSR Patent No. 133063, 1960; Referat. Zhur. Khim. 1961, 19L 138.
148. B. M. Sheiman and V. K. Kuskov, Izvest Vysshikh Uchebn.

Zavedenii, Khim. i Khim. Tekhnol. 3:876, 1960; Referat. Zhur. Khim. 1961, 18J 102.

149. W. Baker and O. M. Lothian, J. Chem. Soc. p. 274, 1936.

150. E. C. Armstrong, R. L. Bent, A. Loria, J. R. Thirtle, and A. Weissberger, J. Am. Chem. Soc. 82:1928, 1960; Referat. Zhur. Khim. 1961, 2J 106.

151. L. N. Nikolenko, Summaries of Reports at the Scientific and Technical Conference on Problems of the Synthesis and Use of Organic Dyes, Ivanovo, State Press for Chemical and Technical Literature, 1961, p. 11.

152. T. Kawai, T. Simizu, and H. Chiba, J. Pharm. Soc. Japan 76:660, 1956; C. A. 51:1067, 1957.

153. J. L. Finar, Organic Chemistry 1:596, 1960.

154. R. Stroh, J. Ebersberger, H. Haberland, and W. Hahn, Angew. Chem. 69:125, 1957.

155. G. Ecke, J. Napolitano, and A. Kolka, J. Org. Chem. 21:711, 1956.

156. L. N. Nikolenko and K. K. Babievskii, Zhur. Obshchei Khim. 25:2231, 1955.

157. L. N. Nikolenko and K. K. Babievskii, Zhur. Obshchei Khim. 28:238, 1958.

158. E. N. Matveeva, E. I. Kirillova, and N. P. Lazareva, Collection of Annotations, Moscow, Scientific Research Institute for Technico-economic Research in Chemistry, State Committee of the Soviet Ministry, USSR, 1961, p. 25.

159. T. Zincke, Ann. 400:27, 1913.

160. German Patent 733031, 1938.

161. Canadian Patent 495049, 1953; Referat. Zhur. Khim. 1955, 53565.

162. Great Britain Patent 711122, 1954; Rubb. Abs. 32:463, 1954; Referat. Zhur. Khim. 1956, 33951.

163. Great Britain Patent 744875, 1956.

164. U. S. Patent 2822404, 1958; Referat. Zhur. Khim. 1960, 90754.

165. U. S. Patent 2999394, 1959; C. A. 53:18532, 1959; Referat. Zhur. Khim. 1959, 29732.

166. U. S. Patent 2862909, 1958; C. A. 53:76706, 1959; Referat. Zhur. Khim. 1961, 7P 230.

167. Swiss Patent 348163, 1960; Zbl. 1961, 6706.

168. German Democratic Republic Patent 17984, 1959; Referat. Zhur. Khim. 1961, 9L 146.

169. U. S. Patent 2877210, 1959; C. A. 53, 15622, 1959; Referat. Zhur. Khim. 1960, 37185.

170. Great Britain Patent 774875, 1956.

171. U. S. Patent 2968630, 1960.

172. U. S. Patent 2944086, 1960; Referat. Zhur. Khim. 1961, 16M 325.

173. German Federated Republic Patent (D.A.S.) 1044808, 1958; Zbl. 1960, 13868.

174. U. S. Patent 2862976, 1958; Referat. Zhur. Khim. 1961, 7L 138.

175. U. S. Patent 2816945, 1957; Referat. Zhur. Khim. 1959, 59330.

176. U. S. Patent 2919294, 1959; C. A. 54:8739, 1960.

177. U. S. Patent 2858342, 1958; Referat. Zhur. Khim. 1961, 9L 139.

178. U. S. Patent 2625521, 1953; Ekspress-Inform. SVM, No. 10:132, 1961.

179. M. E. Mc Greal, V. Neiderl, and J. Niederl, J. Am. Chem. Soc. 61:345, 1939.

180. U. S. Patent 2883365, 1959.

181. E. Perotti and G. Castelfrauchi, Chim. Ind. 42:854, 1960; Res. Rub. Plast. No. 5:529, 1961.

182. U. S. Patent 2628212, 1953.

183. U. S. Patent 2877209, 1959; C. A. 53:11874; Referat. Zhur. Khim. 1960, 20336.

184. U. S. Patents 2883364 and 2883365, 1959; C. A. 53:15621, 1959; Referat. Zhur. Khim. 1960, 86938; Zbl. 1961, 3838.

185. J. Mleziva, M. Lidarik, and S. Stary, Plaste Kautschuk 8:171, 1961; See also Khim. i Tekhnol. Polimerov No. 10:59, 1961.

186. USSR Patent No. 129814, 1960; C. A. 55:4443, 1961.

187. Great Britain Patent 842209, 1960; C. A. 55:4443, 1961.

188. Great Britain Patent 883391, 1961; Chem. Age 86:2206, 1961.

189. German Patent 467728, 1928; Zbl. 1929, 3129.

190. L. A. Kheifits, E. A. Simanovskaya, V. I. Belov, P. V. Ivanov, E. S. Shapiro, and M. Ya. Brainer, Maslob. Zhir. Prom. No. 6:35, 1957.

191. L. A. Kheifits, E. A. Simanovskaya, and V. N. Belov, Neftekhimiya 3:284, 1958.

192. L. A. Kheifits, L. M. Shulov, E. V. Broun, and V. N. Belov, Zhur. Obshchei Khim. 30:672, 1960.

193. U. S. Patent 2831897, 1958; C. A. 52:16293, 1958.

194. Great Britain Patent 790096, 1958; C. A. 52:15044, 1958.

195. Great Britain Patent 719101, 1954; Referat. Zhur. Khim. 1956, 5188.

196. U. S. Patent 2792428, 1953; Zbl. 1961, 17414.

197. K. Auwers, Ber. 40(II):2524, 1907.

198. Seto Shoji and Horiuchi Hikaru, J. Chem. Soc. Japan, Ind. Chem. Sect. 62:1921, 1959; Referat. Zhur. Khim. 1961, 2J 34.

199. U. S. Patent 2526755, 1960; C. A. 45:1338, 1951.

200. U. S. Patent 2976325, 1961; C. A. 55:16484, 1961.

201. U. S. Patent 2967774, 1961; C. A. 55:7872, 1961.

202. Canadian Patent 509243, 1955; Referat. Zhur. Khim. 1956, 63086.

203. Great Britain Patent 758973, 1956; Referat. Zhur. Khim. 1959, 62849.

204. U. S. Patent 2819329, 1958; Referat. Zhur. Khim. 1960, 71785.
205. U. S. Patent 2841627, 1958; Referat. Zhur. Khim. 1960, 29124.
206. U. S. Patent 2905737, 1959; Referat. Zhur. Khim. 1961, 14P 377; C. A. 1961, 2464.
207. U. S. Patent 2929849, 1960; Referat. Zhur. Khim. 1961, 14L 160.
208. U. S. Patent 2929848, 1960; C. A. 54, 14195, 1960.
209. H. L. Bender, A. G. Farnham, J. W. Guyer, F. N. Apel, and T. B. Gibb, Ind. Eng. Chem. 44:1619, 1952.
210. S. R. Finn, G. J. Lewis, and N. J. Megson, J. Soc. Chem. Ind. 69:129, 1950.
211. F. Ullman and K. Brittner, Ber. 42:2539, 1909.
212. Akira Misonou, Jkuei Ogata, and Tsutomu Kuwata (University of Tokyo), Yukagaku 6:325, 1957; C. A. 55, 4406, 1961.
213. A. C. Davis, B. T. Hayes, and R. F. Hunter, J. Appl. Chem. No. 7:312, 1953; Referat. Zhur. Khim. 1954, 28756; 1957, 521.
214. H. Kämmerer and H. Lenz, Kunststoffe 51:26, 1961.
215. P. A. Jenkins, J. Chem. Soc. 1957, 2729; C. A. 51:16476, 1957.
216. Great Britain Patent 417944, 1934; C. A. 29:1536, 1935.
217. U. S. Patent 2822413, 1958; Referat. Zhur. Khim. 1960, 67737.
218. U. S. Patent 2976324, 1961; C. A. 55:16484, 1961.
219. U. S. Patent 2776998, 1957; Referat. Zhur. Khim. 1959, 76929.
220. U. S. Patent 2849516, 1958; Referat. Zhur. Khim. 1960, 7310.
221. Great Britain Patent 796285, 1958; C. A. 53:5748, 1959.
222. A. M. Kuliev and I. M. Orudzheva, Transactions of the First Conference of Zakavkaz'ye University, Baku, 1959, p. 111; Referat. Zhur. Khim. 1961, 12M 225.
223. Austrian Patents 203209, 203210, 203211, 1957; Zbl. 1960, 3056.
224. Great Britain Patent 712390, 1954; Referat. Zhur. Khim. 1956, 37562.
225. Great Britain Patent 8588909, 1961; C. A. 55, 15420, 1961.
226. Great Britain Patent 856447, 1960; Referat. Zhur. Khim. 1961, 19P 197.
227. Austrian Patents 202773, 202774, 1957; Zbl. 1960, 3056.
228. Great Britain Patent 731129, 1955; C. A. 50:5746, 1956.
229. F. Dunning and B. Dunning, J. Am. Chem. Soc. 53:3466, 1931.
230. German Patent 563643, 1932.
231. U. S. Patent 2481619, 1949.
232. W. Hawkins, M. Worthington, and W. Matreyek, J. Appl. Polymer Sci. 3:277, 1960; Referat. Zhur. Khim. 1961, 19P 125.
233. U. S. Patent 2982756, 1960; Offic. Gaz., U. S. Pat. Office, 766, 224, 1961.
234. French Patent 1159299, 1958.
235. U. S. Patent 2887604, 1959; Referat. Zhur. Khim. 1960, 37186.
236. U. S. Patent 2393889, 1946.

237. Great Britain Patent 712100, 1954.

238. Great Britain Patent 753740, 1956.

239. U. S. Patent 2734808, 1956.

240. Great Britain Patent 760315, 1956.

241. Great Britain Patent 771063, 1957.

242. Great Britain Patent 774345, 1957.

243. Great Britain Patent 804113, 1958; Referat. Zhur. Khim. 1960, 78329.

244. Great Britain Patent 812467, 1959.

245. Great Britain Patent 817142, 1959; Referat. Zhur. Khim. 1961, 9L 166.

246. U. S. Patent 2902466, 1959; C. A. 53:23048, 1959; Referat. Zhur. Khim. 1961, 8P 519.

247. U. S. Patent 2883362, 1959; C. A. 53:15620, 1959.

248. U. S. Patent 2929797, 1960; C. A. 54:23403, 1960; Referat. Zhur. Khim. 1961, 18P 271.

249. German Federated Republic Patent 1077667, 1960; Referat. Zhur. Khim. 1961, 18L 132.

250. U. S. Patent 2969394, 1961; C. A. 55:16480, 1961.

251. Great Britain Patent 856286, 1960; C. A. 55:16480, 1961.

252. Great Britain Patent 860923, 1961; C. A. 55:16480, 1961.

253. U. S. Patent 2692288, 1954.

254. U. S. Patent 2897177, 1959; C. A. 54:412, 1960; Referat. Zhur. Khim. 1961, 2P 326.

255. R. G. Rice and E. J. Kohn, J. Am. Chem. Soc. 77:4052, 1955.

256. C. Ainsworth, J. Am. Chem. Soc. 78:1635, 1956.

257. J. Horyna and C. Cerny, Collection Czech. Chem. Commun. 21:906, 1956.

258. J. Ball and J. Bourgeois, Chim. Ind. 68:522, 1952.

259. Great Britain Patent 794073, 1958.

260. U. S. Patent 2875175, 1959.

261. U. S. Patent 2776994, 1957.

262. U. S. Patent 2419739, 1947.

263. U. S. Patent 2530769, 1950.

264. Great Britain Patent 687532, 1953.

265. German Federated Republic Patent 859973, 1953.

266. Great Britain Patent 797138, 1958; C. A. 53:2676, 1959; Referat. Zhur. Khim. 1959, 48040.

267. Great Britain Patent 675585, 1953; C. A. 47:1420, 1953.

268. U. S. Patent 2664407, 1953; Referat. Zhur. Khim. 1956, 5191.

269. Org. Syn. 28:49, 1948.

270. U. S. Patent 2748100, 1956; Referat. Zhur. Khim. 1958, 23429.

271. U. S. Patent 2846435, 1958; C. A. 53:1802, 1959.

272. Australian Patent 167072, 1956; Referat. Zhur. Khim. 1957, 13431.

273. D. Craig, J. Am. Chem. Soc. 60:1458, 1938.

274. H. W. Kilbourne and G. R. Wilder, Rubber Chem. Technol. 32:1155, 1959; Referat. Zhur. Khim. 1960, 71765.

275. A. L. Midzhoyan, A. L. Yaroyan, and A. S. Azopyan, Izvest. Akad. Nauk Arm.SSR, Khim. Nauk. 13:287, 1960; Referat. Zhur. Khim. 1961, 21J 162.

276. U. S. Patent 2988533, 1960.

277. USSR Patent No. 132640, 132814, 1960.

278. E. N. Matveeva and F. Yu. Rachinskii, Plast. Mass. No. 2:12, 1961.

279. U. S. Patent 2419354, 1947.

280. B. Hunter, Ind. Eng. Chem. 46:1524, 1957.

281. B. Peters, Rev. Gen. Caoutchouc 34:1233, 1957.

282. U. S. Patent 2866807, 1959; C. A. 53:12240, 1959.

283. U. S. Patent 2877259, 1959.

284. R. Dumon, Ind. Plastiques Mod. (Paris) 12:16, 1960.

285. U. S. Patent 2733226, 1956; Referat. Zhur. Khim. 1957, 49427.

286. Great Britain Patent 803081, 1958; C. A. 53:12748, 1959.

287. U. S. Patent 2867594, 1959.

288. French Patent 1176735, 1959.

289. U. S. Patent 2841607, 1958.

290. German Federated Republic Patent 950152, 1956.

291. Canadian Patent 517031, 1955.

292. U. S. Patent 2816876, 1957.

293. U. S. Patent 2717902, 1955; Referat. Zhur. Khim. 1957, 17282.

294. Great Britain Patent 803557, 1958.

295. USSR Patent No. 141584, 1961.

296. Great Britain Patent 873697, 1961; Rubb. Abs. 39:6035, 1961.

297. W. Kunz, Ber. 27:2559, 1894.

298. L. Anschütz and N. Walbrecht, J. Prakt. Chem. 183:65, 1932.

299. A. E. Arbuzov and F. G. Valitova, Izvest. Akad.Nauk SSSR, Otdel. Khim. Nauk p. 529, 1940.

300. V. M. Plets, Organic Phosphorus Compounds, Moscow, State Press for the Defense Industry, 1940.

301. A. E. Arbuzov, Collected Works, Moscow, Academy of Sciences USSR Press, 1952.

302. P. A. Kirpichnikov, A. S. Kuz'minskii, L. M. Popova, and V. P. Spiridonova, Transactions of Kirov Kazan' Chemicotechnological Institute, No. 30, 1961.

Chapter IV

AGING AND STABILIZATION OF POLYOLEFINS

Among polymer materials, polyolefins –polyethylene, polypropylene, and copolymers of ethylene and propylene – are acquiring special significance at the present time. These polymers possess high mechanical strength, low density, flexibility at low temperatures, high impact strength, moisture stability, splendid electrical insulation properties, and a number of other properties. However, just like most other high-molecular compounds, polyolefins are subject to oxidative-destructive processes under the influence of atmospheric conditions, increased temperature, light, aggressive media, and a number of other factors. During the process of aging, polyolefins lose their elasticity, become brittle, crack, lose their mechanical strength, dielectric properties, in most cases change color, etc.

In spite of this, the literature contains a limited number of works devoted to the study of the aging or destruction of polyolefins. This apparently is a result of the fact that the processes of destruction of low-molecular analogs of polyolefins are comparatively well studied, and researchers almost entirely transfer the known principles to the processes of destruction of polyolefins. We cannot agree with N. Grassi [1], who, in his monograph "Chemistry of Processes of Polymer Destruction," writes "... A fundamental study of the reactions leading to the destruction of polymers would be of invaluable aid; however, economic considerations make it necessary to eliminate the existing shortcomings of the polymers as rapidly as possible. A stabilization of polymers is frequently accomplished without any clear idea of the nature of the inhibited reaction or of the mechanism of the action of the successfully selected stabilizer."

The data available in the literature, mostly published in the last decade, pertain chiefly to the aging of high-pressure polyethylene, and to a lesser degree to polyolefins produced on complex catalysts. There are practically no data on the destruction of copolymers.

Even the first works [1-6] showed that during the aging process the polyolefin is mainly subject to oxidative destruction, which is accelerated

TABLE 3. Variation of the Characteristic Viscosity of Polypropylene
During Reprocessing and Use at Increased Temperatures

Initial powder	Granulation, 210°C	Casting under pressure, 220°C	Use at 150°C for	
			10 hr	30 hr
4.0	3.08	2.8	1.0	0.5

under the action of light. Processes of purely thermal decomposition
play a lesser role, and the polymer in the absence of oxygen is stable in
practice for prolonged periods of time at temperatures up to 300°C, while
oxidative processes proceed at appreciable rates at temperatures close
to 100°C. As a result of this, the investigation of the aging of polyolefins
has developed mainly along the line of the study of oxidation and destruction processes under the action of ultraviolet illumination.

This chapter cites data characterizing the processes of oxidation and
photodestruction of polyolefins of the series polyethylene—copolymers
of ethylene and propylene—polypropylene, and methods of their stabilization against the destructive action of oxygen and light.

I. OXIDATIVE DESTRUCTION OF POLYOLEFINS

The investigation of the oxidation of polyolefins is conducted mainly
under conditions approaching the conditions of their reprocessing, although the oxidation, for example, of polyethylene already proceeds at
room temperature [7]. The usual methods of reprocessing are pressing
at temperatures 10-15° higher than the temperature of transition to the
viscous-fluid state, casting under pressure at 200-250°C, and extrusion.
Conducting the reprocessing in air sharply reduces the properties of the
material. Thus, for example, the characteristic viscosity of isotactic
polypropylene drops sharply, as can be seen from Table 3, during the
process of granulation, casting, and further use at 150°C in air.

One of the methods for testing polymers for stability to oxidation is
rolling at high temperatures in an atmosphere of air. This method of
testing, which to a certain degree combines various methods of reprocessing, showed even in the first studies that in addition to a deterioration of the mechanical properties and a drop in the molecular weight in
the polymer, an increase in the concentration of carbonyl groups is observed [1], and this leads to a proportional increase in the dielectric
losses [2].

It is natural that rolling could not be a universal method of investigation. Hence most investigations use varied methods, based on a study
of the absorption of oxygen in a closed system. This makes it possible

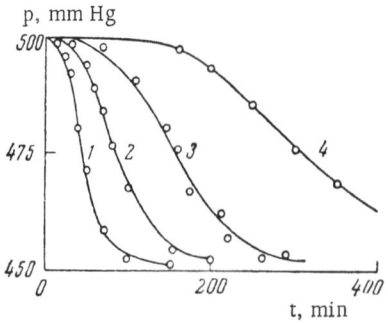

Fig. 45. Variation of the oxygen pressure in the oxidation of polypropylene at a temperature of 150° (1), 140° (2), 130° (3), and 120°C (4).

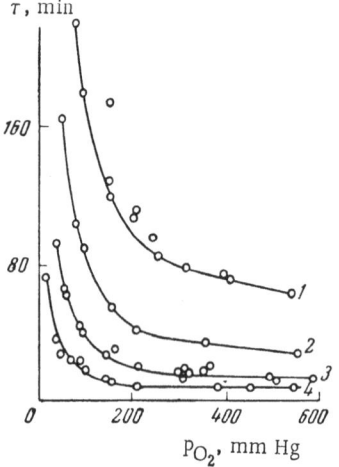

Fig. 46. Dependence of the induction period of the oxidation of polypropylene on the pressure at a temperature of 120° (1), 130° (2), 140° (3), and 150°C (4).

not only to follow the change in the chemical composition of the polymer, but also to isolate and determine the volatile reaction products.

The absorption of oxygen by the polymer, as Fig. 45 shows, proceeds with an appreciable induction period, when the rate of absorption of oxygen is extremely small [8]. Hence the value of the induction period can be used as a characteristic of the rate of the oxidation process.

The value of the induction period τ is sharply reduced with increasing temperature and change in the oxygen pressure [9]. Such a dependence is presented in Fig. 46. It is characteristic that a sharp variation of the value of τ as a function of the oxygen pressure is observed only at pressures below 150-200 mm Hg. For such pressures the reciprocal of the induction period $1/\tau$ is also proportional to the average rate of the initial stage of the reaction, directly proportional to the pressure ($\tau \cdot P_0$ = const). A further increase in the pressure (to 120 atm in [10]) has very little influence on the variation of the induction period. After the end of the induction period the absorption of oxygen occurs according to a curve possessing a sharply pronounced autocatalytic character.

In most cases the physicomechanical properties and molecular weight of the polymer change little during the induction period. A sharp change in the properties is observed in the region of autocatalysts. The nature of the change in properties during the oxidation process becomes understandable from Fig. 47, which cites the dependence of certain quantities on the time of oxidation of low-pressure polyethylene [11]. It is characteristic that the content of carbonyl groups in the polymer increases especially sharply after the end of the induction period. As was shown in [2], it is precisely the increase in the concentration of carbonyl com-

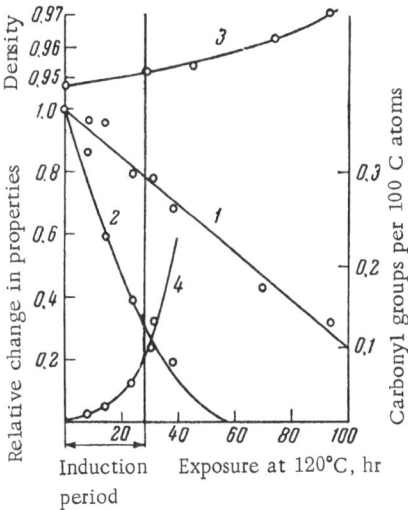

Fig. 47. Variation of the characteristic viscosity (1), relative flexibility of the chains (2), density (3), and content of carbonyl groups in the oxidation of polyethylene in air; T = 120°C.

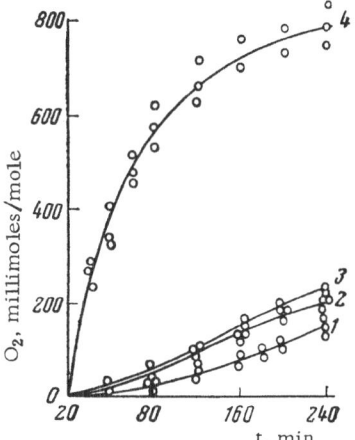

Fig. 48. Absorption of oxygen by various polyolefins at 150°C. 1) High-pressure polyethylene; 2) low-pressure polyethylene; 3) copolymer of ethylene and propylene; 4) polypropylene. Initial oxygen pressure 600 mm Hg.

pounds that leads to a sharp increase in the tangent of the angle of dielectric losses. Thus, the accumulation of $\diagup C = O$ groups to 0.05% by weight leads to a 10-fold increase in tan δ. In view of this, there are indications in the literature that the quantity tan δ can be a measure of the content of carbonyl compounds. More profound oxidation leads to the formation of the −OOH, −OH, −COOH groups [12]; in addition to C = O groups, the formation of various types of C−O groups is observed [13].

The process of polyethylene oxidation is characterized by structuring − the formation of cross-linked structures [12]. Structuring is usually manifested at increased temperatures. Thus, the oxidation of a fusion of polyethylene at 180°C leads to the accumulation of insoluble products: after 30 hr, about 50% of them accumulate; after 61 hr, 90%; after 150 hr, 100% by weight. Under the conditions of structuring, the viscosity of the polymer first drops rapidly during the oxidation process, and then rises as insoluble structures accumulate.

As is shown by spectroscopic investigations, the process of structuring is accompanied by the formation of transverse ether C − O − C bonds. The presence of such bonds is also demonstrated by the fact that the insoluble cross-linked structures are subject to acid and alkaline hydrolysis, as a result of which they begin to dissolve. The absorption spectra in the infrared region show that the C − O − C bonds disappear upon hydrolysis.

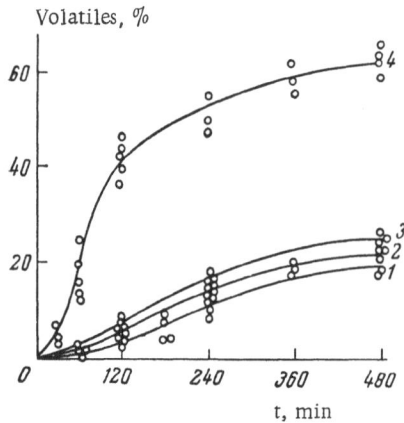

Fig. 49. Accumulation of volatile low-molecular products during the process of oxidation of polyolefins at 150°C. 1) High-pressure polyethylene; 2) low-pressure polyethylene; 3) copolymer; 4) polypropylene.

There are indications [14] that the oxidation of low-pressure polyethylene begins with the amorphous portions of the material, which are more porous and hence more accessible for the penetration of the oxygen in comparison with the crystalline portions. Moreover, as a result of inhomogeneous destruction in the mass of the polymer, a sharp change in the mechanical properties is observed, with the chemical properties practically preserved. It is characteristic that the remelting of samples of the polymer leads to complete restoration of the initial physicomechanical properties. It has been shown for the example of isotactic polypropylene [15] by the method of nuclear magnetic resonance that amorphization of the polymer occurs during the process of oxidative destruction. However, at the present time it is still difficult to say how the degree of crystallinity or amorphousness of the polymer influences its tendency to be oxidized. Considerable influence on the tendency toward oxidation is exerted by the structure of the polymer molecule. Thus, it has been noted [16] that at the early stages the oxidation of branched polymers occurs considerably more rapidly than that of linear polymers. This is explained by the rule known from the chemistry of hydrocarbon oxidation, according to which the tendency of various groups in the hydrocarbon molecule toward oxidation drops in the series $CH > CH_2 > CH_3$.

A clear confirmation of this is an investigation [17] of the comparative oxidation of polyolefins in the series polyethylene–copolymer of ethylene with propylene–polypropylene. In this study, high- and low-pressure polyethylene, the copolymer (CEP) containing 14 mol.% propylene, and polypropylene, in which 77% was made up of isotactic fractions, were used. The absorption of oxygen, quantitative composition of the low-molecular volatile reaction products, and kinetics of the change in the physicochemical properties of the polymer during oxidation were measured. For accumulation of the reaction products, the oxidation process was carried out under conditions of circulation of oxygen, which made it possible to draw off the volatile reaction products from the reaction vessel and concentrate them. The investigation was conducted in the temperature range 120-170°C at an oxygen pressure of 760 mm Hg.

TABLE 4. Properties of Polyolefins after Oxidative Treatment at 150°C

Indices	Polypropylene			Copolymer			Polyethylene					
							High-pressure			Low-pressure		
					Time of oxidation, hr							
	0	4	8	0	4	8	0	4	8	0	4	8
Insoluble portion, %	—	1-2	1.4-2.6	—	9.3-13	37-40	—	3.5-4.3	3.5-6.0	—	23-26	48-58
Characteristic viscosity of soluble portion	2.95	0.1	0.2	2.55	0.07-0.1	0.15-0.2	1.0	0.15-0.2	0.15-0.22	3.30	0.1	0.3
Total oxygen, %	—	19.6-20.6	23.6	—	12.9-13.6	12.7-14.5	—	11.1-12.6	13.3-14.2	—	13.8-15.4	15.2-16.5
Peroxide, active oxygen, %	—	0.26	—	—	0.16	0.04	—	0.5	0.06	—	0.02	0.02
Carbonyl compounds, CO, %	0.03	5.2	3-4	0.02	2-2.5	3.2-3.5	0.1	3.0-4.3	2.4-3.7	0.01	2.8-3.4	2.4-2.9
Carboxyl compounds, COOH, %	—	3.2-4.2	3.34	—	2.7-2.83	1.2-1.4	—	3.3-3.6	4.6-5.4	—	3.0-3.5	2.8-3.0
Unsaturated compounds, bromine number	—	9.8	—	—	6.6	2.8-3.0	—	2.8-3.8	2.7	—	2.6-3.0	2-2.85

The absorption of oxygen by polyolefins for one of the investigated temperatures under comparable conditions is presented in Fig. 48. From the figure it is distinctly evident that the oxidation of polypropylene proceeds incomparably more rapidly than the oxidation of polyethylene and the copolymer. The values of the activation energy of the oxidation process were calculated on the basis of the kinetic curves of the absorption of oxygen at various temperatures. The activation energy proved to be equal to 21.8 kcal/mole for polypropylene, 30.8 kcal/mole for the copolymer, 31.9 kcal/mole for low-pressure polyethylene, and 32.7 kcal/mole for high-pressure polyethylene. This means that the tendency of the polyolefins toward oxidation decreases in the indicated sequence.

As can be seen from Fig. 49, the accumulation of volatile low-molecular reaction products varies in the same sequence at the same temperature. A large amount of volatile products is a characteristic feature of polypropylene and the copolymer, which possess a substantially greater branched character of the molecular chains. It is indicative that at 150°C the weight of a sample of polypropylene begins to drop sharply after only 30-40 min of oxidation, while for low-pressure polyethylene this phenomenon is observed only at 170°C and after 2.5-3 hr. All this gives evidence of the greater tendency of polypropylene toward oxidation and the more profound occurrence of this process.

As can be seen from Table 4, the properties of polyolefins change sharply as a result of oxidative destruction. The content of carbonyl, carboxyl groups, and unsaturated compounds increases. The content of insoluble (cross-linked) structures increases sharply. Structuring is manifested especially strongly in high-pressure polyethylene, least of all in polypropylene. This fact is weighty evidence that oxidation leads to structuring in the case of polyethylene, while it leads mainly to destruction to low-molecular products in the case of polypropylene.

The volatile decomposition products of all the investigated polyolefins, as can be seen from Table 5, possess practically the same qualitative composition. Only the quantitative ratio varies for various polymers. Thus, in four hours of oxidation of polypropylene at 150°C, 15 times as much acid is formed as in the oxidation of low-pressure polyethylene, 13 times as much formaldehyde, and 6 times as much aldehyde. Increasing the temperature leads to an increase in the yield of volatile products, the content of acids and carbonyl compounds mainly increasing. It is characteristic that the basic volatile destruction product is water.

At the present time, attempts are being made to elucidate the detailed mechanism of the process of polymer oxidation. However, as a result of the complexity of the system, and the absence of reliable methods of analyzing high-molecular reaction products, this work is being

TABLE 5. Mixture of Volatile Oxidation Products of Polyolefins at 150°C (in millimoles per mole of the monomer)

Reaction products	Polypropylene		Copolymer		Polyethylene			
					High-pressure		Low-pressure	
	Time of oxidation, hr							
	4	8	4	8	4	8	4	8
Acids, considered on the basis of COOH	110-130	125-135	12-15	16-19	9-12	18-20	7-9	9-12
Esters, considered on the basis of $OCOCH_3$	35-36	35-45	None	2.1-3.2	None	None	None	None
Unsaturated compounds (bromine number)	40-42	37-38	1.1-1.5	None	3.0-3.6	None	2.0-2.6	None
Peroxides, considered on the basis of active oxygen	4-5	2.2-2.5	0.3-0.35	0.3-0.37	0.6-1.0	1.0-1.1	0.5-0.6	0.4-0.5
Carbonyl Compounds:								
Formaldehyde	15-25	28-33	1.6-2.1	3.5-4.0	1.9-2.4	3.4-4.0	1.2-1.8	1.8-2.6
Acetaldehyde	5-8	8-11	1.4-1.8	1.8-2.3	1.4-1.5	1.2-1.8	0.9-1.4	0.5-0.8
Water	670-750	620-650	200-250	380-400	150-200	200-230	190-240	290-305
Carbon dioxide	100-120	—	25-35	—	13-15	—	20-25	—
Carbon monoxide	40-50	—	6-9	—	3-5	—	6-10	—
Hydrogen	17-20	—	—	—	—	—	3.7-4.3	—

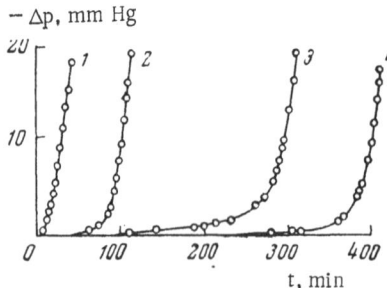

Fig. 50. Influence of antioxidants on the kinetics of the oxidation of poly-propylene. 1) Unstabilized poly-propylene; 2) with an addition of benzidine; 3) with an addition of di-phenylamine; 4) with an addition of neozone D; T = 140°C, initial O_2 pressure 300 mm Hg; concentration of antioxidants 0.003 mole/kg.

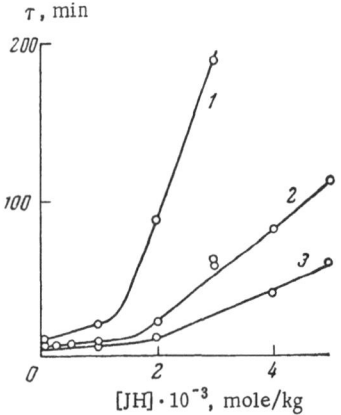

Fig. 51. Dependence of the induction period of polypropylene on the con-centration of 2,2'-methylene-bis-(4-methyl-6-tert-butylphenol) at various temperatures (1 – 190°; 2 – 200°; 3 – 210°C). Initial oxygen pressure 300 mm Hg.

developed extremely slowly. Re-searchers are resting mainly on the analysis of volatile reaction products and on establishing an interrelation-ship among the kinetic behaviors of the latter. In very many cases, research-ers are turning to models, which prac-tically always are liquids, as a result of which the principles found for them cannot always be transferred to oxida-tion in the solid phase. Analysis of the reaction products shows that the prim-ary oxidation product represents per-oxide compounds [1, 2, 18]. Moreover, hydroperoxides are chiefly formed [19].

A detailed quantitative investiga-tion of the kinetics of the accumulation of hydroperoxides and other acid-con-taining oxidation products of polypropyl-ene was the subject of [20], in which it was shown that at the initial stage of the reaction the amount of hydroper-oxides formed corresponds almost ex-actly to the amount of oxygen absorbed by the polymer. Moreover, as can be seen from the work, the kinetic curve of the variation of the rate coincides with the curve of hydroperoxide ac-cumulation. The rates of formation of oxygen-containing volatile reaction products (such as water, aldehydes, acetone, alcohols, etc.) during the oxidation process coincide and are de-scribed by parallel kinetic curves. In [20], the curves of the rates of forma-tion of the reaction products were trans-formed in such a way that the experi-mental data for various products lay on one curve. The rates of formation of the same reaction products in the thermal decomposition of polypropylene hydroperoxide coincide anal-ogously, as is shown in Fig. 12. This is clear evidence that the main volatile oxygen-containing products are formed from hydroperoxides.

Considering the processes of oxidation of polyolefins as a whole, we should indicate that these are radical chain processes, proceeding with branching (see Chapter I).

The kinetics of the formation of the decomposition products of the hydroperoxide of isotactic polypropylene (chief of which is water) in the pure polymer and in the polymer containing a large concentration of cyclohexylphenyl-p-phenylenediamine, was investigated in [21]. Cyclohexylphenyl-p-phenylenediamine, which possesses a positive hydrogen atom in the $-NH-$ group, reacted energetically with the hydroperoxide. It was found that the composition of the products in both cases was qualitatively the same. On the basis of this, the authors conclude that the following reaction is accomplished at the temperatures 120-150°C:

$$ROOH + RH \rightarrow H_2O + \overset{\bullet}{R}O + R^{\bullet} \tag{1}$$

analogously to the reaction of the hydroperoxide with an amine:

$$ROOH + HNR_2' \rightarrow H_2O + R\dot{O} + \dot{N}R_2' \tag{2}$$

II. STABILIZATION OF POLYOLEFINS
AGAINST OXIDATIVE DESTRUCTION

The oxidation of a polyolefin, as we have seen, leads to the formation of a strong oxidizing agent and ROOH, by means of which a branched chain process is effected. Since the reaction proceeds with the participation of free radicals, it is natural to use inhibitors (antioxidants) to suppress the development of chains. Hence, compounds of the type AH, which have a labile hydrogen atom: aromatic amines, substituted (shielded) phenols, mercaptans, etc., have long been used to stabilize polyolefins, just as, in general, for inhibiting the oxidation of hydrocarbons. It is assumed that these inhibitors react with the radicals that carry the oxidation chains, according to the general scheme:

$$R\dot{O}O + AH \rightarrow ROOH + \dot{A} \tag{3}$$

In this case an inhibitor radical, incapable of continuing the oxidation chain, is formed (see Chapter I).

On the other hand, a distinct tendency for the utilization of compounds possessing no labile hydrogen as antioxidants: organic (aliphatic and aromatic) sulfides, disulfides, esters of phosphorous and boric acids, etc., has recently been noted. In the case of compounds of this type, it is assumed that their basic role reduces to a directed decomposition of hydroperoxides, during the process of which free radicals generally are not formed, or intermediate compounds (including radicals), which cannot take part in the branches of the oxidation chain, are formed. At the present time the mechanism of the inhibiting action of such nonhydrogen-containing antioxidants is not exactly known.

Since the antioxidant terminates the kinetic chains during oxidation of a polymer (reacts with ROO˙ or decomposes hydroperoxides, thereby nullifying the possibility of chain development), the external manifestation of its influence on the process consists of lengthening the period during which the reaction proceeds before autocatalytic acceleration; in this case the antioxidant is consumed. At the end of this period, the process of polymer oxidation develops with autoacceleration, at practically the same rates as in the absence of the antioxidant, which is well illustrated by Fig. 50 [22].

Frequently no appreciable absorption of oxygen is observed at the initial stage of the development of the reaction (before autocatalysis), as in the case of the addition of benzidine (curve 2 in Fig. 50) and neozone D (curve 4), or diphenylamine (curve 3), and especially sulfur-containing compounds; slow absorption of oxygen occurs during this period. It is assumed that in the latter case the basic role of the antioxidant reduces to the decomposition of hydroperoxides, which accumulate as a result of the reactions of the steady-state process of development of oxidative chains.

In the first case the induction period of the oxidation, defined in most experimental work as the time during which the oxygen pressure in the system decreases by a quite definite small quantity (1-5 mm Hg), coincides with the beginning of the autocatalytic acceleration of the reaction; in the second case it is always smaller. As a result of this, an inhibition of the oxidation process is observed after the end of the induction period.

The induction period increases with increasing antioxidant concentration. Figure 51 taken from [23], shows the dependence of the induction period on the concentration of 2,2-methylene-bis-(4-methyl-6-tert-butylphenol). It is characteristic that in the region of low concentrations practically no inhibiting action of the compound is observed. Then, when a definite value, called the critical concentration, is reached, the induction period begins to increase sharply. The position of the point of inflection of the curve $\tau = f[AN]$ is displaced into the region of high concentrations with increasing temperature and depends on the chemical nature of the antioxidant. For a number of antioxidants, no critical concentration generally was observed. In the region of high antioxidant concentrations $(10^{-2}-10^{-1}$ mole/kg), the curve of the function $\tau = f[AN]$ approaches a limiting value; sometimes it passes through a maximum and begins to decrease. The decrease in the value of τ at large antioxidant concentrations is explained by certain authors [24] by an initiating action of the inhibitor radicals.

As a result of the fact that the introduction of antioxidants leads to a very great lengthening of the induction period, the physicomechanical properties of the polymer are preserved for a long time, practically until complete consumption of the antioxidant. Hence the basic method of evaluating the effectiveness of stabilizers is the determination of the rate of absorption of oxygen or the induction period of the reaction. We should mention, however, that sometimes the small rate of oxygen absorption and the large induction period still are not responsible for the invariability of the basic physicomechanical indices of the polymer. Thus, it was shown in the work of one of the authors that the induction period of the oxidation of polypropylene at 200°C in the presence of 0.03 mole/kg 2,2'-dithio-bis-(4-methyl-6-tert-butylphenol) lasts for more than 1100 min. At the same time, the characteristic viscosity of the polymer drops from 4.0 to 0.5 after only 50 min of oxidation. Obviously some sort of interaction of the disulfide or its conversion products with the polymer occurs in such cases. This makes it necessary to investigate not only the oxygen absorption, but also the variation of other physical or mechanical indices of the polymer.

Noteworthy is the method of chemiluminescence [25], based on the registration of the weak luminescence that accompanies the process of oxidation of certain polymers, recently used to investigate the oxidation of polypropylene and the comparative effectiveness of antioxidants. The luminescence intensity increases in direct proportion to the oxygen concentration and decreases in the presence of various antioxidants, in proportion to their concentration. It has been proposed that this method be used for a rapid evaluation of compounds as antioxidants.

The use of antioxidants of various chemical classes for stabilizing polyolefins against oxidative destruction possesses specific features and its own limits of applicability, depending on the nature of the object into which the polymer is reprocessed. Hence we shall briefly characterize individual, most widely used classes of antioxidants.

Various Classes of Antioxidants

1. Amines and Stabilizers Based on Them

Practically all the secondary amines and derivatives of para-phenylenediamine used for the stabilization of raw and cured rubbers are suitable for the stabilization of polyolefins [26]. However, in this case we must consider that the amines color the polymer a blackish-brown color. Hence they can be used only in the preparation of technical objects, colored deep shades. The amines most widely used in the practice of stabilization of polyolefins are cited in Table 6. We should indicate that phenyl-β-naphthylamine, phenylcyclohexyl-p-phenylenediamine, and N,N'-di-β-naphthyl-p-phenylenediamine are the most effective.

TABLE 6. Amines Most Frequently Used for the Stabilization
of Polyolefins

Amines	Recommended concentration, moles/kg	Literature
Diphenylamine	0.01	[28, 33, 34]
Phenyl-α-naphthylamine (neozone A)		
Phenyl-β-naphthylamine (neozone D)	0.005	[35]
Mixture of neozone A and di-phenyl-p-phenylenediamine	(0.05 % by weight)	[3]
Diphenyl-p-phenylenediamine	0.003-0.005	[36]
N,N'-phenylcyclohexyl-p-phenylenediamine	0.005-0.008	[29, 36]
N,N'-di-β-naphthyl-p-phenylene-diamine	0.008	[36]

The use of these amines guarantees invariability of the physico-mechanical characteristics during the treatment of polyolefins in air at temperatures up to 250°C.

At the present time, attempts are being made to expand the circle of amines for practical use. Thus, it was shown in [27] and [28] that certain secondary amines of the thiophene series, the synthesis of which is more accessible than that of amines of the benzene series, possess good stabilizing properties (as applied to rubbers and polyolefins). More-over, the replacement of the phenyl radical in the molecule of phenyl-β-naphthylamine by the thenyl radical in a number of cases exerts a more favorable influence on the effectiveness of the amine. For example, 2-thenyl-β-naphthylamine is not inferior to neozone D in effectiveness, and sometimes surpasses it. The effectiveness of the amine is substantially increased if a hydroxyl group is introduced into the molecule. The in-troduction of an OH group into the molecule of diphenylamine [29] in-creases the effectiveness of the latter. An analogous picture is observed with amines of the thiophene series [28].

Characterizing the mechanism of the action of amines, the over-whelming majority of the authors believe that their basic role reduces to the termination of kinetic oxidation chains on account of reaction (3). Actually, stable radicals $(C_6H_5)_2NO^{\cdot}$ have been detected by the method of electron paramagnetic resonance as a result of the interaction with peroxide radicals, formed in the liquid-phase oxidation of a hydrocarbon inhibited with diphenylamine [30]. The interaction of certain secondary amines (Ar_2NH) with peroxide radicals, prepared by oxidative radiolysis,

was investigated by this same method in [31]. It was found that the reaction of ROO' with an amine proceeds according to reactions leading to
the formation of a diarylazotoxide radical:

$$Ar_2NH + ROO' \rightarrow ROOH + Ar_2N' \tag{4}$$

$$Ar_2N' + ROO' \rightarrow RO' + Ar_2NO' \tag{5}$$

Later it was revealed that the interaction of an amine with peroxide
radicals is not limited to these reactions. Diarylazotoxide radicals interact energetically with radicals R' and ROO', which has been confirmed by direct experiments. As the authors assume, this interaction
leads to the formation of stable products – derivatives of diarylhydroxyl-
amine:

$$R' + ON'Ar_2 \rightarrow RONAr_2 \tag{6}$$

$$ROO' + ON'Ar_2 \rightarrow RONAr_2 + O_2 \tag{7}$$

Thus, when amines are used, the oxidation chains can be terminated
not only on account of the reaction of hydrogen transfer according to reaction (4), but also by reactions of disproportionation (5) and cross recombination (7).

It was established quite recently [32] that stable azotoxide radicals
(especially shielded) possess stabilizing effectiveness against oxidative
destruction. It was shown for the example of the oxidation of polypropylene that the diphenylazotoxide radical is superior to diphenylamine in
effectiveness. Thus, reaction (4) proves to be unprofitable in the case
of amines in comparison with reaction (7), since it results in the formation of a branching agent – ROOH.

From this follows a rational method of increasing the effectiveness
of amines as antioxidants: their preliminary oxidation to azotoxide stable radicals and the use of the latter for stabilization purposes.

It was also shown recently that secondary amines (such as diphenylamine and N-phenyl-N'-cyclohexyl-p-phenylenediamine [21]) in the solid
phase are capable of reacting not only with radicals ROO', but also with
polymer hydroperoxide, forming water as the basic reaction product,
thereby greatly accelerating decomposition of the hydroperoxide:

$$ROOH + HNAr_2 \rightarrow H_2O + RO' + Ar_2N' \tag{8}$$

This reaction with diphenylamine proceeds even at room temperature,
which is evidenced by the rapid accumulation of stable radicals $(C_6H_5)_2N'O$.

TABLE 7. Phenols Used Most for the Stabilization of Polyolefins

Name	Literature
p-Hydroxyphenylcyclohexane	[33]
Di-p-hydroxyphenylcyclohexane	[33]
Dicresylolpropane	[2]
Mixtures of alkylated phenols	[34]
2,6-Di-tert-butyl-4-methylphenol	
2,4,6-Tri-tert-butylphenol	[35]
Condensation products of dialkylphenols with formaldehyde	[36]
Reaction products of phenol with styrene	[37]
1,1'-Methylene-bis-(4-hydroxy-3,5-tert-butylphenol)	
2,2'-Methylene-bis-(4-methyl-6-tert-butylphenol)	[40]
2,6-(2-Tert-butyl-4-methyl-6-methylphenol)-p-cresol	[38]
Phenylethylpyrocatechol and phenylisopropylpyrocatechol	[39]
2,2'-Thio-bis-(4-methyl-6-tert-butylphenol)	[41]
4,4'-Thio-bis-(3-methyl-6-tert-butylphenol)	[42]

Thus, secondary amines perform two functions during the oxidation process: they suppress chains and decompose peroxides. We might think that the decomposition of peroxides is not a secondary process, since it results in the formation of an effective inhibitor – the arylazot-oxide radical.

2. Phenols

Phenols are very widely used for stabilizing various polymer materials, thanks to the fact that they practically do not color the material at low temperatures and possess the lowest possible toxicity. The latter circumstance permits their use as stabilizers of food oils.

The basic requirements set for phenols in the stabilization of polymer materials are good compatibility with the polymer and a high boiling point. In the case of polyolefins, the former is achieved mainly as a result of aliphatic substituents in the aromatic ring, the latter by high molecular weight. In view of the latter, as can be seen from Table 7, which cites the substituted phenols used for the stabilization of polyolefins, a tendency of monoolefins to be replaced by bi- and tri-olefins is observed. Not only a decrease in the volatility of the antioxidant, but also an increase in its effectiveness is thereby achieved. We should indicate that the stabilizing effectiveness (according to the data of a number of authors) does not increase in proportion to the increase in the active functional groups of the compound.

An important factor determining the effectiveness of a substituted phenol is the structure of the substituent and its position in the molecule with respect to the OH group. It has been established [43] that phenols with substituents in the 2-, 4-, and 6-positions are the most effective. The introduction of an amino group, as has been indicated, increases the effectiveness of the phenol. The effectiveness of phenols increases especially greatly when the thioether group is introduced into the molecule [28], or in the formation of thio-biphenols [41]. Substituted phenols preserve the physicomechanical properties of the polymer during its reprocessing in air at 200-220°C and during use effectively and for a comparatively long time. However, at high temperatures and in the case of protracted use (especially in the light), the polymer acquires a weak coloration, which changes with time. The color intensity depends on the concentration of radicals of the polymerization catalyst. Coloring of the polymer can be avoided by using phenol sulfides; however, in this case the catalyst radicals exert a strong effect.

Since phenols have long been used as inhibitors, and in varied branches of industry, the problem of the mechanism of their action has been comparatively well studied and described in the literature. This problem is treated in especially great detail in the monograph [44].

It has been established that phenol donates a hydrogen atom from the OH group when it reacts with an active radical. A stable radical is formed in this case. Many phenoxyl-radicals have been synthesized by mild oxidation of phenols [45], which permitted an investigation of their properties and structure (this was done, for example in [25] and [47]). Solutions of phenoxyl radicals are colored, which is apparently partially responsible for the appearance of a color in polymers inhibited by phenols. Many studies are known devoted to the investigation of the conversion product of phenols. These investigations have made it possible to establish that the phenoxyl radicals formed from phenol can react with oxygen, forming peroxides of the type $A - O - O - A$ [48], and in the presence of high concentrations of peroxide radicals $ROO^{\cdot}$ – peroxides of the type $A - O - O - R$ [49]. Peroxides of this nature, just like phenoxyl radicals, represent colored compounds, stable at room temperature.

3. Sulfur-Containing Compounds

Of the compounds containing a sulfur atom in the molecule, mercaptans possess comparatively high inhibiting effectiveness. Hence, the use of mercaptobenzimidazole [50, 51], mercaptobenzothiazole, β-naphthylthiol, and certain aliphatic mercaptans, for example, dodecyl mercaptan [41], is recommended for polyolefins. However, to guarantee the desired stability to oxidation under the conditions of reprocessing of the polymer at temperatures of 200-250°C, a comparatively high concentration of such stabilizers (of the order of 1% by weight) is required, which greatly limits their use. Aliphatic mercaptans are viscous liquids; this hinders their introduction into the polymer. Mercaptobenzimidazole possesses limited compatibility with the polymer. It has been shown that the introduction of mercaptobenzimidazole into polypropylene to the extent of only 0.5% by weight has no effect on the properties of the polymer. At higher concentrations, the physicomechanical properties of polypropylene are sharply changed as a result of the poor solubility of the mercaptans.

With respect to the mechanism of their inhibiting action, mercaptans apparently are included among the group of compounds whose participation in the inhibition process begins with the stripping of a hydrogen atom. Thus, it has been established that the replacement of a hydrogen atom in the SH group of mercaptobenzimidazole by a methyl radical leads to a sharp drop in the inhibiting effectiveness in the process of oxidation of polypropylene, while an analogous substitution in the NH group has a considerably weaker effect.

In view of this the authors propose that the mechanism of the inhibiting action of mercaptans can be expressed by the scheme:

$$ROO^{\cdot} + R'SH \rightarrow ROOH + R'S^{\cdot} \tag{9}$$

$$2R'S \rightarrow R'SSR' \tag{10}$$

$$ROO^{\cdot} + R'S^{\cdot} \rightarrow \text{stable products} \tag{11}$$

The possibility that the processes occur in the indicated direction is confirmed by the generally known formation of disulfides from mercaptans during oxidative processes, as well as by the appreciable interaction of mercaptans with free radicals of the diphenylazotoxide type [52], leading to the destruction of the latter. In the work it was also shown that the replacement of the hydrogen atom in the SH group by an aliphatic radical leads to the appearance of such an interaction even in the presence of hydroperoxides, when the reaction proceeds especially rapidly.

In a number of cases aromatic thioethers can be used for the stabilization of polyolefins; however, in most cases their stabilizing effectiveness is low. Of the best investigated compounds, diphenyl sulfide, phenylbenzyl sulfide, di-β-naphthyl sulfide, methyl-β-naphthyl sulfide [41], sulfides of the thiophene series [28], etc., are known. In most cases amino- or phenol sulfides such as 2,2'-thio-bis-(4-methyl-6-tert-butyl-phenol), thio-bis-β-naphthol, thio-bis-(N-phenyl-2-naphthylamine) [41], etc., are used.

In most studies devoted to the investigation of the inhibiting action of organic sulfides, it is assumed that their basic role reduces to the decomposition of hydroperoxides. This concept was formulated in [53], where it was shown that sulfides react with the peroxides formed during oxidation and form sulfoxides:

$$RSR + R'OOH \rightarrow \underset{\underset{O}{\downarrow}}{RSR} + R'OH \tag{12}$$

The latter, also possessing inhibiting properties, are further oxidized by peroxides to sulfones:

$$RSOR + R'OOH \rightarrow RSO_2R + R'OH \tag{13}$$

The formation of sulfoxides and sulfones has been demonstrated experimentally. It has been revealed that reaction (12) proceeds more rapidly than the reaction of hydroperoxide formation during oxidation. This leads to a suppression of the chain branching.

The kinetics of the decomposition of a number of hydroperoxides under the influence of organic sulfides has been studied [54]. In many works devoted to this problem, it is assumed that the interaction of the sulfide with hydroperoxides proceeds without the formation of free radicals. However, it was established in [52] by the method of electron paramagnetic resonance that in the presence of hydroperoxides, most sulfides or other sulfur-containing compounds capable of reacting with hydroperoxides react with stable diarylazotoxide radicals. The authors assume that the reaction of the sulfide with the peroxide leads to the formation of an active radical of the thiyl type, which recombines with the azotoxide radicals.

There are indications [41] that a number of organic disulfides can be used as stabilizers of polyolefins. Many of them (for example, 1-dodecylsulfide [41], or 2,2'-dithio-bis-benzimidazole [56]) are more effective than the corresponding mercaptans. As a result of this, we might assume that the mechanism of the inhibiting action of disulfides reduces to their dissociation into two thiyl radicals, followed by termination of the oxidation chain according to reaction (11).

Of the other sulfur-containing compounds, certain alkylthiurammono- and disulfides [29, 55] and esters of thiodipropionic acid [57] can be used.

In most cases sulfur-containing compounds change the color of polyolefins only to a negligible degree, as a result of which they can be used to stabilize light-colored objects. The interest of researchers and practical workers in these compounds has increased in recent years in connection with the development of new types of elastomers on the basis of polyolefin copolymers.

4. Other Antioxidants

In addition to the classes of compounds considered, epoxy compounds [58]; azomethyne derivatives of o- and p-aminophenols [59], aromatic hydrazones [60], organic zinc compounds [61], tin dibutyl mercaptide, salts of metals of groups II and IV of the Periodic System and organic acids containing from 8 to 24 carbon atoms in the molecule [62], alkaline salts of mercaptans, in particular, mercaptobenzimidazole [28], and a number of others have been recommended and tested as stabilizers of polymers. Most of these recommendations were taken from data on the stabilization of other classes of polymers and are not specific for polyolefins.

In view of this, antioxidants prepared on the basis of phosphorous acid are of the greatest interest for polyolefins. At the present time methods are known for synthesizing a large number of simple and mixed esters of phosphorous acid and various phenols, possessing high effectiveness [62].

In [63] it was shown for the example of inhibition of the oxidation of polypropylene that of the phosphites, esters of pyrocatecholphosphorous acid, such as the 2,6-di-tert-butyl-4-methylphenyl ester of pyrocatecholphosphorous acid and the phosphite of α-naphthol and pyrocatechol, possess the greatest effectiveness. Simple esters of phosphorous acid [for example, trinonylphenyl phosphite (polygard) or tri-p-tert-butylphenyl phosphite] are always less effective than the mixed esters. The high inhibiting effectiveness of phosphites permits their use for the protection of readily oxidized polypropylene in the formation of fibers from a melt [64]. As was shown by one of the authors, good inhibiting properties are possessed by the phosphite of the reaction product of phenol with styrene. This is a thick liquid, straw-yellow in color; in a concentration of 0.5% by weight it guarantees conservation of the physicomechanical characteristics of various types of polyolefins under varied conditions of reprocessing. The product does not color the polymers, is nontoxic as a result of which it can be used in the stabilization of plastics used for everyday purposes.

According to the mechanism of their action, phosphites belong to the class of compounds that decompose hydroperoxides. According to [65] and [66], phosphites react energetically with various hydroperoxides. In this case an alcohol possessing the structure of the initial peroxide is formed. It has been demonstrated that the phosphite is oxidized to phosphate as a result of the reaction, in view of which the general gross scheme of the process can be expressed by the equation

$$R'OOH + P(OR)_3 \rightarrow R'OH + (RO)_3PO \qquad (14)$$

TABLE 8. Values of the Induction Periods of the Oxidation of Polypropylene during the Use of Antioxidants and Their Mixtures (Temperature 200°C, Oxygen Pressure 200 mm Hg)

Sulfur-containing compounds	Induction period of individual compounds, min	Compounds containing the amino group					
		DPA	H-DPA	N-D	N-A	H-ND	TDHQ
		Induction period of individual compound, min					
		25	100	680/350	180	150	10
		Induction periods of mixtures, min					
Phenylbenzyl sulfide	20	25	–	–	–	–	–
2,2'-tert-Butyl-4-methylphenol sulfide	350•	475 (1:4)	–	–	–	375 (1:3)	–
Tetramethyl-thiurammonium sulfide	19	260 (1:1,5)	–	–	–	550 (1:3)	–
Tetramethyl-thiuram disulfide	45	90 (1:1)	290 (1:2)	–	–	320 (1:3)	260 (4:1)
2,2'-Diphenyl-diamine disulfide	200	250 (1:4)	–	–	–	–	–
4,4'-Diphenyl-diamine disulfide	95	170 (1:1)	270 (1:1)	–	–	–	–
Mercaptobenzimidazole	75/65•	100 (1:1)	300 (1:4)	700	280• (1:3,5)	620 (1:19)	250• (1:1)
Mercaptobenzothiazole	20/15•	60 (1:1)	220 (5:1)	60• (1:1)	180•	150	65 (1:1)
Dodecyl mercaptan	25/10•	65• (1:2)	175 (1:2)	285•	–	–	75 (1:1)
α-Naphthol	130/75•	125	–	800	280•	–	–

Note. The values of the induction period noted by an asterisk (•) pertain to experiments in which the concentrations of the individual compounds or the total concentrations of their mixtures were 0.05 mole/kg of the polymer. In all the remaining cases the concentration was 0.1 mole/kg. DPA = diphenylamine; H-DPA = p-hydroxydiphenylamine; N-D = phenyl-β-naphthylamine (neozone D); N-A = phenyl-α-naphthylamine (neozone A); H-ND = p-hydroxyphenyl-β-naphthylamine; TDHQ = 2,2,4-trimethyldihydroquinoline.

The detailed mechanism of this process has not been elucidated. In [66] the possibility of formation of an intermediate complex $(RO)_3P\text{---}\overset{\displaystyle H}{O}\text{---}O\text{---}R'$, decomposition of which leads to the final reaction product, was assumed. In [63] a considerable critical concentration of phosphite was noted in the inhibition of polymer oxidation, which permits us to assume the possibility of reaction of phosphites with peroxide radicals as well.

5. Mutual Intensification of the Effectiveness of Antioxidants

The division of inhibitors of oxidation processes into two large
groups – those that terminate chains and those that decompose peroxides – long ago led to the idea of using mixtures of compounds of these
two groups. In this case we might expect that a mixture of two compounds will possess higher effectiveness than the individual compounds.
This was established for the example of the inhibition of processes of
oxidation of oils [54, 67, 68].

Mixtures of compounds of different classes have found wide use in
the stabilization of food oils [44]. It has been established [29] that when
mixtures of antioxidants are used, three cases of manifestation of their
inhibiting effectiveness are possible: mutual intensification of their action, usually called the phenomenon of synergism in the literature, additive addition of the effectivenesses, determined mainly by the effectiveness of the strongest antioxidant, and weakening of the effectiveness of
the strong antioxidant. The method of taking diagrams of the binary mixtures, according to which the value of the induction period is measured
as a function of the molar composition of the mixture, has been proposed in [69] for rapid establishment of the tendency of mixtures of antioxidants to manifest synergism.

In the case of intensification of the effectiveness of antioxidants, the
functional dependence of the induction period on the molar composition
of the mixture of antioxidants is expressed by a curve with a maximum
(Fig. 52). The maximum effectiveness is observed in the case of a strictly determined molecular ratio of the components of the mixture, which
depends on the chemical nature of the compounds used. In most cases
the molar ratio of the components of the mixture at which the induction
period possesses the maximum value is close to 1 : 1 (Table 8). In certain cases a maximum is observed at ratios of about 1 : 20, as in mixtures of p-hydroxyphenyl-β-naphthylamine with mercaptobenzimidazole.

The value of the induction period at the maximum for effective compositions is several times as great as the induction period in the use of
the individual compounds, taken in a concentration equal to the total concentration for the mixture. The induction period decreases with decreasing total concentration (see Fig. 52); the position of the maximum
is thereby conserved. The induction period is also substantially reduced when the temperature is increased.

In [29] the intensification of the effectiveness of amines and aminophenols when they were mixed with certain sulfur-containing compounds
was investigated. The data cited in Table 8 show that the maximum
achievable value of the induction period depends on the chemical nature

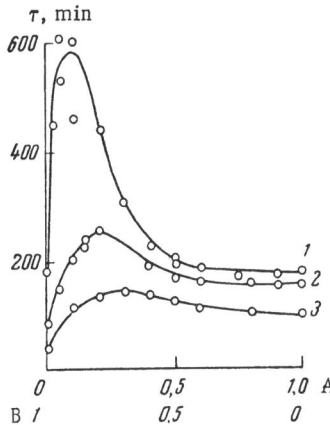

Fig. 52. Variation of the induction period of the oxidation of polypropylene at 200°C as a function of the molar composition of a mixture of p-hydroxy-phenyl-β-naphthylamine (A) with mercapto-benzimidazole (B). Total concentrations of mixtures in moles/kg: 1) 0.1; 2) 0.05; 3) 0.025.

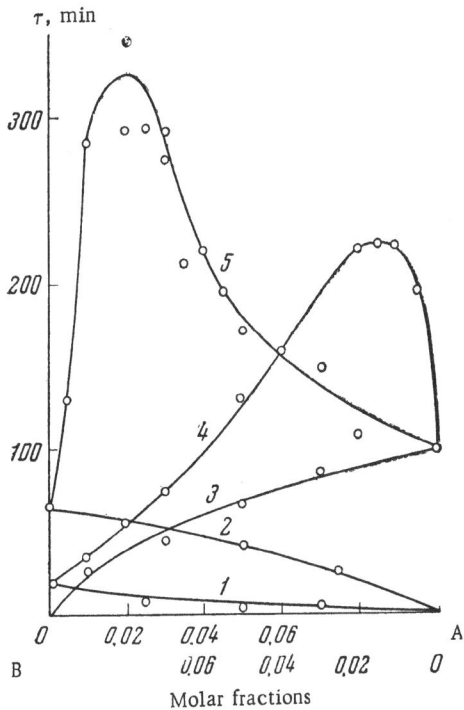

Fig. 53. Influence of the chemical nature of the second component in a mixture of antioxidants. 1) Mercaptobenzothiazole; 2) mercapto-benzimidazole; 3) p-hydroxydiphenylamine; 4) mixture of 1 and 3; 5) mixture of 2 and 3. Total concentration of the mixture 0.1 mole/kg. Scale A: hydroxydiphenylamine; scale B: mercaptobenzimidazole and mercaptobenzothiazole

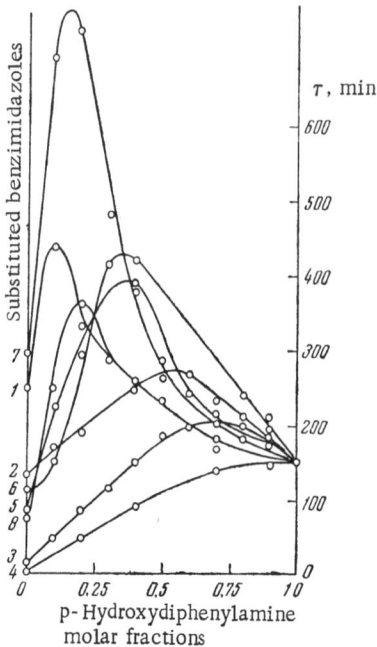

Fig. 54. Influence of the replacement of hydrogen atoms in the =NH and =SH groups of mercaptobenzimidazole on the effectiveness of mixtures with p-hydroxydi-phenylamine.

	Curve No.	1	2	3	4	5	6	7	8
(structure)	R_1	H	CH$_3$	H	CH$_3$	H	CH$_3$	H	H
(structure)	R_2	H	H	CH$_3$	CH$_3$	Na	Na	(structure)	(structure)

of the compound. Thus, a mixture of tetramethylthiuram disulfide with diphenylamine in a 1 : 1 ratio at 200°C and a total concentration of 0.1 mole/kg guarantees stability of the polymer for 90 min. Replacement of diphenylamine by p-hydroxydiphenylamine increases this time to 290 min. In a number of cases changing the chemical composition of the second component also changes the position of the maximum. This case is cited in Fig. 53 for mixtures of p-hydroxydiphenylamine with mer-captobenzimidazole and mercaptobenzothiazole.

Just as in the case of individual compounds, in mixtures of two anti-oxidants replacement of the active hydrogen in the functional group by an alkyl radical leads to a reduction of the effectiveness of the mixtures.

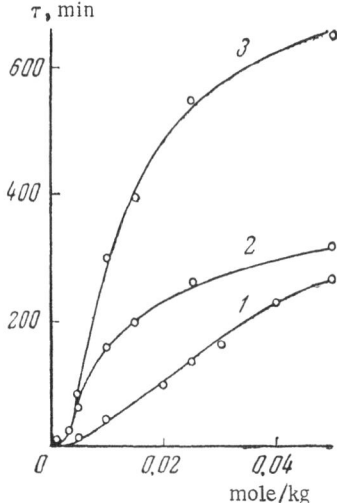

Fig. 55. Dependence of the induction period on the concentration of 2,6-di-tert-butyl-4-methylphenyl ester of pyrocatecholphosphorous acid (1), 2,2'-thio-bis-(4-methyl-6-tert-butylphenol) (2), and their mixture (3) in a 2 : 3 ratio.

Such an influence was very distinctly detected on the effectiveness of mixtures of p-hydroxydiphenylamine with substituted mercaptobenzimidazole.

From Fig. 54, on which these data are cited, we can also see that replacement of the H-atom in the sulfhydryl group by a radical containing a sulfur atom or NH group guarantees conservation of the effectiveness and in a certain cases even leads to an intensification of the effectiveness.

The dependence of the induction period on the summary concentration, taken in the molar ratio corresponding to the maximum, is expressed, as is shown in Fig. 55 (curve 3), by an S-shaped curve with a characteristic break after the critical concentration and a break in the region of large concentrations [63].

In addition to increasing the effectiveness, mixing two antioxidants leads in a number of cases to a weakening of the effectiveness of the strongest antioxidant. This is observed when mercaptobenzothiazole or tetramethylthiuram disulfide is used in mixtures with certain secondary amines [56] or amino sulfides [28]. One of these examples is shown in Fig. 56. It is characteristic that in this case the use of mercaptobenzimidazole leads to an intensification of the effectiveness. The weakening of the effectiveness (antagonism) in the mixing of antioxidants is of no less significance that the intensification effect. This is especially important for rubbers and elastomers of polyolefins, where various compositions are used. The investigation of antagonism is also important in the presence of impurities in the polymer and when antioxidants are combined with photostabilizers.

Mixed compositions of mercaptobenzimidazole and alkylated thiurams with phenol sulfides and certain aminophenols [29, 69], phosphites with phenol sulfides [63], phenols with sulfides, and especially with thiodipropionic acid [70] are of the greatest practical significance.

In spite of the great effects achieved in the use of mixtures of antioxidants, and their increasing practical utilization, up to the present time there have been practically no studies in the literature devoted to the investigation of the mechanism of the mutual intensification or weak-

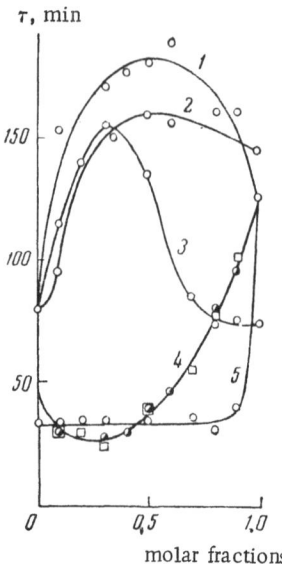

Fig. 56. Dependence of the induction period in the oxidation of polypropylene on the molar composition of mercaptobenzimidazole with compounds A (1), B (2), and C (3), tetramethylthiuram disulfide (4) [O] with compound A; □) with compound B], and mercaptobenzothiazole with compound C (5). T = 200°C.

A. $C_4H_9SCH_2$—⟨⟩—CH_2NH—

B. C_2H_5S—⟨⟩—CH_2NH—

C. H_3C—O—⟨⟩—S—⟨⟩—CH_2NH—

ening of the action. Most authors believe that one of the antioxidants terminates the kinetic chains, while the other decomposes hydroperoxides. This representation is based mainly on an investigation of the ratio of the individual antioxidants to peroxide compounds and has not been confirmed by investigations of the chemism of the oxidation process in the presence of mixtures of stabilizers. There is some basis for believing that the mechanism of synergism will vary depending on the chemical nature of the compounds used.

In [71-75], devoted to the investigation of the inhibition of the oxidation of plant and animal oils, the phenomenon of synergism is explained by the fact that the synergist (certain organic acids) regenerates the inhibitor (mainly of the phenol type) from the products of its oxidative transformation. The duration of the work of the inhibitor and, consequently, the induction period as well are thereby lengthened.

It has been shown that the role of the synergist can be reduced to an elimination of the initiating action of the radicals obtained in the oxidation of the inhibitor. This question was discussed in greater detail in Chapter I.

For mixtures of amines and certain sulfur-containing compounds (especially thiurams), the possibility of the formation of intermediate compounds, possibly of the type of sulfenamides with S – N bonds, is also not excluded. Thus, in an investigation of the interaction of diarylazotoxide radicals with tetramethylthiuram disulfide [52] by the EPR method, the formation of radicals, the g-factor of whose EPR spectrum corresponded to radicals containing –S–N =, was discovered. An increase in the stabil-

TABLE 9. Influence of Sulfur-Containing Compounds (0.1%)
on the Oxidation of Pure Polyethylene and Polyethylene
Containing Carbon Black [41]

Content	Time of absorption (hr) of 10 cm^3 of oxygen per g of polyethylene	
	Without additive	With 3% carbon black
Polyethylene (control)	6	35
Polyethylene + 1% elementary sulfur	–	120
Thioethers		
4,4'-Thio-bis-(3-methyl-6-tert-butylphenol)	600	750
Thio-bis-(β-naphthol)	240	730
Thio-bis-(N-phenyl-β-naphthyl-amine)	190	260
Methylthio-2-naphthalene	6	55
Disulfide		
Diphenyl disulfide	6	120
3-Tolyl disulfide	6	520
1-Dodecyl disulfide	8	350
Polymeric 1,10-decanedithiol	110	540
Thiols		
2-Toluenethiol	6	200
2-Naphthylthiol	6	900
Mercaptobenzothiazole	35	380
1-Dodecyl mercaptan	6	160

izing effectiveness of the trimer 2,2,4-trimethyldihydroquinoline after
its treatment with mercaptobenzimidazole before introduction into the
polymer was established. As it was found, the treatment leads to a loss
of paramagnetic properties by the dihydroquinoline.

As can be seen from the above, the mechanism of the action of binary
or more complex mixtures of antioxidants has as yet been far from elucidated and is in need of a detailed investigation. These investigations
are all the more essential in that, in addition to the selection of the most
important synergic pairs, they will aid in finding rational means of synthesizing highly effective stabilizers.

Of special interest are studies [76-78] devoted to the investigation of
the intensification of the effectiveness of certain antioxidants by carbon
blacks, in connection with the problem of stabilizing various technical
objects, mainly tubes, insulation of pipes and cables [79].

TABLE 10. Variation of the Characteristic Viscosity of Polypropylene during the Process of Reprocessing and Subsequent Oxidative Aging at 150°C

Item No.	Stabilizers	Concentration, % by weight	Induction period at 200°C, min	After granulation	After casting under pressure	Characteristic viscosity•			
						After aging of duration			
						50 hr	100 hr	150 hr	200 hr
1	Initial polypropylene		6-8	3.1	2.8	—	—	—	—
2	2,2'-Methylene-bis-(4-methyl-6-tert-butylphenol)	0.3	220	3.2	3.1	2.5	—	1.2	—
3	4-Methyl-6-isobornyl-phenol	0.5	30	2.8	2.6	1.6	—	—	—
4	Phenylcyclohexyl-p-phenylenediamine	0.2	340	4.0	3.7	3.7	3.7	3.2	3.2
5	Di-β-naphthyl-p-phenylenediamine	0.3	405	4.0	3.9	3.6	3.0	3.2	3.4
6	2,2'-Thio-bis-(4-methyl-6-tert-butyl-phenol)	0.3	440	3.74	3.24	2.4	2.4	1.8	2.0
7	Phosphite of product of phenol with styrene	0.5	95	3.96	3.96	3.25	3.2	3.0	2.5
8	α-Naphthyl ester of pyrocatecholphosphorous acid	0.2	220	4.1	4.0	3.8	—	—	—
9	Mixtures No. 6 and No. 8	0.5	925	4.1	4.0	4.0	4.0	—	4.0
10	Mixture of mercapto-benzimidazole and p-hydroxyphenyl-β-naphthylamine	0.5	450	4.0	3.9	3.9	3.9	3.9	3.8

•Characteristic viscosity of polypropylene powder 4; after aging for 10 hr, 1; after aging for 30 hr, 0.5.

Carbon blacks are widely used as light absorbers in the protection of materials against the action of light. However, in polyolefins they are also antioxidants. The best in antioxidizing properties is finely dispersed (particle size 150-250 A) channel black.

The effectiveness of carbon black is determined by the degree of oxidation of its surface, on which OH, CO, quinoid, and other groups are to be found, which permits us to consider carbon black as high-molecular phenols or quinones.

The stabilizing effectiveness of channel black is high. Thus, 3% of it in polyethylene at 140°C protects the polymer for a longer time than 2,6-di-tert-butyl-4-methylphenol.

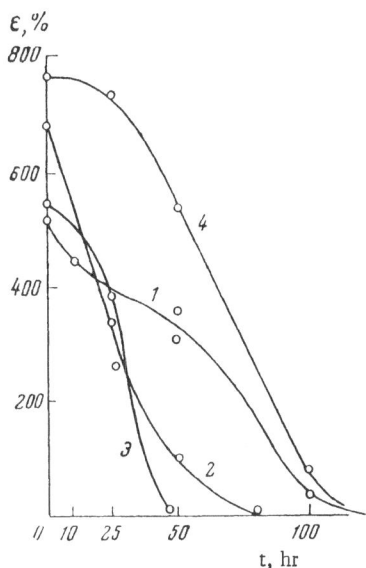

Fig. 57. Variation of the relative lengthening of polyolefin films during the process of irradiation by the light of a PRK-4 lamp in air. 1) High-pressure polyethylene; 2) low-pressure polyethylene; 3) polypropylene; 4) copolymer of ethylene with propylene; T = 20°C.

Various antioxidants behave differently in mixtures with carbon black. Most substituted phenols and certain amines (for example, diphenyl-p-phenylenediamines) become still less effective in the presence of carbon black (phenomenon of antagonism). However, thioesters prepared from these compounds are not only themselves stronger antioxidants, but also manifest synergism in mixtures with carbon black.

In [41] a substantial increase in the effectiveness was demonstrated when a number of thioethers, benzyl-aryl sulfides, thiols, disulfides, and certain polymeric sulfur-containing compounds were introduced into polyethylene containing 2-3% channel black. As an example, Table 9 indicates the values of the effects achievable in this case.

It is characteristic that the practical absence of autocatalysis during oxidation is observed for a number of sulfur-containing compounds in conjunction with carbon black. For example, at 140°C the induction period of the oxidation of polyethylene in the presence of 0.1% 2-thionaphthol was equal to 10 hr, while in the presence of 3% channel black it was 30 hr. In the case of the joint action of these antioxidants, slow oxidation began after 1000 hr; autocatalysis was not observed even after 2000 hr. An analogous picture is observed for combinations of carbon black with thiurams [90].

The mechanism of synergism in pairs of carbon black and sulfur-containing compounds is quite unknown. It is assumed only that the structure of the carbon black plays an important role here. Carbon blacks subjected to pyrolysis entirely lose their protective action and their ability to manifest synergism with antioxidants. It has been shown [76] that polyacenes (model of deactivated carbon black), for example, tetracene, pentacene, and perylene possess high effectiveness in conjunction with phenol sulfides and thiols. Thus, the addition of 0.1% perylene and 0.1% β-naphthyl mercaptan greatly decelerates the oxidation of polyethylene at 140°C. In this case autocatalysis is not observed for more than 2000 hr.

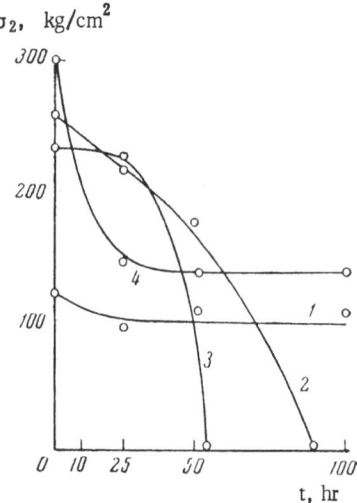

σ_2, kg/cm^2

Fig. 58. Variation of the breaking strength of polyolefin films during the process of photodestruction. 1) High-pressure polyethylene; 2) low-pressure polyethylene; 3) polypropylene; 4) copolymer of ethylene and propylene.

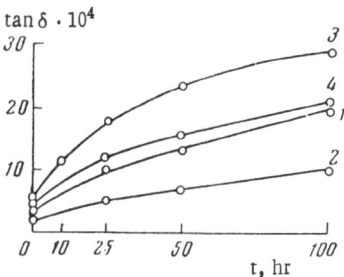

tan δ · 10^4

Fig. 59. Variation of tan δ of polyolefins during the process of photo-destruction. 1) High-pressure polyethylene; 2) low-pressure polyethylene; 3) polypropylene; 4) copolymer of ethylene and propylene.

6. Comparative Effectiveness of Stabilizers of Polyolefins

In selecting stabilizers for practical application, an important factor is a knowledge of the effectiveness of compounds of various classes. Unfortunately, there are no data on series of antioxidants in the published literature as applied to concrete examples. Comparative data on the effectiveness of a limited number of antioxidants and their mixtures for polyethylene and polypropylene are cited only in [29, 41, 63, 64, 79, 80]. However, even in these works the researchers limited themselves to a measurement of the induction period of the oxidation process, without investigating widely the change in properties of the polymer. The remaining literature data, especially the patent data, pertain to individual compounds, which in the absence of a single approach to the evaluation of effectiveness makes it difficult to draw a correct conclusion on the advantages of certain stabilizers over others. Such a situation at the present time, when the demands of the polymer industry with respect to stabilizers are growing, poses the question of the development of single, standardized methods for evaluating stabilizers according to groups or types of polymers.

Broad testing of a number of stabilizers of industrial significance has been conducted for polypropylene. The variation of the physicomechanical and other properties of polypropylene during the process of aging in air at 150°C has been investigated. The variation of the characteristic viscosity of a solution of the polymer in tetralin in the presence of various antioxidants during the aging process is indicative. These data are cited in Table 10, from which we can see that the polymer already begins to be destroyed during reprocessing,

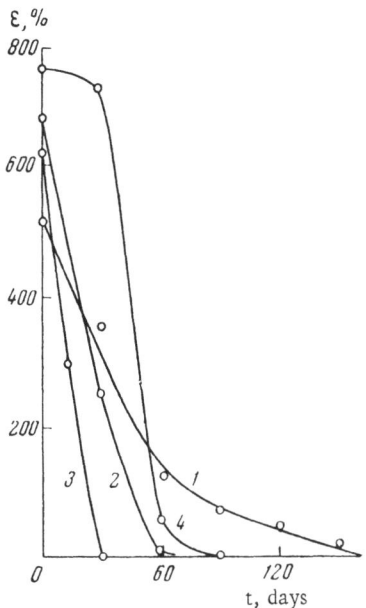

Fig. 60. Variation of the relative lengthening during breaking of polyolefin films in the process of atmospheric aging. 1) High-pressure polyethylene; 2) low-pressure polyethylene; 3) polypropylene; 4) copolymer of ethylene and propylene.

even in the presence of certain antioxidants. The advantages of mixtures of stabilizers are also distinctly evident. In spite of the fact that 2, 2'-thio-bis-(4-methyl-6-tert-butylphenol) is a strong antioxidant, a mixture of it with the α-naphthyl ester of pyrocatechol-phosphorous acid is substantially more effective. The polymer inhibited by phenol sulfide alone was mechanically destroyed after 220 hr under conditions of aging. Inhibition with mixture No. 9, on the other hand, led to a conservation of the properties of the polymer for more than 450 hr. Each of the antioxidants cited in Table 8 is of industrial significance and can be used for the stabilization not only of polypropylene, but also of other polyolefins.

As can be seen from the material cited, at the present time there are good stabilizers inhibiting the thermooxidative aging of polyolefins; however, the searches for new antioxidants are continuing.

III. DESTRUCTION OF POLYOLEFINS UNDER THE ACTION OF LIGHT

The properties of polyolefins change greatly under the influence of ultraviolet radiation. The action of light is manifested especially strongly in an atmosphere of oxygen. It was shown in [81] that irradiation of polyolefin films in air at room temperature by the light of a mercury-quartz lamp (PRK-4) leads to a comparatively rapid deterioration of the physicomechanical properties. The relative lengthening drops sharply (Fig. 57); the breaking strength changes (Fig. 58); the value of the tangent of the angle of dielectric losses increases (Fig. 59). During the process of photoaging, the polymer cracks, becomes brittle, acquires a color. It was shown in the same work that the variation of the physicomechanical properties of polyolefins under conditions of atmospheric aging proceeds according to the same laws as in the case of artificial ultraviolet irradiation. Only the rate of the process changes. As an example, we cite in Fig. 60 the variation of the relative lengthening of films during breaking in the process of prolonged atmospheric aging. During atmospheric aging the maximum rates of decomposition of the polymer correspond to the months of the greatest solar radiation.

It is characteristic that polyolefins can be arranged in the following series with respect to stability to ultraviolet irradiation, just as in thermal oxidation: polypropylene–low-pressure polyethylene–copolymer–high-pressure polyethylene. Branching of the polymer chains also reduces the stability in photodestruction.

At the present time there is a comparatively large number of works in the literature studying the mechanism of the photooxidative destruction of polyolefins. Unfortunately, most of them are devoted to high-pressure polyethylene, in view of which it does not seem possible to make any comparison between individual types of this class of polymers. However, the general picture of the photooxidation possesses much in common with the thermal process of oxidation and differs from the latter mainly in the stage of initiation.

Ultraviolet irradiation intensifies the oxidation of the polymer, the rate of oxygen absorption increasing with the light intensity [82] and with increasing temperature [83].

The influence of light (mercury arc or mercury-quartz lamps) leads to a progressive increase in the concentration of various oxygen-containing groups in the structure of the polymer. An analysis of the infrared absorption spectra during the process of irradiation of polyethylene indicates an increase in the absorption in the region of 1180 cm^{-1} (carboxyl group), in the region of 1700 cm^{-1} (carbonyl group), and in the region of 3300 cm^{-1} (hydroxyl group) [81]. In addition, an increase in the absorption in the region of 940 cm^{-1} is observed, which corresponds to unsaturated bonds. The changes in the infrared spectrum of polyethylene as a result of photooxidation can be seen in Fig. 61. The concentration of carbonyl groups changes especially sharply. Figure 62 [13] presents typical kinetic curves of the increase in C = O groups, obtained in the case of irradiation of samples of high- and low-pressure polyethylene by the light of a PRK-2 lamp. Noteworthy is the difference of the kinetic curves for high- and low-pressure polyethylene. In the latter case the increase in the C = O groups approaches a limit, while for high-pressure polyethylene, new groups continue to be formed at an increasing rate. This difference is apparently explained by differences in the degree of crystallinity of high- and low-pressure polyethylene. It is known that the crystallinity of low-pressure polyethylene is higher; it is assumed that the crystal packing hinders penetration of oxygen into the sample.

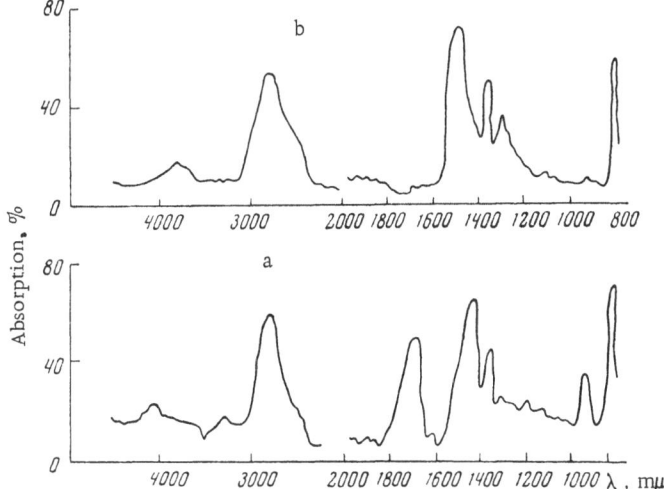

Fig. 61. Influence of ultraviolet irradiation of polyethylene on the structure of the infrared absorption spectrum. a) Before irradiation; b) after 89 hr irradiation by UV light.

A study of the absorption of light in the region corresponding to the absorption of carbonyl groups [84] led to the conclusion that the chain reaction of photooxidation begins with the absorption of light precisely by the carbonyl groups, which are always present in linear polymers. Such groups can appear in the polymer molecule even during the process of polymerization in the presence of oxygen. When light is absorbed by the carbonyl group of aldehydes of ketones (wavelengths shorter than 3300 A), the latter decompose into macroradicals. In an investigation of the photolysis of polyethylene [85], carbon monoxide, water, acetaldehyde, and acetone were detected in the gas phase, and vinyl groups in the polymer.

Such a composition of the products confirms the decomposition of the carbonyl groups of the compounds.

In view of this it is assumed that the decomposition proceeds according to the schemes:

$$R-CO-CH_2-CH_2-CH_2-R + h\nu \longrightarrow R-CO-CH_3 + CH_2=CH-R \qquad (15)$$

$$R-CH_2-CH_2-C\overset{O}{\underset{H}{\diagdown}} + h\nu \begin{cases} \longrightarrow R'-CH=CH_2 + CH_3C\overset{O}{\underset{H}{\diagdown}} \\ \\ \longrightarrow R-CH_2-\overset{\cdot}{C}H_2 + \overset{\cdot}{C}\overset{O}{\underset{H}{\diagdown}} \end{cases} \qquad (16)$$

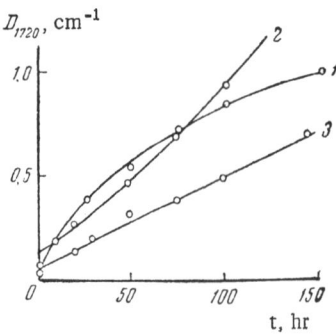

Fig. 62. Kinetics of the increase in carbonyl groups (> C = O) in the photodestruction of polyethylene. 1) Low-pressure polyethylene; 2) high-pressure polyethylene; 3) high-pressure polyethylene stabilized with a mixture of phenyl-β-naphthylamine, diphenyl-p-phenylenediamine, and carbon black.

The macroradical formed to reaction (16) interacts with oxygen, leading to a normal chain reaction and the formation of hydroperoxides. According to the data of [85], the quantum yield of carbonyl groups in the photolysis of polyethylene is no greater than 0.1. This means that the kinetic chains in photooxidation are not very long. In view of this, termination of the chains by means of antioxidants is hindered in practice, since in the case of short chains and a high rate of initiation, effective inhibition is impossible. The formation of unsaturated compounds during photolysis [reactions (15) and (16)] facilitates further oxidation of the polyolefin.

In spite of the fact that the peroxide compounds formed during photodestruction, in the opinion of certain researchers [86], are more stable than the corresponding formation of aldehydes and ketones. According to the opinion expressed in [82], aldehydes are formed in the decomposition of secondary hydroperoxides, ketones in the decomposition of tertiary hydroperoxides. In the presence of oxygen, aldehydes are oxidized to acids under the action of light.

Under atmospheric conditions, the photooxidation of polyolefins is promoted by the presence of many oxidizing agents contained in air: ozone, SO_2, NO_2, H_2O_2, free radicals, etc. Ozone enters from the upper layers of the atmosphere or is formed as a result of the photooxidation of SO_2 and NO_2; the formation of free radicals is possible in the photolysis of water

$$H_2O + h\nu \rightarrow H^{\cdot} + O^{\cdot}H \tag{17}$$

or hydrogen peroxide

$$H_2O_2 + h\nu \rightarrow 2\ O^{\cdot}H \tag{18}$$

Hydrogen peroxide can be formed, for example, according to the reaction

$$H_2O + O_3 + h\nu \rightarrow O_2 + H_2O_2 \tag{19}$$

The process of photodestruction of polyolefins is extremely complex, and at present there are only tentative data with respect to its mechanism.

IV. INCREASING THE LIGHT STABILITY OF POLYOLEFINS

The range of light waves 300-400 mμ is most dangerous for poly-olefins. Hence, the practical problem of photostabilization reduces mainly to the introduction into the polymer of compounds that would absorb precisely this part of the spectrum. The mechanism of the action of light absorbers is very complex, has received little study, and is apparently based on the absorption of light by the light absorber and further conversions of the energy of electronic excitation of the molecules of the light absorber to other forms of energy. Light absorbers can be divided into three groups, indicated in [82], according to the mechanism of transformation of absorbed energy of light quanta.

After absorbing light quanta, the compounds can give off the energy of electronic excitation in the form of radiation of the same frequency (fluorescence); in this case no light-stabilizing action is observed.

Excited light-absorber molecules can also transfer energy through collision. A shorter time is required for this than for fluorescence. This type of energy transfer leads to a sensitizing influence of light absorbers, and hence to an increase in the rate of destruction of the polymer.

Certain antioxidants – secondary amines, in particular, phenyl-β-naphthylamine [85] – absorb light in a region close to the absorption of carbonyl groups. However, they intensify photooxidation. This is explained by the fact that the energy of electronic excitation of the amine molecule is expended on a chemical reaction of the type:

$$
\begin{aligned}
& R_1NHR_2 + h\nu \rightarrow [R_1NHR_2]^* \\
& [R_1NHR_2]^* + O_2 \rightarrow R_1\dot{N}R_2 + HO\dot{O} \\
\text{or}\quad & [R_1NHR_2]^* + RCH_2C\underset{H}{\overset{O}{\diagup\!\!\!\backslash}} \rightarrow R_1NHR_2 + \\
& + \left[RCH_2C\underset{H}{\overset{O}{\diagup\!\!\!\backslash}}\right]^* \rightarrow R_1NHR_2 + \dot{R} + \dot{C}H_2C\underset{H}{\overset{O}{\diagup\!\!\!\backslash}}
\end{aligned}
\qquad (20)
$$

which results in the formation of active radicals, capable of carrying on the chain of polymer oxidation. The replacement of a hydrogen atom in the amino group substantially reduces the ability of the amines for sensitization.

Tests for a large number of different compounds capable of absorbing ultraviolet have shown that a tendency for sensitization is also possessed by certain compounds that possess no labile hydrogen in the molecule. They include, for example, benzophenone, which is characterized by localization of the absorbed light energy in the carbonyl group. This

TABLE 11.   Effective Light Stabilizers Based on Benzophenone

Type of stabilization	Structural formula	Compound
Hydroxy derivatives of benzophenone	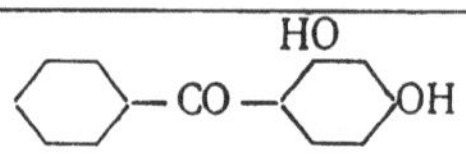	2,4-Dihydroxybenzophenone
	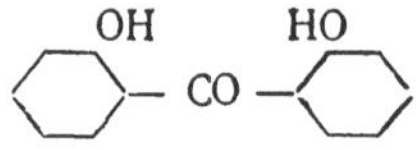	2,2'-Dihydroxybenzophenone
	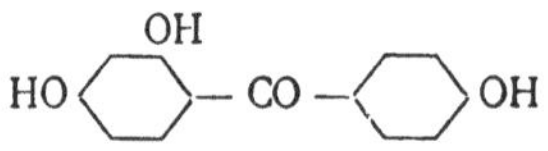	2',4',4-Trihydroxybenzophenone
	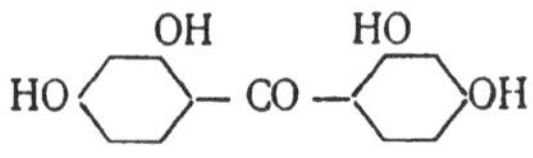	2,2',4,4'-Tetrahydroxybenzophenone
Monohydroxyalkoxy derivatives of benzophenone	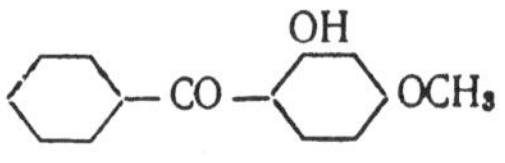	2-Hydroxy-4-methoxybenzophenone
	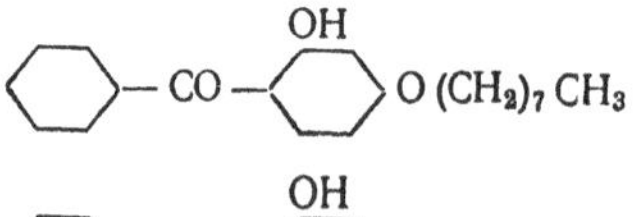	2-Hydroxy-4-octyloxybenzophenone
	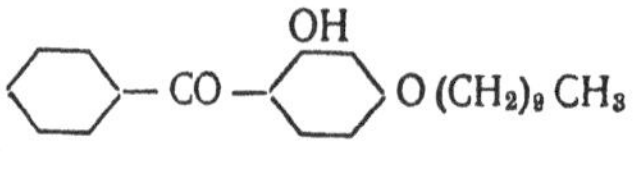	2-Hydroxy-4-decyloxybenzophenone
	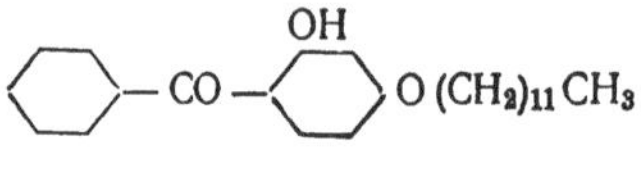	2-Hydroxy-4-dodecyloxybenzophenone
Monohydroxydialkoxy derivatives of benzophenone	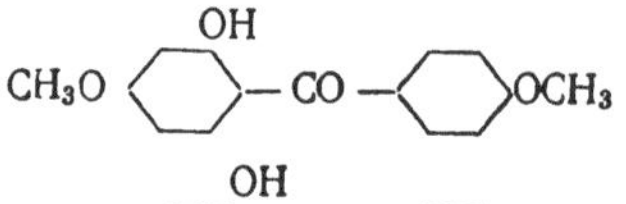	2-Hydroxy-4,4'-methoxybenzophenone
	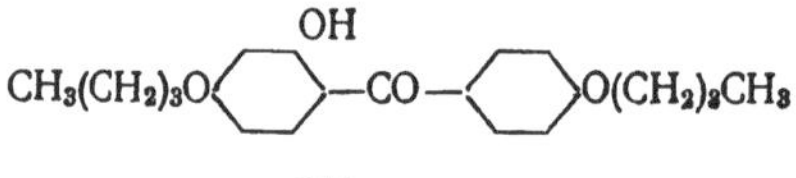	2-Hydroxy-4,4'-butoxybenzophenone
	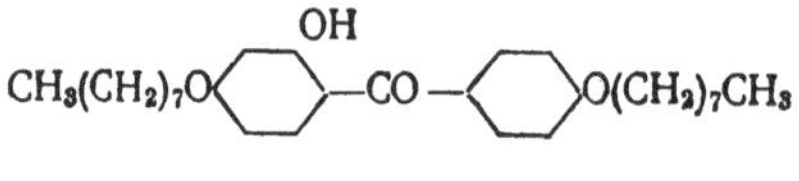	2-Hydroxy-4,4'-octyloxybenzophenone
Dihydroxyalkoxy derivatives of benzophenone	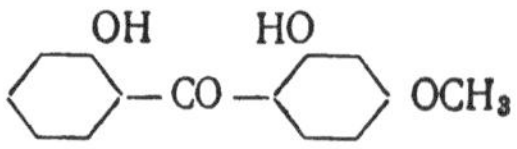	2,2'-Hydroxy-4-methoxybenzophenone
	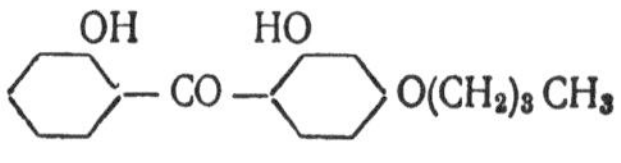	2,2'-Hydroxy-4-butoxybenzophenone
	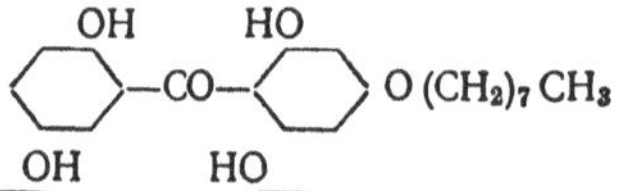	2,2'-Hydroxy-4-octyloxybenzophenone
	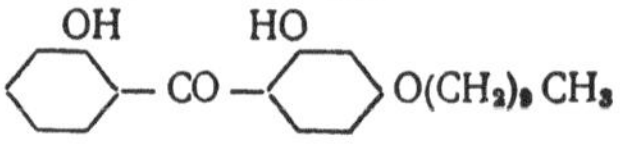	2,2'-Hydroxy-4-decyloxybenzophenone

Table 11 (conclusion)

Type of stabilization	Structural formula	Compound
Dihydroxydialkoxy derivatives of benzophenone	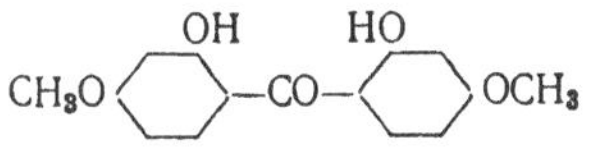 CH₃O⟨ring⟩(OH)—CO—(HO)⟨ring⟩OCH₃	2,2'-Hydroxy-4,4'-methoxybenzophenone
Modified hydroxybenzophenones	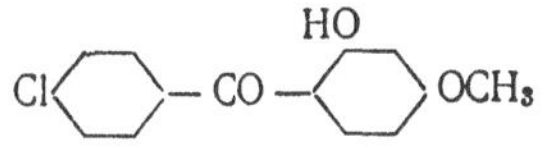 Cl⟨ring⟩—CO—(HO)⟨ring⟩OCH₃	2-Hydroxy-4-methoxy-5-chlorobenzophenone
	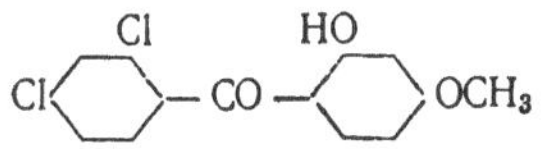 Cl⟨ring⟩(Cl)—CO—(HO)⟨ring⟩OCH₃	2-Hydroxy-4-methoxy-2',4'-chlorobenzophenone
	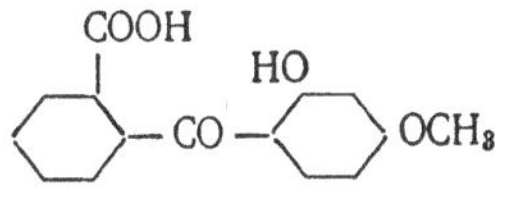 ⟨ring⟩(COOH)—CO—(HO)⟨ring⟩OCH₃	2-Hydroxy-4-methoxy-2'-carboxybenzophenone
	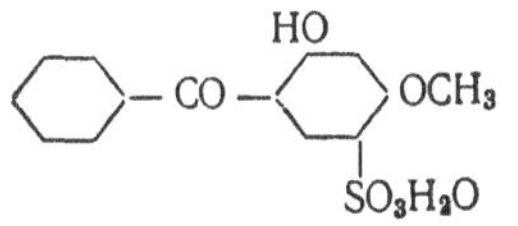 ⟨ring⟩—CO—(HO)⟨ring⟩OCH₃ (SO₃H₂O)	2-Hydroxy-4-methoxy-5-sulfobenzophenone

energy is then transferred through collision to the polymer molecules, which leads to the formation of free radicals and further destruction. The formation of benzophenone radicals has been experimentally demonstrated during the irradiation of solutions of benzophenone. Such radicals dimerize; the dimer can be isolated in crystalline form [85].

The sensitizing action of many antioxidants must be considered in those cases when the polymer should be protected not only from the action of light, but also from oxidative destruction.

Finally, the absorption of light can give rise to the minimum level of electronic excitation of the light-absorber molecule. Such molecules in this case will rapidly give off part of their energy in the form of thermal energy and in the form of light quanta. The transfer of energy of excitation to the polymer in low-energy quanta is not dangerous for it. In such a case stabilization against the action of light is observed. A typical representative of this type of light absorber is hydroxybenzophenone, as well as some of its derivatives.

The transfer of light energy in hydroxybenzophenone is accomplished thanks to its ability to be rearranged into a quinoid structure [82, 85], which results in the emission of energy with lower frequencies than the frequency of the absorbed light. Such a rearrangement proceeds in accord with the scheme:

$$+ RH + hv \rightarrow \qquad + RH \rightarrow \qquad + RH + hv' \qquad (21)$$

where $hv' < hv$.

There is a labile equilibrium between the benzene and quinoid structures of benzophenone in this process, which permits the conversion of large amounts of energy.

It is precisely this ability of hydroxybenzophenone to transform absorbed light energy into energy safe for the polymers that has led to the wide practical utilization of its derivatives [84, 87-93].

The most effective light stabilizers based on benzophenone are cited in Table 11. Their recommended concentrations for polyolefins are 0.2-1.5% by weight. Light stabilizers based on benzophenone can be divided into two basic types: mono-2-hydroxybenzophenones and 2,2'-dihydroxybenzophenones.

The production of these compounds is based in most cases on a condensation reaction according to Friedel-Crafts or alkylation of the corresponding benzophenone derivatives [92-94]. Depending on the structure of these compounds, their light absorption properties, compatibility with polyolefins, color, and solubility in organic solvents vary considerably. All the light absorbers of the benzophenone class possess an

o-hydroxyphenolketone group , the hydrogen bond in which (chelate group) is responsible for strong absorption in the region of the near ultraviolet (absorption maximum 340 mμ); 2,2'-dihydroxy derivatives of benzophenone possess a greater conjugation effect in the chelate group. Hence their absorption in the ultraviolet region is more intense [87, 88] and extends almost to the boundary of visible light, evidence of which is given by the ultraviolet absorption spectra of certain benzophenone derivatives, presented in Fig. 63. As a result of this, in most cases the compounds are colored yellow.

When the hydroxyl groups are replaced, steric hinderances for the formation of the chelate group arise, and light absorption in the ultraviolet region is considerably reduced (the absorption maximum at 340 mμ disappears).

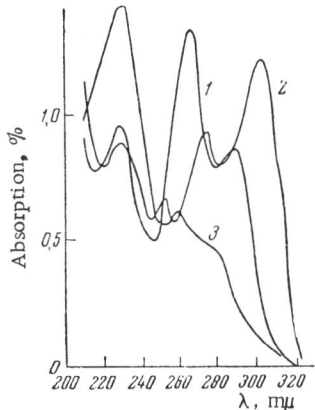

Fig. 63. Ultraviolet absorption spectra of solutions of benzophenone derivatives in dichloroethane. 1) 2-Hydroxy-2-methoxybenzophenone; 2) 2,2'-hydroxy-4-methoxybenzophenone; 3) 2,4-dimethoxybenzophenone.

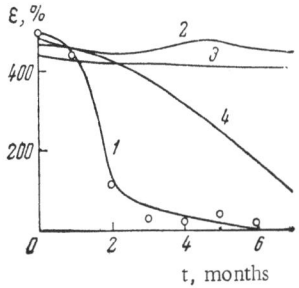

Fig. 64. Influence of light stabilizers on the variation of the relative lengthening of high-pressure polyethylene film during the process of atmospheric aging. 1) Unstabilized polyethylene; 2) with an addition of 2-hydroxy-4-octyloxybenzophenone; 3) carbon black + diphenyl-p-phenylenediamine; 4) derivative of 4-thiazolidone.

The introduction of an alkoxy group into the 4-position practically does not change the light absorption properties, but substantially improves the compatibility of the benzophenone derivatives with the polyolefins.

It has been established that the best compatibility is possessed by compounds in whose alkoxy groups the number of carbon atoms is equal to 8 to 12. Usually all benzophenone derivatives are insoluble in water and readily soluble in organic solvents. Solubility in water is achieved by introducing the sulfo group into the benzene ring.

The light yellow color of hydroxybenzophenone derivatives at the concentrations used leads to practically no change in the color of the polymer; hence they can be used to stabilize colorless polymer materials. Derivatives of benzoic and salicylic acids and resorcinol [88, 91, 94, 99], 4-thiazolidone [90], and benzotriazole [87, 88, 96, 97] are also used as colorless light stabilizers. However, as applied to polyolefins these compounds are less effective in comparison with hydroxybenzophenone derivatives. They are finding wide use in the stabilization of other classes of polymers. The mechanism of their action has been very little studied.

Good stabilizing action on polyolefins is exerted by finely pulverized channel black [78, 98]. Many researchers explain the light-stabilizing action of carbon black by its ability to absorb the entire range of ultraviolet and visible radiation and to transform the absorbed light energy into a less dangerous form for the polymer – thermal energy. On the other hand, thanks to the structural peculiarities of the crystal lattice, carbon black possessed the ability to block free radicals, initiating thermal and

photooxidation. The best light-stabilizing action is exerted by gas channel black with a particle dispersion of 0.17-0.25 A.

Substances that reflect or scatter ultraviolet and visible radiation
are also recommended for increasing the light stability of polymers [82,
92]. However, such a method is less effective than the use of light absorbers.

For polymers possessing low stability to the action of light, i.e.,
great light absorption and a large quantum yield (such as, for example,
polypropylene), to obtain the maximum effect it is expedient to use mixtures of light absorbers with antioxidants. In this case, not only is the
harmful action of light blocked, but the oxidation reactions are also inhibited.

In a number of cases high absorption proceeds with a low quantum
yield. Here one light absorber usually proves sufficient, since a low
quantum yield is accompanied by short kinetic chains, and the use of an
antioxidant will be useless. Photostabilization of ultraviolet-transparent
polymers can be accomplished by using an antioxidant alone [89]. In
selecting antioxidants for polymers in this case, the antioxidant should
not absorb light in the region of 290-300 mμ and should not be subject
to photolysis. The use of light stabilizers permits a substantial increase in the operating period of objects made from polyolefins, as is
distinctly visible in Fig. 64, which presents curves characterizing the
variation of the relative lengthening of stabilized and unstabilized high-
pressure polyethylene during the process of atmospheric aging. This
figure to some degree also characterizes the difference in the effectiveness of various light stabilizers.

The optimum concentration of light stabilizers is usually determined experimentally and depends on the compatibility of the absorber
with the polymer, its effectiveness, and the thickness of the object.

BIBLIOGRAPHY

1. C. S. Meyer, Ind. Eng. Chem. 44(5):1095, 1952.
2. A. Basoni, Materie Plastiche 19(5):361, 1963.
3. B. S. Biggs and W. Hawkins, Ind. Plastiques Mod. (Paris) 6(10):55, 1954.
4. J. E. Wilson, Ind. Eng. Chem. 47(10):2201, 1955.
5. H. Beachell and P. Nemphes, J. Polymer Sci. 21(97):113, 1956.
6. S. P. Foster and W. W. Spohn, Wire and Wire Products 30(12):1487, 1955.
7. T. G. Borras and J. V. Cavender, J. Polymer Sci. 24(105):138, 1957.

8. V. B. Miller, M. B. Neiman, V. S. Pudov, and L. I. Lafer, Vysokomolekulyarnye Soedineniya 1(11):1696, 1959.

9. V. B. Miller, M. B. Neiman, and Yu. A. Shlyapnikov, Vysokomolekulyarnye Soedineniya 1(11):1703, 1959.

10. V. V. Dudorov, M. B. Neiman, and A. F. Lukovnikov, Plast. Mass. No. 12:3, 1961.

11. B. Grieveson, R. Havard, and B. Wright, Thermostable Polymer Symposium, London, 1960.

12. F. Grafmüller and E. Husemann, Markomol. Chem. 40(3):161, 172, 1960.

13. A. L. D. Gol'denberg, L. N. Pirozhnaya, G. S. Popova, and L. I. Tarutina, Collection: Molecular Spectroscopy, Leningrad University Press, 1960.

14. N. Kavafian, J. Polymer. Sci. 24(107):499, 1957.

15. M. B. Neiman, G. I. Likhtenshtein, Yu. S. Konstantinov, N. P. Karpets, and Ya. G. Urman, Vysokomolelulyarnye Soedineniya 5(11):1706, 1963.

16. H. C. Beachell and G. W. Tarbet, J. Polymer Sci. 45(146):451, 1960.

17. E. N. Matveeva, N. P. Lazareva, and M. Z. Kremen', USSR Patent No. 658346/23, 1960; Byul. Izobretenii, No. 15, 1962.

18. J. D. Burnett and G. Miller, J. Polymer Sci. 15(80):592, 1955.

19. G. Natta, E. Beati, and F. Severini, J. Polymer Sci. 34(127):685, 1959.

20. V. S. Pudov, B. A. Gromov, M. B. Neiman, and E. G. Sklyarova, Neftekhimiya 3(5):543, 1963.

21. V. S. Pudov and M. B. Neiman, Neftekhimiya 2(6):918, 1962.

22. Yu. A. Shlyapnikov, V. B. Miller, M. B. Neiman, E. S. Torsueva, and B. A. Gromov, Vysokomolekulyarnye Soedineniya 2(9):1409, 1960.

23. M. B. Neiman, V. B. Miller, Yu. A. Shlyapnikov, and E. S. Torsueva, Doklady Akad. Nauk SSSR 136(3):647, 1960.

24. A. S. Kuz'minskii, N. N. Lezhnev, and Yu. S. Zuev, Oxidation of Raw and Cured Rubbers, Moscow, State Press for Chemical Literature, 1957.

25. G. E. Ashby, J. Polymer Sci. 50(153):99, 1961.

26. L. G. Angert and A. S. Kuz'minskii, The Role and Use of Antioxidants in Raw and Cured Rubbers, Moscow, State Press for Chemical Literature, 1957.

27. L. G. Angert, L. A. Gol'dfarb, G. I. Gorushkina, A. I. Zenchenko, A. S. Kuz'minskii, and B. P. Fedorov, Zhur. Priklad. Khim. 32:408, 1959.

28. B. P. Fedorov, F. M. Stoyanovich, A. F. Lukovnikov, T. A. Bulgakova, and P. I. Levin, Vysokomolekulyarnye Soedineniya (in press).

29. M. S. Khloplyankina, A. F. Lukovnikov, and P. I. Levin, Vysoko-molekulyarnye Soedineniya 5(2):195, 1963.
30. J. R. Thomas, J. Am. Chem. Soc. 82(22):5955, 1960.
31. A. T. Koritskii and A. F. Lukovnikov, Doklady Akad. Nauk SSSR 147:5, 1126, 1962.
32. G. I. Likhtenshtein, N. A. Sysoeva, and M. B. Neiman, Tr. po Khim. i Khim. Tekhnol. No. 2:320, 1962.
33. Zbl. 121, N 1, 11, 115, 1951; Zbl. 122, N 25, 11S, 1951.
34. U. S. Patent 2675366, 1954.
35. Z. A. Rogovin and T. V. Druzhinina, Nauchn. Dokl. Vysshei Shkoly, No. 1:131, 1958.
36. U. S. Patent 2820775, 1958.
37. E. N. Matveeva, E. I. Kirillova, and N. P. Lazareva, Byul. Izo-bretenii, No. 9, 1961; USSR Patent No. 138026, 1961.
38. Great Britain Patent 758973, 1954.
39. E. N. Matveeva, N. P. Obol'yaninova, M. Z. Kremen', and N. P. Lazareva, USSR Patent No. 140987; Byul. Izobretenii, No. 17, 1961.
40. D. R. Stevens and A. C. Dubbs, U. S. Patent 2570402, 1951.
41. W. L. Hawkins, V. L. Lanza, B. B. Loeffler, W. Matreyek, and F. H. Winslow, J. Appl. Polymer Sci. 1:43, 1959; Rubber Chem. Technol. 32(4):1171, 1959.
42. Mod. Plastics, No. 12:46, 1956.
43. D. R. Stevens and W. A. Gruse, U. S. Patent 2263582, 1941.
44. N. M. Emanuél' and Yu. N. Lyaskovskaya, Inhibition of Processes of Fat Oxidation, Moscow, State Press for the Food Industry, 1961.
45. E. Müller, K. Ley, K. Scheffler, and R. Mayer, Ber. 91:2682, 1958.
46. A. L. Buchachenko, Ya. S. Lebedev, and M. B. Neiman, Zhur. Strukt. Khim. 2(5):558, 1962.
47. A. L. Buchachenko and M. B. Neiman, Doklady Akad. Nauk SSSR 139(4):916, 1961.
48. C. D. Cook, D. A. Kuhn, and P. Flann, J. Am. Chem. Soc. 78:2002, 1956.
49. A. F. Bickel and E. C. Kooyman, J. Chem. Soc. 1953, 3211.
50. U. S. Patent 2727879, 1951.
51. French Patent 1159958, 1956.
52. M. S. Khloplyankina, A. L. Buchachenko, A. F. Lukovnikov, and P. I. Levin, Summaries of Reports at the Conference on the Aging and Stabilization of Polymers, Moscow, Academy of Sciences, USSR Press, 1961.
53. G. H. Denison, Ind. Eng. Chem. 36:477, 1944; G. H. Denison and P. C. Condit, Ind. Eng. Chem. 37(11):1102, 1945.
54. G. W. Kennerly and W. L. Patterson, Ind. Eng. Chem. 48:1917, 1956.

55. Canadian Patent 498511, 1953.

56. N. T. Notley, Trans. Faraday Soc. 58(1):66, 1962.

57. Zbl. 121, N 1, 115, 1951.

58. German Federated Republic Patent 1026953, 1955.

59. F. Yu. Rachinskii, T. G. Potapenko, E. N. Matveeva, and M. Z. Kremen', Plastmassy No. 2:12, 1961; Byul. Izobretenii, No. 20:18, 1960.

60. F. Yu. Rachinskii, E. N. Matveeva, M. Z. Kremen', T. G. Potapenko, and N. P. Lazareva, Byul. Izobretenii, No. 20, 1961.

61. Great Britain Patent 779807, 1957.

62. P. A. Kirpichnikov, Summaries of Reports at the Conference on the Aging and Stabilization of Polymers, Moscow, Academy of Sciences USSR Press, 1961.

63. P. I. Levin, P. A. Kirpichnikov, A. F. Lukovnikov, and M. S. Khloplyankina, Vysokomolekulyarnye Soedineniya 5(8):1152, 1963.

64. N. V. Mikhailov, L. G. Tokareva, P. A. Kirpichnikov, and A. G. Popov, Summaries of Reports at the Conference on the Aging and Stabilization of Polymers, Moscow, Academy of Sciences USSR Press, 1961.

65. C. Walling and R. Rabinowitz, J. Am. Chem. Soc. 81:1243, 1959.

66. D. B. Denney, W. F. Goodyear, and B. Goldstein, J. Am. Chem. Soc. 82:1393, 1960.

67. K. U. Ingold and J. E. Puddington, Ind. Eng. Chem. 51:185, 1959.

68. K. I. Ivanov, E. D. Vilyanskaya, and A. A. Luzhetskii, Teploenerg. No. 11:34, 1960.

69. P. I. Levin, A. F. Lukovnikov, M. B. Neiman, and M. S. Khloplyankina, Vysokomolekulyarnye Soedineniya 3(8):1243, 1961.

70. French Patent 1245606, 1955; 1247236, 1960; Great Britain Patent 851670, 1959; Austrian Patent 210625, 1960.

71. C. Golumbic and H. A. Mattill, J. Am. Chem. Soc. 63:1279, 1941.

72. C. Golumbic, Oil and Soap, JAOCS 19:181, 1942.

73. C. Golumbic, Oil and Soap, JAOCS 20:105, 1943.

74. V. P. Calcins and H. A. Mattill, J. Am. Chem. Soc. 66:239, 1944.

75. H. A. Mattill, Oil and Soap, JAOCS 22:1, 1945.

76. W. L. Hawkins and M. A. Worthington, Chem. Ind. No. 32:1023, 1960.

77. W. L. Hawkins, M. A. Worthington, and F. H. Winslow, Rubber Age 88(2):279, 1960; Ekspress-Inform. CBM, No. 8, 1961, Reference 107.

78. W. L. Hawkins, Rubber Plast. Week. 142(8):291, 293, 299, 1962.

79. W. L. Hawkins and F. H. Winslow, Plastics Inst. (London) Trans. J. 29:81, 88, 1961.

80. Yu. A. Shlyapnikov, V. B. Miller, M. B. Neiman, E. S. Torsueva, and B. A. Gromov, Vysokomolekulyarnye Soedineniya 2(9):1409, 1960.

81. E. N. Matveeva, A. A. Kozodoi, and A. L. Gol'denberg, Summaries of Reports at the Conference on the Aging and Stabilization of Polymers, Moscow, Academy of Sciences USSR Press, 1961.
82. P. Maltese, Materie Plastische 23(2):107, 1957.
83. V. Ya. Efremov, M. B. Neiman, B. V. Rozynov, and Yu. E. Vilents, Plast. Mass. No. 9:4, 1962.
84. A. Pross and P. Black, J. Soc. Chem. Ind. 69(4):113, 1950.
85. A. Burgess, Chem. & Ind. No. 4:78, 1952; Natl. Bur. Std. (U.S.) Cir. No. 525:149, 1953.
86. A. S. Kuz'minskii, Oxidation of Raw and Cured Rubbers, Moscow, State Press for Chemical Literature, 1957.
87. W. S. Penn, Rubber Plast. Week. No. 1:23, 1962.
88. C. P. Vale, Chem. Ind. No. 9:8, 1961.
89. F. R. Hansen, Plastics World No. 11:14, 1961.
90. J. W. Tamblyn, G. S. Newland, and M. T. Watson, Plastics Technol. 4:427, 455, 1958.
91. R. A. Coleman and G. A. Weicksel, Mod. Plastics 8:117, 1959.
92. N. A. Lapin, E. N. Matveeva, and V. Smirnova, Zhur. Obshchei Khim. 30:2377, 1960.
93. N. A. Lapin, E. N. Matveeva, and L. Kazhutkina, Zhur. Obshchei Khim. 32:367, 1962.
94. U. S. Patent 2839418, 1958.
95. U. S. Patent 2853521, 1959.
96. SPE Journal 13(2):23, 1957.
97. Mod. Plastics No. 2:103, 1959.
98. C. G. Cottfried, J. Appl. Polymer Sci. 5:17, 612, 1961.

Chapter V

AGING AND STABILIZATION OF POLYFORMALDEHYDE

1. THERMAL AND THERMOOXIDATIVE DESTRUCTION

Polyformaldehyde – a high-molecular polyoxymethylene, possessing and average degree of polymerization of 1000-4000 and a melting point of 175-180°C, possesses high mechanical, dielectric, and technological properties.

However, a vital shortcoming of this polymer for reprocessing into objects is its extreme instability; at temperatures below the melting point (~100°C) the polymer decomposes readily, liberating monomeric formaldehyde. Hence the reprocessing of polyformaldehyde can be accomplished only after preliminary stabilization of the product.

In the works of Kern and Cherdron [1-3], it was shown that the thermal destruction of polyformaldehyde occurs according to an ionic mechanism, and that a vital influence on this process is exerted by the nature of the terminal groups. Thus, the blocking (esterification) of the terminal hydroxyl groups considerably increases the stability of the polymer to thermal decomposition, preventing depolymerization from the ends of the chain. The presence of oxygen accelerates the decomposition reaction, which is apparently related to the oxidation of formaldehyde to formic acid, which also gives rise to supplementary acidolytic decomposition of the polyoxymethylene chains.

Later Dudina and Enikolopyan [4-12] expressed considerations in favor of the possible occurrence of the process of thermal destruction of polyformaldehyde according to ionic and radical mechanisms. Moreover, the authors established that in the thermal decomposition of the polymer, the natural gaseous product is monomeric formaldehyde. The participation of oxygen in the thermal oxidation of polyformaldehyde is manifested only in an initiation of depolymerization, which proceeds according to the "laws of chance" for the polymer with blocked terminal groups, and in the fact that the oxidative direction of the reaction is absent in this case.

However, these works contain no direct evidence in favor of the occurrence of radical processes during the thermal destruction of polyformaldehyde. However, the considerable experimental material currently available in the field of study of the thermal destruction of various polymers (polyolefins, polyamides, epoxide resins, etc.) gives evidence in favor of the occurrence of these processes according to a radical mechanism.

By analogy with other polymers (polypropylene, epoxide resins, etc.) [13-15], the thermal decomposition of which is a radical reaction, M. B. Neiman and associates [16] proposed that the thermal decomposition of polyformaldehyde can also proceed according to a radical mechanism. Two types of free radicals can be produced in the initial stage of the process:

$$\sim O - CH_2 - O - CH_2 - O^\cdot \tag{1}$$

$$\sim O - CH_2 - O - \overset{\cdot}{C}H_2 \tag{2}$$

Formaldehyde is produced when the radicals formed decompose.

From the energy standpoint, cleavage of a bond such that it is accompanied by the formation of a double bond is the most favorable.

Isomerization of the radicals formed is also possible. Thus, it has been shown in a number of studies [14, 15] that isomerization of the polymer radicals formed occurs in the thermal destruction of polypropylene and polyethylene oxide.

In [16] the hypothesis was advanced that the isomerization of the radicals (1) and (2) can proceed according to the following scheme, forming formic acid and methyl formiate:

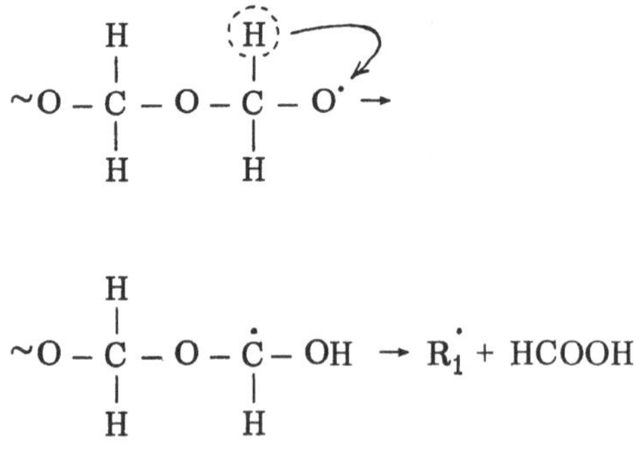

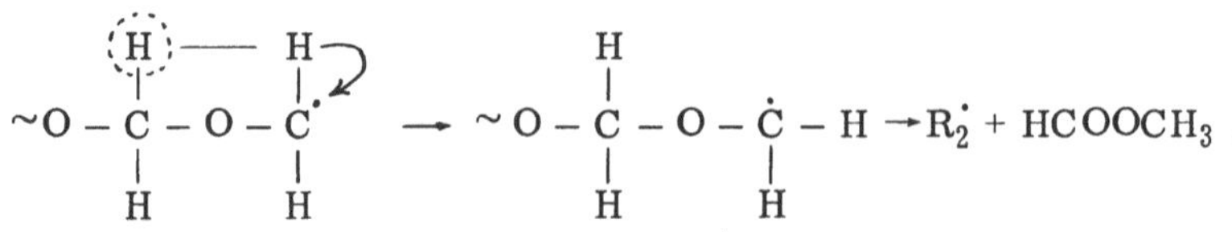

If isomerization occurs on account of a jumping of hydrogen from the farther carbon atoms, then methanol can be formed:

(1) 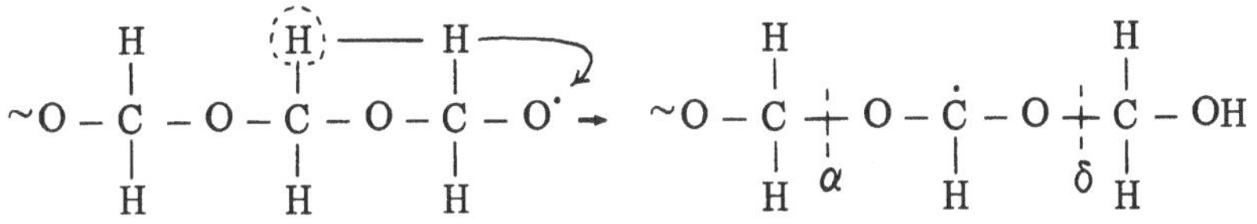
$$\sim O - \underset{\underset{H}{|}}{\overset{\overset{H}{|}}{C}} - O - \underset{\underset{H}{|}}{\overset{\overset{(H)}{|}}{C}} - O - \underset{\underset{H}{|}}{\overset{\overset{H}{|}}{C}} - O^{\cdot} \rightarrow \sim O - \underset{\underset{H}{|}}{\overset{\overset{H}{|}}{C}} \underset{\alpha}{+} O - \underset{\underset{H}{|}}{\overset{\overset{\cdot}{|}}{C}} - O \underset{\delta}{+} \underset{\underset{H}{|}}{\overset{\overset{H}{|}}{C}} - OH$$

This radical can decompose in the following way:

a) 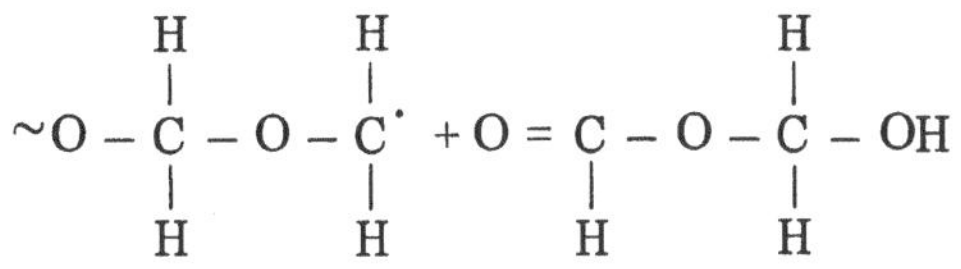
$$\sim O - \underset{\underset{H}{|}}{\overset{\overset{H}{|}}{C}} - O - \underset{\underset{H}{|}}{\overset{\overset{H}{|}}{C}}^{\cdot} + O = \underset{\underset{H}{|}}{\overset{\overset{H}{|}}{C}} - O - \underset{\underset{H}{|}}{\overset{\overset{H}{|}}{C}} - OH$$

b) 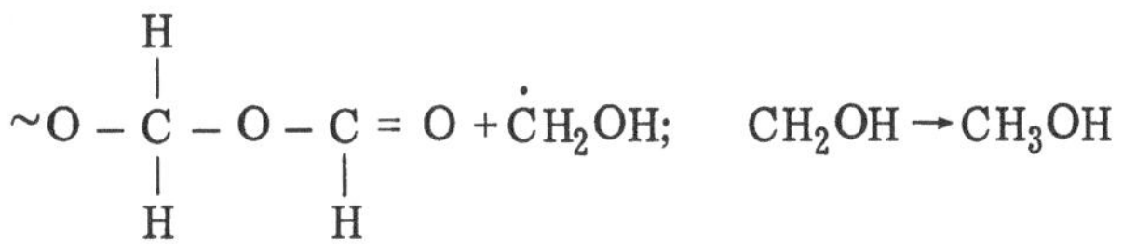
$$\sim O - \underset{\underset{H}{|}}{\overset{\overset{H}{|}}{C}} - O - \underset{\underset{H}{|}}{\overset{}{C}} = O + \dot{C}H_2OH; \quad CH_2OH \rightarrow CH_3OH$$

In addition, more complex molecules of the $-C-O-C-O-C-$ type with various functional groups on the ends can also be formed.

Thus, it can be assumed that if methyl formiate, formic acid, methanol, and certain other substances of more complex structure are found in the thermal decomposition products of polyformaldehyde, this will be important evidence in favor of a radical mechanism of the thermal decomposition reaction.

The thermal destruction of polyformaldehyde with acetylated terminal groups was conducted on a circulation setup at a temperature of 300°C. Since the basic product of the thermal destruction of polyformaldehyde is monomeric formaldehyde, a successful performance of the analysis necessitated its separation from the other decomposition products, the amounts of which were two to three orders of magnitude smaller than the amount of the formaldehyde formed. For this a method was developed for separating formaldehyde, utilizing its property of spontaneous polymerization to Eu-polyoxymethylene at the temperature of dry ice. The yield of formaldehyde was about 98% of the weight of the destructured polymer. After its removal, the decomposition products were subjected to chromatographic separation.

A chromatogram of the destruction products of polyformaldehyde is presented in Fig. 65. The following products were preliminary chromatographed: methyl formiate, methanol, water, and formic acid, and the quantitative dependences of the area of the peak on the amount of the substance introduced into the chromatograph were obtained. When the destruction products were identified, the following were found: methyl formiate, methanol, and water. The yield of methyl formiate was ~1%.

In addition to chromatographic separation and identification with a mass spectrometer, the IR spectra of the destruction products of polyformaldehyde were taken. For this purpose the destruction products were dissolved in carbon tetrachloride.

On the IR spectrum of the decomposition products, presented in Fig. 66, the absorption bands corresponding to pure methyl formiate are plotted with a dotted line. The coincidence of the frequencies and intensity ratio of the bands obtained confirms the presence of methyl formiate in the destruction products. The intense absorption lines corresponding to 1100 and 1030 cm^{-1} on the spectrum of the destruction products of polyformaldehyde are also observed in simple ethers and correspond to the valence vibrations of the ether bonds in compounds of the type $-C-O-C-O-C-$.

Thus, methyl formiate, methanol, hydrogen, and methane were detected in the destruction products of polyformaldehyde using various methods.

It was shown in the cited work that under these conditions methyl-formiate is not formed according to a Tishchenko reaction

$$2CH_2O = HCOOCH_3$$

as was indicated in a number of studies of the acidolysis of low-molecular formaldehyde polymers [3, 17].

Consequently, the formation of methyl formiate and methanol is possible only as a result of isomerization of the radicals arising during destruction, which confirms the radical nature of the thermal decomposition of polyformaldehyde. Formic acid could not be detected in chromatographic, mass spectral, and polarographic analyses of the destruction products of polyformaldehyde.

The thermal oxidation of polyformaldehyde with acetylated terminal groups was studied on a special static setup, in which all parts of the reaction vessel could be kept at the same temperature, so as to avoid poly-

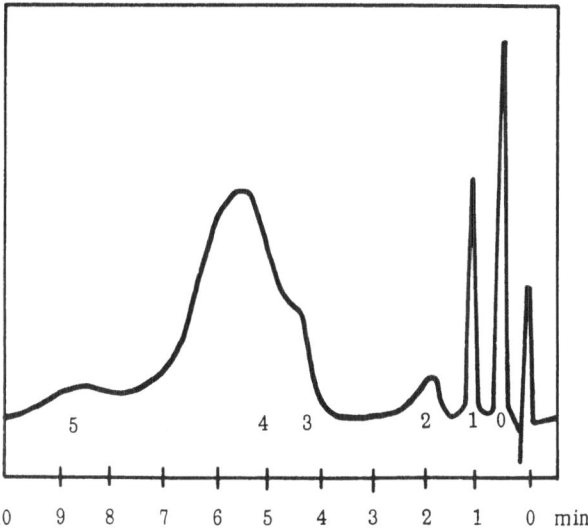

Fig. 65. Chromatogram of the destruction products of polyformaldehyde.   1) Methyl formiate; 2) methanol; 3) water.

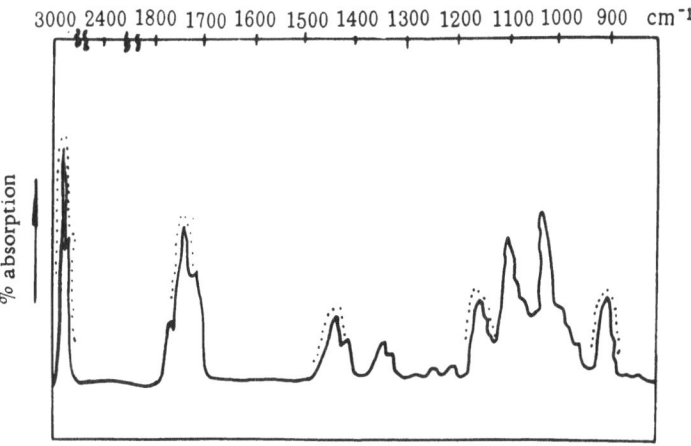

Fig. 66. IR spectrum of the destruction products of polyformaldehyde. The
absorption bands corresponding to pure methyl formiate are plotted with a
dotted line.

merization of the formaldehyde liberated upon oxidation [18, 19].   Parallel with the experiments on gas evolution, experiments were conducted
on the loss in weight of the polyformaldehyde at constant temperature, and
it was established that the curves of the increase in pressure characterize the kinetics of the process, just like the curves of the loss in weight.

A comparison of the data on the kinetics of gas evolution and loss in
weight during the oxidation of polyformaldehyde is presented in Fig. 67.

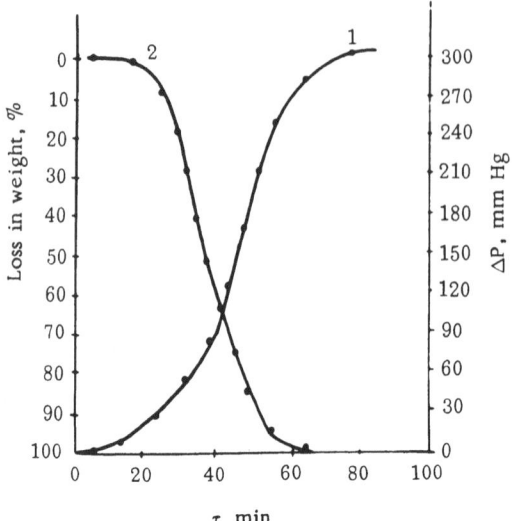

Fig. 67. Kinetic curves of the gas evolution (1) and loss in weight (2) in the thermooxidative destruction of poly-formaldehyde. T = 200°C.

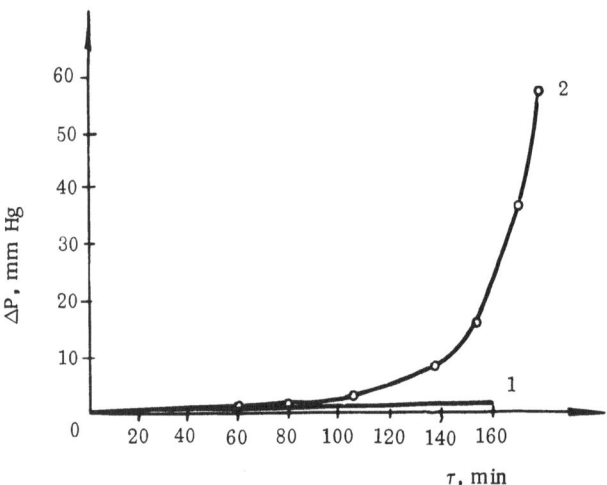

Fig. 68. Curves of increase in pressure in thermal (1) and thermo-oxidative destruction of polyformaldehyde (2). T = 160°C.

If there is no oxygen in the reaction vessel, then at 145°C the in-crease in the pressure on account of the decomposition of polyformalde-hyde occurs extremely slowly (Fig. 68, curve 1). In the presence of oxygen, on the other hand (P_{O_2} = 600 mm Hg; curve 2), this process is considerably accelerated, from which we can conclude that oxygen stim-ulates the thermal decomposition of the polymer.

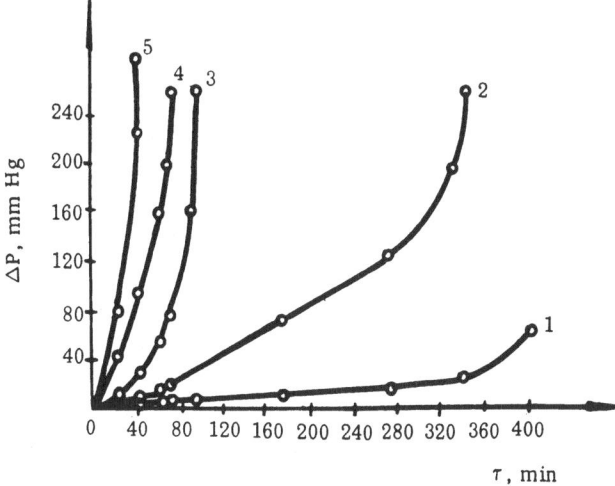

Fig. 69. Kinetic curves of the oxidation of polyformaldehyde.
P_{O_2} = 200 mm Hg. 1) 160°; 2) 190°; 3) 200°; 4) 210°; 5) 220°C.

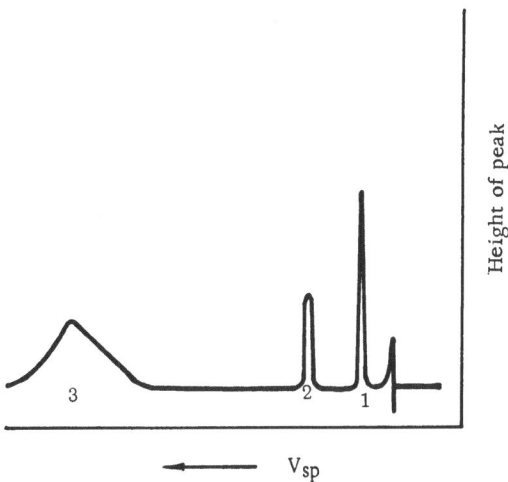

Fig. 70. Chromatogram of the oxidation product of
polyformaldehyde. 1) Gas; 2) methyl formiate; 3)
water.

The induction period τ, during which the pressure increases comparatively slowly, and a period of rapid occurrence of the reaction can
be distinctly seen on the curve. When the partial pressure of oxygen is
increased, or when the temperature is raised, the value of the induction
period decreases. The kinetic curves of the oxidation of polyformaldehyde are presented in Fig. 69. The induction period in the oxidation of
polyformaldehyde is appreciably lengthened when oxidation inhibitors are
introduced.

Attempts to determine hydroperoxides in the oxidation of polyformaldehyde by an iodometric method did not give positive results. Apparently, in this case hydroperoxides do not accumulate in analytically determinable amounts. In [11], hydroperoxides also were not detected in the oxidation of polyformaldehyde.

It was shown in [20, 21] that two types of radicals are formed when polyformaldehyde is irradiated: the stable radical $\sim O - \overset{\cdot}{C}H - O \sim$ and short-lived radicals, formed as a result of cleavage of the polymer chains. It was noted in this work that the reaction of the polyformaldehyde radicals with oxygen is not accompanied by the formation of peroxide radicals, in spite of an acceleration of their destruction under the influence of oxygen, and it was established that the destruction of polyformaldehyde radicals in the presence of oxygen is described by a second-order equation.

To explain the second order of the destruction of radicals in oxygen, the authors proposed a kinetic scheme proceeding through the formation and decomposition of peroxide radicals and showed that the decomposition of the peroxide radical occurs at a great rate.

An analysis of the destruction products was conducted to refine the mechanism of the oxidation of polyformaldehyde.

On the basis of the hypothesis that oxygen initiates the thermal decomposition of polyformaldehyde, we also had the right to expect the formation of such substances as methyl formiate, methanol, etc., produced in the thermal destruction of the polymer, in this case as well.

The oxidation of polyformaldehyde was conducted on a circulation setup at temperatures of 150-170° and 190°C and various initial oxygen pressures. At these temperatures thermal decomposition of the polymer is practically absent. The destruction products were frozen out in liquid oxygen. An analysis of the products was conducted on a chromatograph and on a mass spectrometer. The sample was preliminarily preheated for a long time under vacuum at a temperature of 80-90°C to remove the sorbed water. Figure 70 presents a chromatogram of the oxidation products of polyformaldehyde. The negligible amounts of the formaldehyde formed in the oxidation were removed according to the procedure described above. The basic oxidation products of polyformaldehyde at temperatures of 150 and 170°C (in addition to formaldehyde) are water and methyl formiate.

Formic acid was not detected in these experiments. The amount of water in the oxidation products increases with increasing initial oxygen pressure and temperature. Thus, the yield of water in the oxidation of the polymer at a temperature of 150°C and a pressure of 125 mm Hg com-

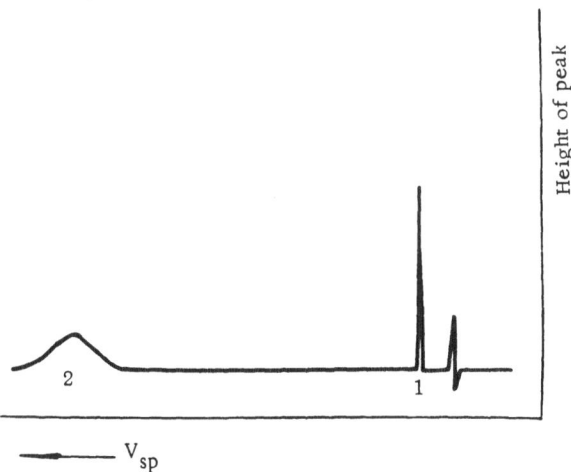

Fig. 71. Chromatogram of the products obtained in the heating
of preliminary oxidized polyformaldehyde: 1) Gases; 2) water.

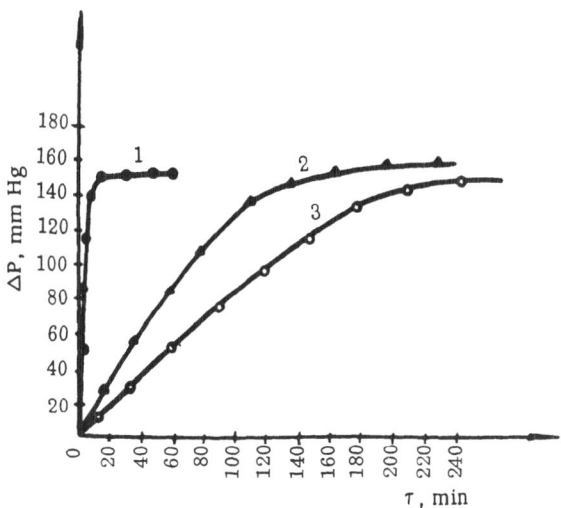

Fig. 72. Kinetics of the absorption of polyformaldehyde by
polyamides: 1) Polyamide 68; 2) polycaproamide; 3) polyamide 54; T = 200°C; P_{CH_2O} = 450 mm Hg.

prised 5.4×10^{-6} mole; when the process was conducted at the same
temperature and an oxygen pressure of 375 mm Hg, the yield of water
was 1.5×10^{-5} mole per 100 mg of the initial sample.

Heating of the initial sample in a stream of helium, conducted under
analogous conditions, showed the absence of water in the destruction
products of the polymer. However, in the case of preheating of preliminarily oxidized samples of polyformaldehyde in a stream of helium at a

temperature of 150°C, water was again detected in the destruction products (Fig. 71).

The formation of methyl formiate indicates isomerization of the radicals in the oxidation of polyformaldehyde. These same processes also occur in the thermal decomposition of the polymer. Water arises during decomposition of oxygen-containing groups formed in the oxidation of the polymer, which confirms the intrinsically oxidative direction of the thermooxidative destruction of polyformaldehyde.

The presence of induction periods and autoacceleration in the oxidation of polyformaldehyde, the inhibition of the reaction by small additions of inhibitors, and the occurrence of processes of radical isomerization during oxidation permit us to assume that the thermooxidative destruction of polyformaldehyde is a radical chain process.

2. STABILIZATION OF POLYFORMALDEHYDE

Kern and Cherdron [1, 2] were the first to establish that to obtain stable polyformaldehyde, one must block (esterify) the terminal groups of the polymer and thereby prevent depolymerization from the ends of the chains. As Dudina and Enikolopyan [7] have shown, replacement of the terminal hydroxyl groups by acetyl groups increases the activation energy of polymer decomposition from 26 to 32 kcal/mole.

In addition to blocking of the terminal groups, Kern and Cherdron proposed that the monomeric formaldehyde evolved during decomposition be bonded by suitable additives (polyamides, urea, etc.), in order to prevent its oxidation to formic acid, which promotes acidolytic cleavage of the polymer macromolecules.

Enikolopyan and Vardanyan [4] proposed that the oxidative destruction of polyformaldehyde be prevented not only by bonding the monomeric formaldehyde liberated, but also by tying up the formic acid formed.

Kern and Cherdron [2] propose the introduction into the polymer not only of additions of formaldehyde acceptors, but also of antioxidants, which, as has been shown on other polymers, satisfactorily solve the problem of stabilization against oxidation. However, the use of antioxidants was not substantiated in the indicated studies, if we consider that the decomposition of polyformaldehyde proceeds according to an ionic mechanism according to the data of these authors. However, it is known that stabilizing additives of the type of phenols, amines, etc., decelerate processes that proceed only through the formation of free radicals.

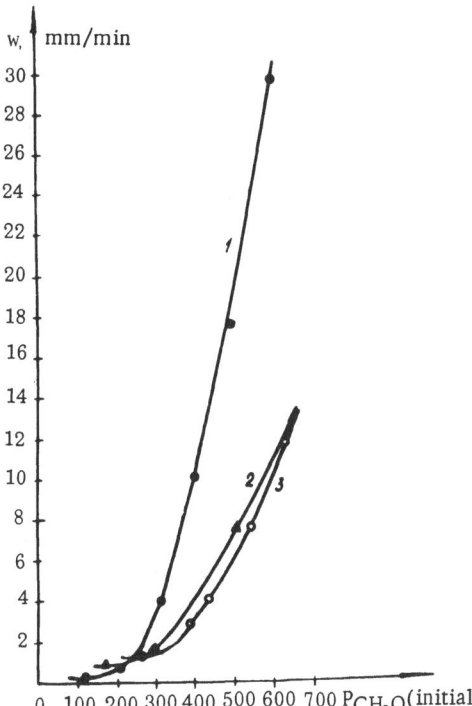

Fig. 73. Dependence of the rate (w) of absorption of formaldehyde on its initial pressure for polyamides: 1) Polyamide 68; 2) polyamide 54; 3) polycaproamide. T = 200°C.

3. BONDING OF FORMALDEHYDE BY POLYAMIDES

Various polyamides are usually used as acceptors of formaldehyde, the liberation of which accelerates the decomposition of polyformaldehyde. There are no comparative data in the literature pertaining to the kinetics and mechanism of the bonding of formaldehyde by polyamides of various structures.

To solve this problem, a special procedure was developed in [22], permitting the evaluation of the effectiveness of polyamide resins as formaldehyde acceptors. The authors utilized the reaction of direct absorption of formaldehyde by the polyamide resin. It might be assumed that the most reactive with respect to formaldehyde are polyamide resins that also prove to be the best acceptors when added to the polymer. α-Polyoxymethylene, the decomposition of which under the experimental conditions ended in the first three minutes, was used as the source of formaldehyde. The experiments were conducted on a static setup, guaranteeing complete thermostatic control of all the parts of the reaction vessel. The effectiveness of the resins was evaluated according to the drop in the formaldehyde pressure in the system on account of its absorption by the polyamide.

The following were investigated: polyamide, polycaproamide, polyamide 68 (polyhexamethylenesebacamide), and polyamide-54 (condensation product of 1 : 1 caproamide and hexamethyleneadipamide).

Figure 72 presents the kinetic absorption curves of polyformaldehyde by various polyamides. The greatest rate of absorption of polyformaldehyde is noted for polyamide 68, the saturation limit of which is established in the first 15 min from the beginning of the experiment. The lowest rate of absorption is noted for polyamide 54, for which the equilibrium state is achieved 20 min after the beginning of the experiment.

The rate of achievement of the maximum absorption of formaldehyde, as is shown on Figs. 73 and 74, increases sharply with increasing initial pressure of formaldehyde and temperature. Thus, when the initial formaldehyde pressure is raised from 300 mm to 600 mm Hg, the rate of its absorption for polyamide 68 increases sevenfold, and five- to sixfold for capron and polyamide 54. However, the final volume of the formaldehyde absorbed by the resins (Fig. 73) falls with increasing temperature.

Such a dependence was also observed in [23], where the decrease in the absorption of formaldehyde by the polyamide resin was partially explained by an increase in the number of cross-links in the sample on account of the interaction of methylol derivatives.

In a study of the kinetics of the absorption of formaldehyde, it was of interest to elucidate the possible transformations that occur in polyamides.

It has been shown in a number of studies [23, 24] that when formaldehyde reacts with a polyamide, methylol groups are formed on the nitrogen atom according to the reaction

$$R - NH - CO - R_1 + CH_2O \rightarrow R - \underset{\underset{CH_2OH}{|}}{N} - CO - R_1$$

which at high temperatures (above 100-150°) are capable of reacting with one another and with the hydrogen of the amino group, forming methylene ether and methylene bonds in the polyamides [25]:

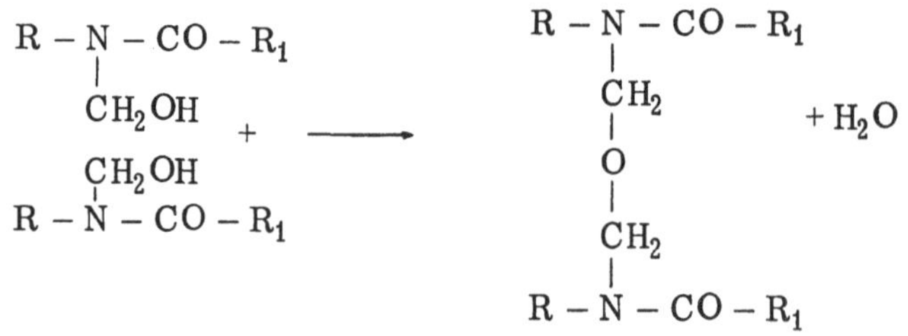

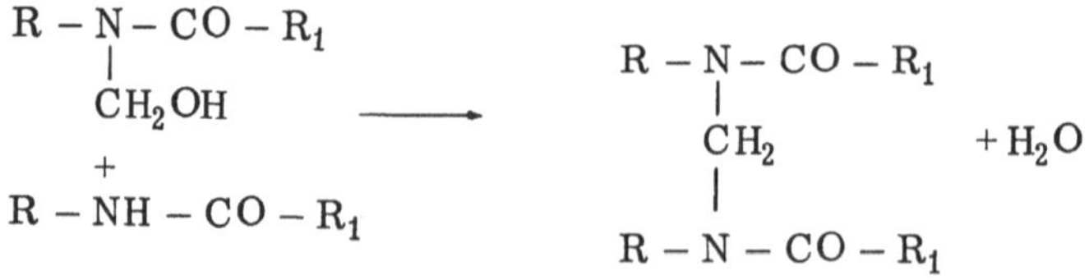

$$\begin{array}{c} R - \underset{\underset{CH_2OH}{|}}{N} - CO - R_1 \\ + \\ R - \underset{\overset{|}{CH_2OH}}{N} - CO - R_1 \end{array} \longrightarrow \begin{array}{c} R - \underset{\underset{CH_2}{|}}{N} - CO - R_1 \\ \underset{\overset{|}{O}}{} \\ \underset{\overset{|}{CH_2}}{} \\ R - \underset{}{N} - CO - R_1 \end{array} + H_2O$$

$$\begin{array}{c} R - \underset{\underset{CH_2OH}{|}}{N} - CO - R_1 \\ + \\ R - NH - CO - R_1 \end{array} \longrightarrow \begin{array}{c} R - \underset{\underset{CH_2}{|}}{N} - CO - R_1 \\ \underset{}{} \\ R - \underset{}{N} - CO - R_1 \end{array} + H_2O$$

$$
\begin{array}{c}
R - N - CO - R_1 \\
| \\
CH_2 \\
| \\
O \\
| \\
CH_2 \\
| \\
R - N - CO - R_1
\end{array}
\quad\longrightarrow\quad
\begin{array}{c}
R - N - CO - R_1 \\
| \\
CH_2 \qquad + CH_2O \\
| \\
R - N - CO - R_1
\end{array}
$$

A study of the infrared spectra of hardened methylol-polyamide [26] showed that the appearance of three-dimensional structures in such resins is related to the formation of methylene bridges.

A determination of the solubility of polyamides, conducted in [22] at various stages of their reaction with formaldehyde at a temperature of 200°C, showed that a sharp drop in solubility, which reaches 50%, occurs 10 min after the beginning of the experiment for polycaproamide and after 60 min for polyamide 54. After 200 minutes from the beginning of the experiment, the solubility of these resins is reduced to 10-20% (Fig. 75). The data obtained permit us to conclude that the drop in the solubility of the resins in the reaction with formaldehyde is a result of the formation of three-dimensional structures in them.

Even at the early stages of absorption of formaldehyde, no methylol derivatives could be detected in these resins. This is apparently due to the rapid formation of methylene bonds in the resins. In polyamide 68, methylol derivatives were detected (up to 3% of the weight of the polymer 15 min after the beginning of the experiment) at all stages of the absorption of formaldehyde. The content of methylol groups was determined polarigraphically. It corresponded to the amount of free polyformaldehyde liberated in the decomposition of methylol groups in alkaline medium [27]. However, even at the maximum absorption of formaldehyde, polyamide 68 remains entirely soluble in tricresol under these conditions.

Thus, all the polyamides used are capable of bonding formaldehyde; however, the rate of these processes varies. As was indicated above, the bonding of formaldehyde leads to various chemical conversions in polyamides.

The possibility of transferring the results obtained with formaldehyde to polyformaldehyde required verification. Actually, in the oxidation of polyformaldehyde with additions of 1-2% of the corresponding polyamides, analogous results were obtained.

All the brands of polyamides studied, in conjunction with an antioxidant, manifest considerable effectiveness in the stabilization of polyformaldehyde.

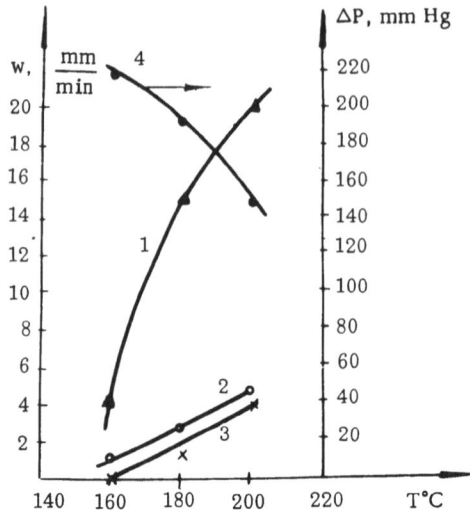

Fig. 74. Dependence of the rate w (curves 1, 2, 3) and the maximum of the absorption (curve 4) of form-aldehyde by polyamides on the temperature: 1, 4) Polyamide 68; 2) polyamide 54; 3) polycaproamide. P_{CH_2O} = 600 mm Hg.

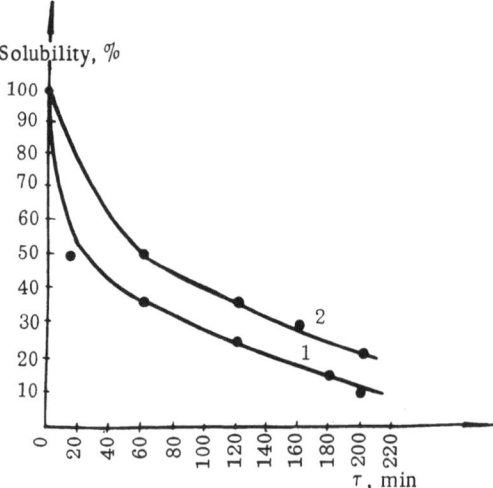

Fig. 75. Kinetics of the change in solubility of poly-amides after their interaction with formaldehyde. 1) Polycaproamide; 2) polyamide 54.

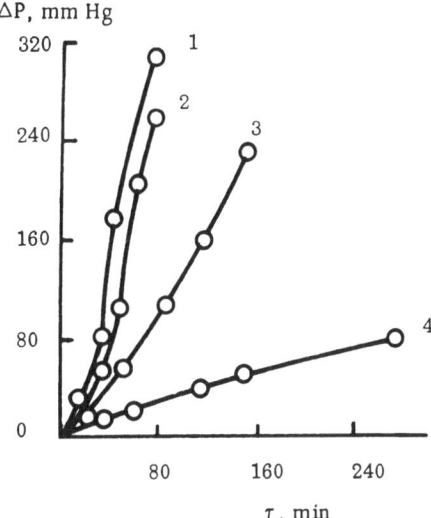

Fig. 76. Curves of the increase in pressure in the oxidation of polyformaldehyde (200°C, P_{O_2} = 200 mm Hg) without additions (1) and with additions of polyamide (2), anti-oxidant (3), and a mixture of them (4).

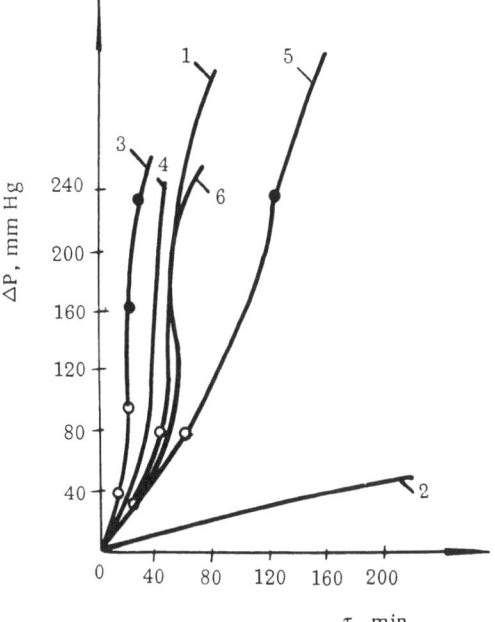

Fig. 77. Effectiveness of the action of inhibitors of the type of phosphites and sulfides, introduced into the poly-mer (with an addition of 2% polyamide) in a 0.5% con-centration; T = 200°C; P_{O_2} = 200 mm Hg. 1) Polymer without additives; 2) stable polymer; 3) ionol pyro-catechol phosphite; 4) polygard; 5) SAO-6; 6) polysulfide.

4. INHIBITION OF THE PROCESS OF OXIDATION
OF POLYFORMALDEHYDE BY INHIBITORS,
DERIVATIVES OF AROMATIC AMINES AND PHENOLS,
PHOSPHITES, AND SULFUR-CONTAINING COMPOUNDS

Experiments we conducted have shown that the addition of a poly-amide alone or an inhibitor alone is relatively ineffective, somewhat reducing the gas evolution in the oxidation of polyformaldehyde. Only the joint introduction of a polyamide and an inhibitor proves far more effec-tive than the introduction of the polyamide and antioxidant individually (Fig. 76). The effectiveness of the inhibitor was evaluated in [22] ac-cording to the amount and rate of gas evolution in the thermooxidative destruction of polyformaldehyde [18] on a static setup.

The principles obtained confirm the complex mechanism of the pro-cess of oxidation of polyformaldehyde. Although the introduction of in-hibitors that terminate chain oxidation processes is sufficient to de-celerate (inhibit) decomposition processes for a number of polymers (polyolefins, polyamides, etc.), for polyformaldehyde, additives·are needed which, on the one hand, might inhibit chain oxidation processes, and, on the other, would prevent acceleration of the decomposition of the polymer, by tying up the monomeric formaldehyde.

The use of derivatives of phenols, aramines, urea, thiourea, and hydrazines is proposed in the available patent literature [29-35] for the stabilization of polyformaldehyde.

The action of inhibitors: phenol derivatives, aromatic amines, sulfur- and phosphorus-containing compounds was studied in conjunction with polyamide.

A study of the effectiveness of the inhibitors showed [19] that phos-phites and sulfides (Fig. 77) exert no appreciable inhibiting action on the thermooxidative destruction of polyformaldehyde. Aramine (Fig. 78) and phenol (Fig. 79) derivatives are extremely effective in the stabiliza-tion of this polymer (data are cited in comparison with the stable poly-mer).

The most effective of the oxidation inhibitors studied are: inhibitor 22-46 − 2, 2'-methylene-bis-(4-methyl-6-tert-butyl)phenol, santovar "0" − 2, 5-di-tert-butyl-hydroquinone, diphenylamine, p-hydroxyneozone, etc. These inhibitors, in conjunction with polyamide, considerably lengthen the induction period and reduce the gas evolution in the oxida-tion of polyformaldehyde.

In evaluating the effectiveness of the action of additives that increase the stability of polyformaldehyde, the optimum ratios of polyamide and

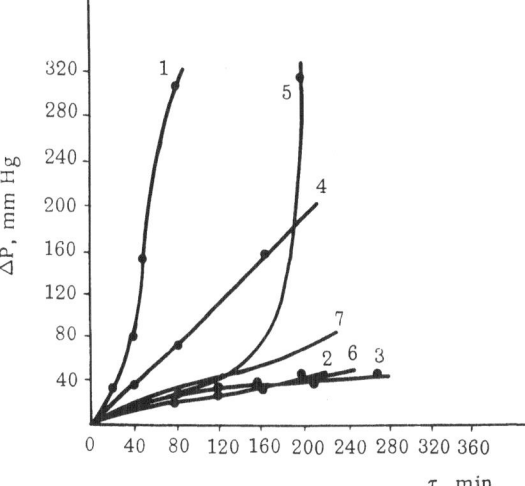

Fig. 78. Effectiveness of the action of inhibitors of the aramine type, introduced into the polymer (with an addition of 2% polyamide) in a 0.5% concentration. T = 200°C; P_{O_2} = 200 mm Hg. 1) Polymer without additives; 2) stable polymer; 3) p-hydroxyneozone; 4) inhibitor 40-10 (N-phenyl-N'-cyclohexyl-p-phenylenediamine); 5) mercaptobenzimidazole; 6) diphenylamine; 7) polyazophenylene.

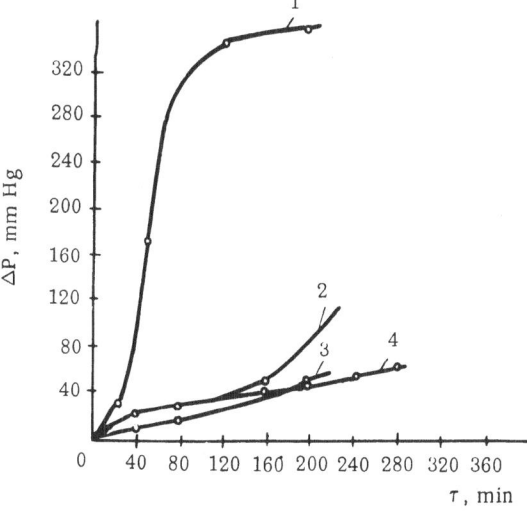

Fig. 79. Effectiveness of the action of inhibitors: phenol derivatives, introduced into the polymer (with an addition of 2% polyamide) in a 0.5% concentration. T = 200°C; P_{O_2} = 200 mm Hg. 1) Polymer without additives; 2) santovar "0" (2,5-di-tert-butyl-hydroquinone); 3) stable polymer; 4) inhibitor 22-46 [2,2'-methylene-bis-(4-tert-butyl-6-methyl)-phenol].

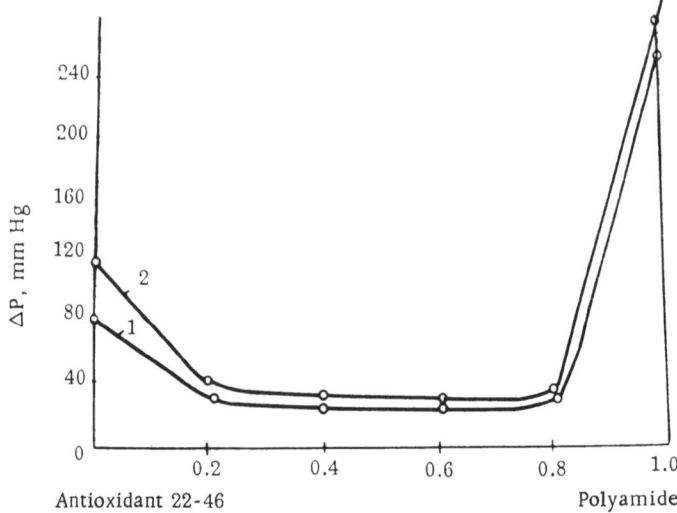

Fig. 80. Dependence of the gas evolution in the thermooxidative destruction of polyformaldehyde on the ratio of stabilizing additives (polyamine
and inhibitor 22-46 at a summary ratio of 2.5%). T = 200°C; P_{O_2} = 200
mm Hg. 1) 80 min after the beginning of the process; 2) 100 min after
the beginning of the process.

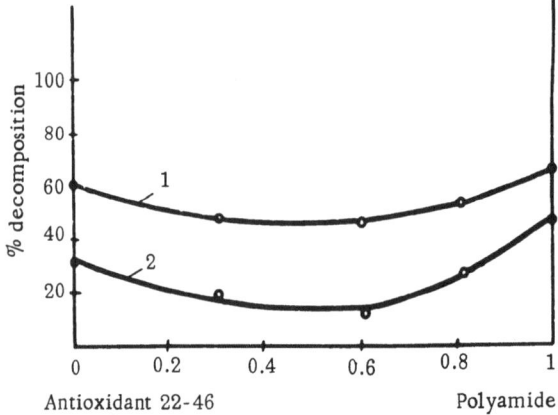

Fig. 81. Dependence of the loss in weight of polyformaldehyde
on the ratio of stabilizing additives (polyamide and inhibitor
22-46 at a summary ratio of 2.5%). P_{O_2} = 200 mm Hg. 1) T
= 230°C; 2) T = 240°C.

antioxidant were established. Various ratios of the polyamide and antioxidant 22-46 were used in summary concentrations of 2.5 and 1.5% with
respect to polyformaldehyde. The variation of the pressure in the system on account of gas evolution in the oxidation of polyformaldehyde in
the presence of additives taken in various ratios is presented in Fig. 80.

The optimum ratio of the polyamide and antioxidant can be considered to be 0.6:0.4. However, within a broad interval of ratios (0.4–0.8 with respect to polyamide), deviation from the optimum has almost no effect. An analogous picture is also observed in the establishment of the optimum ratios of polyamide and antioxidant according to the loss in weight in the oxidation of polyformaldehyde (Fig. 81).

5. INHIBITION OF THE OXIDATION OF POLYFORMALDEHYDE BY RADICAL-TYPE INHIBITORS

The inhibiting action of stable nitrogen oxide radicals of the following structures on formaldehyde was studied in [22]:

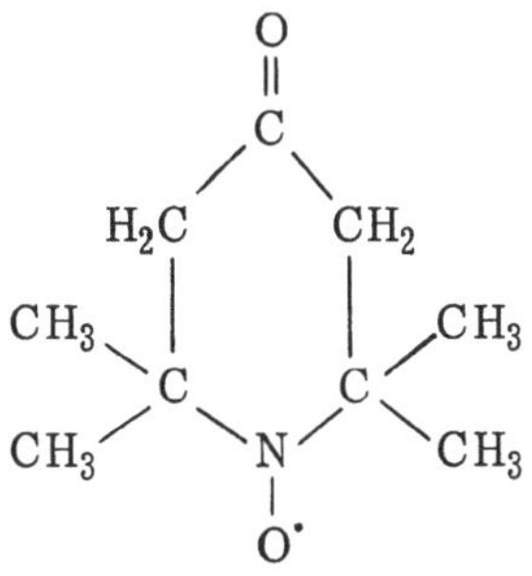

I

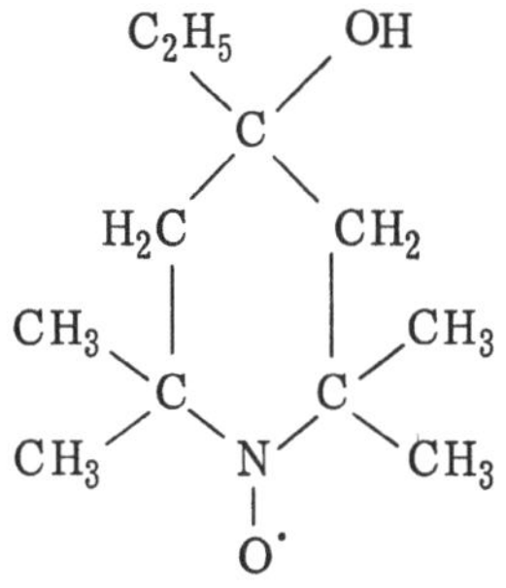

II

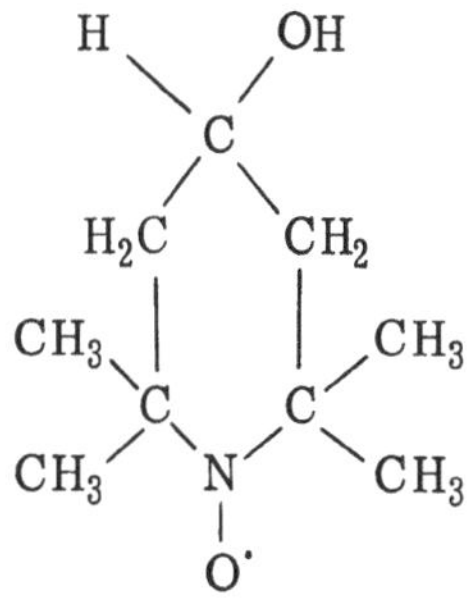

III

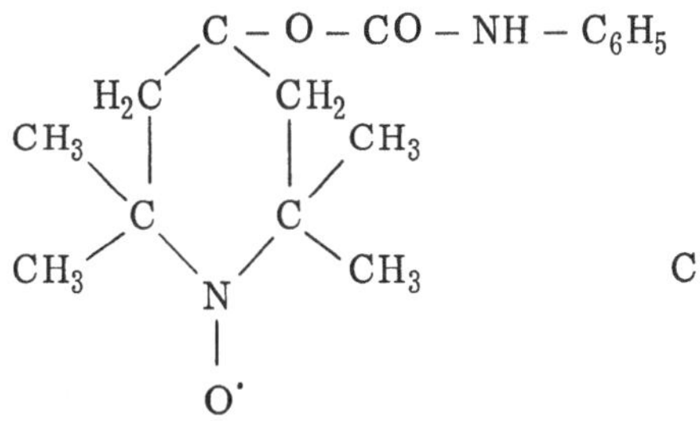

IV

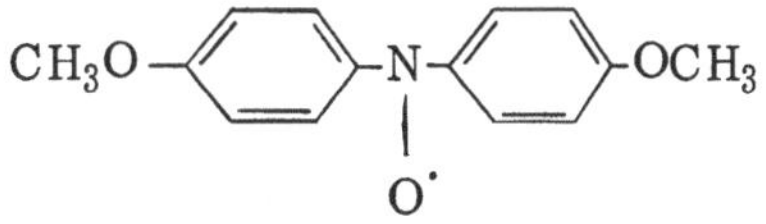

V

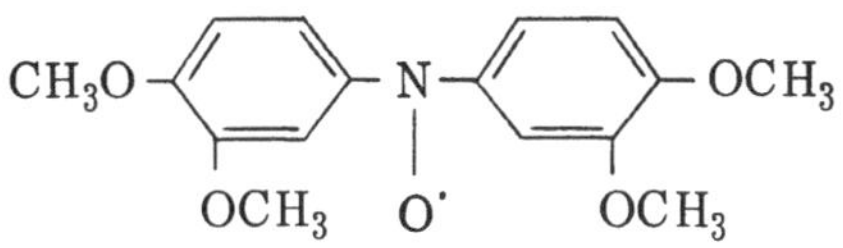

VI

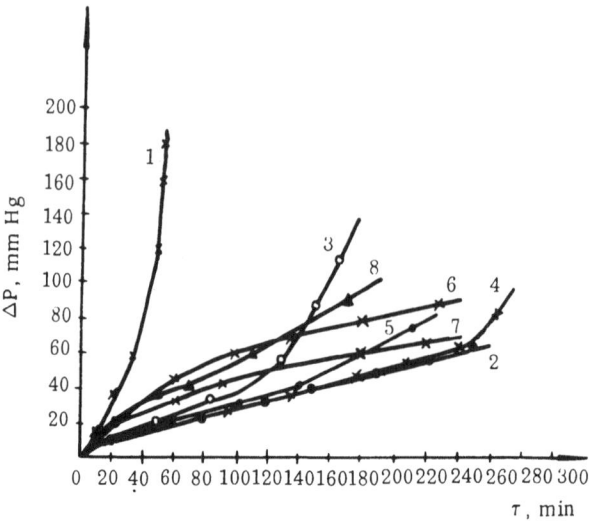

Fig. 82. Effectiveness of the action of stable radical inhibitors introduced into the polymer (1) (with an addition of polyamide 54 in 1.5%
concentration) in a 1% concentration. 2) Stabilizer 22-46; radical inhibitors: 3) I; 4) II; 5) III; 6) IV; 7) V; 8) VI.   T = 200°C; P_{O_2} = 200
mm Hg.

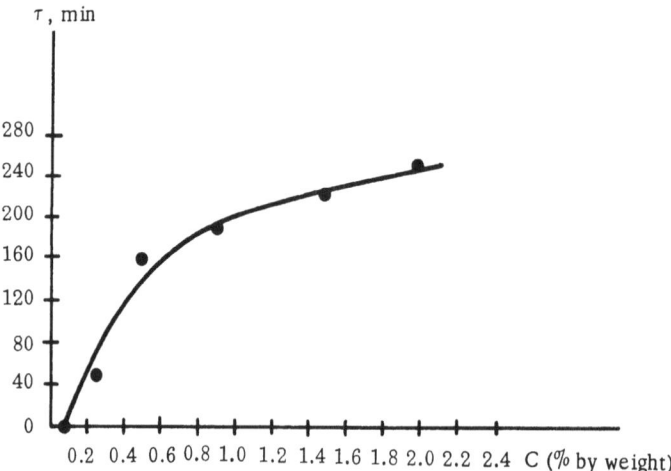

Fig. 83. Dependence of the induction period τ on the weight concentration (C) of the stabilizing additive (radical inhibitor II and polyamide 54). P_{O_2} = 200 mm Hg; T = 200°C.

Figure 82 presents the kinetic curves of gas evolution in the thermal oxidation of polyformaldehyde with additions of radical inhibitors (conc. 1%) in conjunction with polyamide 54 (conc. 1.5%). We can see from the figure that the oxidation of polyformaldehyde in the presence of radical inhibitors is characterized by the presence of considerable induction periods. Radical II is the most effective of the investigated stable radicals.

Data on the gas evolution in the oxidation of polyformaldehyde with an addition of the known oxidant 22-46 are also cited for comparison of the effectiveness. The radical inhibitors proved especially effective at lower oxidation temperatures (180°C).

A study of the dependence of the duration of the induction period in the oxidation of polyformaldehyde on the weight concentration of the stabilizing additive showed that at first the induction period increases with increasing concentration of the stabilizing additive, but then it remains practically constant (Fig. 83).

The maximum duration of the induction period corresponds to 2.0-2.2% of the stabilizing resin, which agrees with the optimum ratio of the stabilizing additive, including the usual inhibitor of the phenol or amine type.

As was also shown for polyamides, a break is detected on the curve (Fig. 83). This indicates that the dependence of the induction period on the initial inhibitor concentration is not directly proportional, but is more complex in character; the inhibitor is apparently consumed not only for chain termination, but also for side processes (volatilization, oxidation, initiation). An analogous picture is also observed in the stabilization of polyformaldehyde by phenols (Fig. 84).

6. CONSUMPTION OF NITROGEN OXIDE STABLE RADICALS IN THE THERMAL OXIDATION OF POLYFORMALDEHYDE

In [22] the kinetics of the consumption of the radical inhibitor during the induction period was investigated. For this purpose, the kinetics of the decomposition of polyformaldehyde and the simultaneous reduction of the inhibitor concentration in the polymer were studied on a specially designed instrument. Thermal oxidation was conducted in the resonator of an EPR spectrometer at 200°C. The variation of the radical concentration was determined approximately according to the change in amplitude of the first derivative of the EPR absorption line, while decomposition of polyformaldehyde was determined according to the increase in pressure in the system. As can be seen from Fig. 85, the end of the induction period coincides with the disappearance of the EPR signal, i.e., the radical inhibitor is entirely consumed during the induction period.

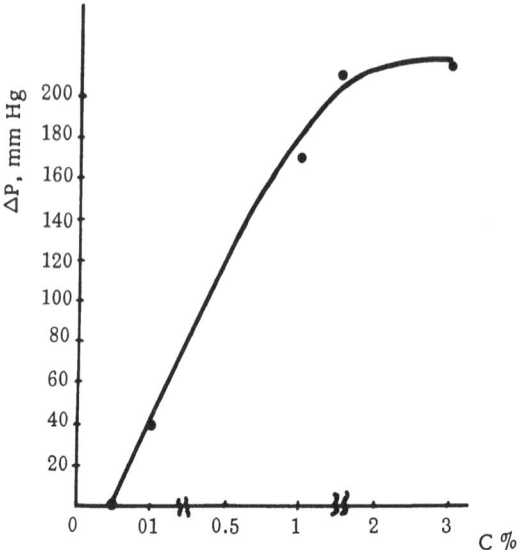

Fig. 84. Dependence of the induction period (τ) on the weight concentration (C) of the stabilizing additive (poly-amide 54 and inhibitor 22-46). P_{O_2} = 200 mm Hg; T = 200°C.

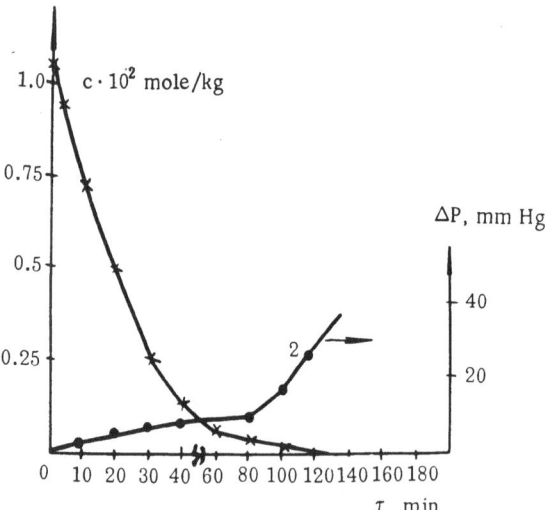

Fig. 85. Consumption of the radical inhibitor II during the induction period (curve 1) and kinetics of gas evolution (curve 2) in the thermooxidative destruction of polyformal-dehyde with an addition of polyamide 54 (c = 0.3%) and in-hibitor (c = 0.2%). P_{O_2} = 200 mm Hg; T = 200°C.

The kinetics of the destruction of the radical inhibitor is described by a first-order equation, i.e., its consumption is apparently related not only to termination of the oxidation chains. A substantial role in this case is also played by evaporation of the inhibitor from the polymer at high temperatures.

It was successfully shown with the aid of the EPR method that radical II is very stable under vacuum and in inert silicone liquid at 200°C. A study of the evaporation of the radical inhibitor II from the polymer at 200°C in air showed that by the end of the experiment this value comprises only 6-10% of the initial radical concentration.

We can see from these data that the radical inhibitors are consumed mainly according to a reaction related to inhibition of the thermal oxidation of polyformaldehyde.

The inhibition of the oxidation of polyformaldehyde by radical-type stabilizers is also indirect evidence in favor of the occurrence of the process of destruction of polyformaldehyde according to a radical mechanism.

BIBLIOGRAPHY

1. V. Kern and H. Cherdron, Makromol. Chem. 40:101, 1960.
2. V. Kern, H. Cherdron, and V. Jaacks, Angew. Chem. 73(6):177, 1961.
3. H. Cherdron, T. Köhr, and V. Kern, Makromol. Chem. 52:48, 1962.
4. N. S. Enikolopyan and M. S. Vardanyan, Zh. Vses. Khim. Obshch. im. Mendeleeva 7(2):194, 1962.
5. L. A. Dudina and N. S. Enikolopyan, Vysokomolek. Soed. 4(6):869, 1962.
6. L. A. Dudina and N. S. Enikolopyan, Vysokomolek. Soed. 5(6):861, 1963.
7. L. A. Dudina and N. S. Enikolopyan, Vysokomolek. Soed. 5(7):986, 1963.
8. L. A. Dudina and N. S. Enikolopyan, Vysokomolek. Soed. 5(8):1135, 1963.
9. L. A. Dudina, L. V. Karmilova, and N. S. Enikolopyan, Vysokomolek. Soed. 5(8):1160, 1963.
10. L. A. Dudina, L. A. Agayants, L. V. Karmilova, and N. S. Enikolopyan, Vysokomolek. Soed. 5(8):1245, 1963.
11. L. A. Dudina, L. V. Karmilova, and N. S. Enikolopyan, Doklady Akad. Nauk SSSR, 150(2):309, 1963.
12. L. A. Dudina, A. A. Berlin, L. V. Karmilova, and N. S. Enikolopyan, Doklady Akad. Nauk SSSR 150(3):580, 1963.

13. M. B. Neiman, L. I. Golubenkova, B. M. Kovarskaya, A. S. Strizhkova, I. I. Levantovskaya, M. S. Akutin, and V. D. Moiseev, Vysokomolek. Soed. 1(1):1531, 1959.

14. V. D. Moiseev, M. B. Neiman, and A. I. Kryukova, Vysokomolek. Soed. 1(1):1552, 1964.

15. V. D. Moiseev and M. B. Neiman, Vysokomolek. Soed. 3(3):1383, 1961.

16. A. B. Blyumenfel'd, M. B. Neiman, and B. M. Kovarskaya, Doklady Akad. Nauk SSSR 154(3):631, 1964.

17. A. L. Kammick and A. R. Boevee, J. Chem. Soc. 123:228, 1963.

18. V. R. Alishoev, M. B. Neiman, and B. M. Kovarskaya, Plastmassy No. 7:11, 1962.

19. V. R. Alishoev, M. B. Neiman, B. M. Kovarskaya, and V. V. Gur'yanova, Vysokomolek. Soed. 5(5):644, 1963; 4(12), 1962.

20. A. L. Buchachenko and M. B. Neiman, Vysokomolek. Soed. 3(9):1285, 1961.

21. M. B. Neiman, T. S. Fedoseeva, G. V. Chubarova, A. L. Buchachenko, and Ya. S. Lebedev, Vysokomolek. Soed. 5(9):1339, 1963.

22. B. M. Kovarskaya, M. B. Neiman, V. V. Gur'yanova, and E. G. Rozantsev, Vysokomolek. Soed. 6(9), 1964.

23. T. B. Lewis and R. T. Reynolds, Chem. & Ind. J. 45:958, 1952.

24. K. N. Vlasova and L. A. Rodivilova, Vestnik Tekhn. i Ekonom. Informatsii No. 5, 1958.

25. L. A. Rodivilova, K. N. Vlasova, and G. S. Petrov, Izv. Vysshikh Uchebn. Zavedenii, Khimiya i Khim. Tekhnologiya No. 4, 1958.

26. S. M. Raitburd, L. A. Rodivilova, K. N. Vlasova, A. E. Shabadash, and L. A. Igonin, Plastmassy No. 7:20, 1960.

27. D. Walker, Formaldehyde [Russian Translation], State Press for Chemical Literature, Moscow, 1957.

28. U. S. Patent 27689994, 1956.

29. British Patent 770717, 1957; 748856, 1956.

30. Israeli Patent 10397, 1958.

31. U. S. Patent 2871220, 1959.

32. U. S. Patent 2920059, 1960.

33. German Federal Republic Patent 1076363, 1960.

34. U. S. Patent 2810708, 1957.

Chapter VI

AGING AND STABILIZATION OF POLYVINYL CHLORIDE AND COPOLYMERS OF VINYL CHLORIDE

Polyvinyl chloride and vinyl chloride copolymers have found widespread use in various fields of industry and the national economy, thanks to a number of valuable chemical and physicomechanical properties. However, both the homopolymer and the copolymers of vinyl chloride possess an extremely vital shortcoming – they break down comparatively easily under the action of physical (heat, radiation) and chemical (oxygen, ozone) agents and thereby lose their valuable qualities.

The decomposition and stabilization of vinyl chloride homopolymer have been the subject of many studies. However, the mechanism of the breakdown and stabilizing action of additives has not yet been strictly demonstrated, and many questions remain open to dispute. The investigation of the breakdown and stabilization of vinyl chloride copolymers has essentially only begun in the last few years.

I. FACTORS DETERMINING THE MECHANISM AND RATE OF DECOMPOSITION OF POLYVINYL CHLORIDE

From the literature devoted to the study of the decomposition of polyvinyl chloride under the influence of heat, various types of radiations, oxygen, and ozone, it is known that the basic routes of decomposition are dehydrochlorination, oxidation, destruction, and structuring.

It is currently the custom to consider that the decomposition of polyvinyl chloride proceeds both according to an ionic-molecular and according to a free-radical mechanism. However, until comparatively recently this process was considered in the overwhelming majority of investigations mainly from the standpoint of classical organic chemistry; moreover, preference was given to the ionic mechanism, and the factors determining the decomposition of the polymer as a result of free-radical chain reactions were not taken into consideration. This attitude was based both on the electronic concepts developed by organic chemistry and on the concrete results of experiments, obtained in the study of the properties of polyvinyl chloride.

It has been shown that polyvinyl chloride possesses a primarily linear structure, with predominant mutual arrangement of the units in the chain "head-to-tail" [1]:

$$\sim CH_2 - CHCl - CH_2 - CHCl - CH_2 - CHCl \sim$$

In a study of the thermal and thermooxidative dehydrochlorination of polyvinyl chloride, the opinion has been expressed that allyl activation of the chlorine atoms next to the double bonds occurs; the question of the mechanism of the primary event of decomposition of the polymer remained not entirely clear [2].

The fact that the rate of decomposition becomes considerable at relatively low temperatures gave a basis for expressing the opinion that labile structures, which disturb the regularity of the alternation of the $-CH_2-$ and $-CHCl-$ groups and are responsible for the primary event of decomposition, are present in the polyvinyl chloride molecules [3]. The possibility of the formation of such structures in the polyvinyl chloride macromolecules is due to the nature of the reactions that take place during the process of radical polymerization of vinyl chloride. The composition of the macromolecules can include not only fragments of the initiator, but also branches, as well as double bonds on the ends of the polymer chain; the appearance of branching is the result of reactions of chain transfer through the polymer, while that of unsaturated terminal groups results from reactions of disproportionation and chain transfer through the monomer [4].

The presence of tertiary carbon atoms in polyvinyl chloride, at the sites of branching of the polymer chain, and the negative influence of branching on the stability of the polymer have been confirmed experimentally [5-7]. The negative influence of unsaturated terminal groups on the stability of the polymer has also been confirmed experimentally. The determination of the rates of thermal and thermooxidative dehydrocylorination of polyvinyl chloride, produced by polymerization at temperatures from 20 to 70°C, has shown that the low-molecular polymer is less stable than the high-molecular polymer [3, 8].

Considering the fact that the rate of decomposition of the polymer produced at various temperatures can also be influenced by other structures, the concentration of which depends on the conditions of polymerization, a stricter evidence of instability of the terminal unsaturated groups lies in the results obtained in an investigation of various fractions of the polymer sample. For fractions with molecular weights from 12,500 to 78,000, the rate of thermal dehydrochlorination is directly dependent on the reciprocal of the molecular weight, $1/M$ (Fig. 86) [9]. Other evidence of the negative influence of unsaturation on the rate of polymer decomposition is the decrease in the rate of dehydrochlorination after mild chlorination (Fig. 87) [10].

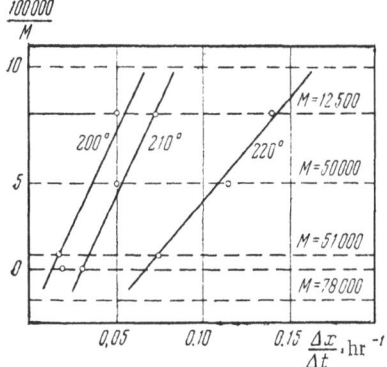

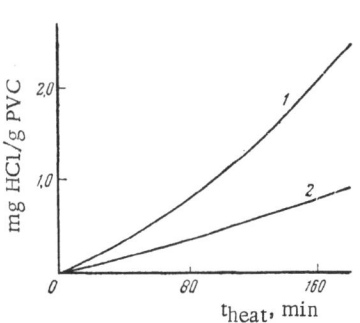

Fig. 86. Influence of the molecular weight (M) on the rate of dehydrochlorination of polyvinyl chloride at various temperatures in a stream of nitrogen. x is the fraction of chlorine split out of the total amount in the polymer.

Fig. 87. Influence of mild chlorination of polyvinyl chloride (PVC) on the rate of splitting out of HCl at 150°C in a stream of nitrogen. 1) Before chlorina-tion; 2) after chlorination.

The observed phenomena were satisfactorily explained by the prin-ciples established for the chemical transformations of low-molecular compounds containing analogous structures. The lability of the chlorine atoms at the sites of branching of the polymer chain, as well as those next to unsaturated terminal bonds may be the cause of the splitting out of the first molecules of hydrogen chloride. The formation of polyene structures, producing coloration of the polymer, can be explained by allyl activation of the halogen atoms situated next to the newly formed unsaturated bonds. Acceleration of dehydrochlorination in the presence of oxygen was explained by the easy oxidizability of the polyene struc-tures and α-carbon atoms, situated at the isolated or conjugated double bonds, as well as by the increase in the lability of the chlorine atoms when oxygen-containing functional groups are formed in the polymer chain (Figs. 88 and 89). Nonetheless, a number of peculiarities of the decomposition of polyvinyl chloride, related, in particular, to oxidation processes, could not be satisfactorily explained by the ionic character of the reactions in the decomposition of the polymer.

Studies in which the decomposition of low-molecular and high-molec-ular halogen-containing hydrocarbons, in particular, polyvinyl chloride, is considered as a chain process, proceeding according to a free-radical mechanism, have been developed quite recently [11-14]. In one of them [12] the inadequacy of representations of an ionic mechanism of the de-composition of polyvinyl chloride was reasoned on the basis of two ar-guments. The first argument was the absence of the proper substantia-tion for considering reactions of thermal decomposition of alkyl halides from the standpoint of the principles established for nucleophilic sub-

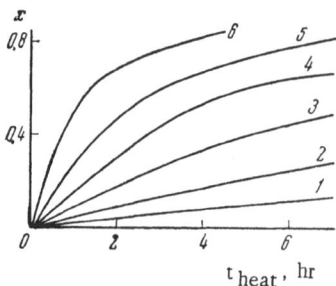

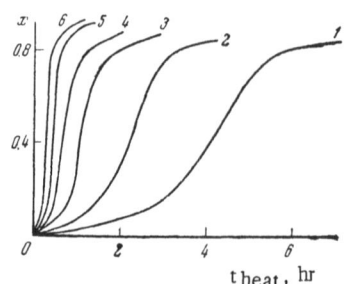

Fig. 88. Rate of dehydrochlorination of PVC with molecular weight M = 51,000 in a stream of nitrogen at various temperatures: 1) 200°; 2) 210°; 3) 220°; 4) 230°; 5) 240°; 6) 250°C; x is the fraction of chlorine split out of the total amount in the polymer.

Fig. 89. Rate of dehydrochlorination of PVC with molecular weight M = 51,000 in a stream of nitrogen at various temperatures: 1) 200°; 2) 210°; 3) 220°; 4) 230°; 5) 240°; 6) 250°C; x is the fraction of chlorine split out of the total amount in the polymer.

stitution reactions. Data characterizing the high intrinsic thermal stability of allyl chloride [15-17] are cited as an example illustrating the erroneous nature of conclusions of allyl activation of the chlorine atoms during the thermal dehydrochlorination of polyvinyl chloride. The second argument pertains to the development of the dehydrochlorination chain and the mechanism of the formation of polyene structures. It is proposed that the formation of the system of conjugated bonds responsible for the appearance of the color of the polymer and, consequently, including no less than seven conjugated double bonds, is impossible as a result of allyl activation of the chlorine atoms on account of the increase in the energy of conjugation of the system when each successive hydrogen chloride molecule is split out. We can scarcely agree with such an assertion, since the development of polyene structures, as will be shown later, in some cases can lead not to chain termination, but, on the contrary, to a further splitting out of hydrogen chloride. In accord with the results of investigations pertaining to the reactions of hydrocarbon oxidation and to the reactions of halogenation and decomposition of the chlorinated hydrocarbons [8-20], the following mechanism of the thermooxidative decomposition of polyvinyl chloride is proposed:

Dehydrochlorination in the presence of labile chlorine atoms in the polymer:

$$\sim CH_2-CHCl-CH_2 \sim \to \sim CH_2-\dot{C}H-CH_2 \sim + Cl^{\cdot}$$
$$\sim CH_2-CH-CH_2 \sim + Cl^{\cdot} \to \sim CH=CH-CH_2\sim + HCl$$

Dehydrochlorination in the presence of free radicals in the system:

$$R^{\cdot} + \sim CHCl-CH_2-CHCl \sim \to RH + \sim CHCl-\dot{C}H-CHCl \sim$$
$$\sim CHCl-\dot{C}H-CHCl \sim \to Cl^{\cdot} + \sim CH=CH-CHCl \sim$$

Oxidation:

$$R^{\cdot} + \sim CH_2-CHCl-CH_2 \sim \to RH + \sim CH_2-\overset{\cdot}{C}Cl-CH_2 \sim$$

$$\sim CH_2-\overset{\cdot}{C}Cl-CH_2 \sim + O_2 \to \sim CH_2-\underset{\underset{\displaystyle OO^{\cdot}}{|}}{C}Cl-CH_2 \sim$$

$$\sim CH_2-\underset{\underset{\displaystyle OO^{\cdot}}{|}}{C}Cl-CH_2 \sim + RH \to \sim CH_2-\underset{\underset{\displaystyle OOH}{|}}{C}Cl-CH_2 \sim + R^{\cdot}$$

$$\sim CH_2CCl-CH_2 \underset{\underset{\displaystyle OOH}{|}}{\sim} \to \sim CH_2-\underset{\underset{\displaystyle O^{\cdot}}{|}}{C}Cl-CH_2 \sim + {}^{\cdot}OH$$

Destruction in β-cleavage of the alkoxy radical:

$$\sim CHCl-CH_2-CCl-CH_2-CHCl-CH_2 \underset{\underset{\displaystyle OOH}{|}}{\sim} \to \sim CHCl-CH_2-\underset{\underset{\displaystyle O^{\cdot}}{|}}{C}Cl-CH_2-CHCl-CH_2 \sim + {}^{\cdot}OH$$

$$\sim CHCl-CH_2-\underset{\underset{\displaystyle O^{\cdot}}{|}}{C}Cl-CH_2-CHCl-CH_2 \sim \to \sim CHCl-CH_2-\underset{\underset{\displaystyle \|}{O}}{C}Cl + {}^{\cdot}CH_2-CHCl-CH_2 \sim$$

$$\sim {}^{\cdot}CH_2-CHCl-CH_2 \sim \to CH_2{=}CH-CH_2 \sim + Cl^{\cdot}$$

Destruction in cleavage of ketones:

$$\sim CHCl-CH_2-\underset{\underset{\displaystyle O^{\cdot}}{|}}{C}Cl-CH_2-CHCl-CH_2 \sim + RH \to R^{\cdot} + \sim CHCl-CH_2-\underset{\underset{\displaystyle OH}{|}}{C}Cl-CH_2-CHCl-CH_2 \sim$$

$$\sim CHCl-CH_2-\underset{\underset{\displaystyle OH}{|}}{C}Cl-CH_2-CHCl-CH_2 \sim \to HCl + \sim CHCl-CH_2-\underset{\underset{\displaystyle \|}{O}}{C}-CH_2-CHCl-CH_2 \sim$$

$$\sim CHCl-CH_2-\underset{\underset{\displaystyle \|}{O}}{C}-CH_2-CHCl-CH_2 \sim \to \sim CHCl-\overset{\cdot}{C}H_2 + {}^{\cdot}CH_2-CHCl-CH_2 \sim + CO$$

Structuring in the reaction of two macroradicals:

$$\left. \begin{array}{l} \sim CHCl-CH_2-\overset{\cdot}{C}H-CH_2-CHCl \sim \\[4pt] \underset{\underset{\displaystyle O^{\cdot}}{|}}{} \\[4pt] \sim CHCl-CH_2-\overset{\cdot}{C}Cl-CH_2-CHCl \sim \end{array} \right\} \to \begin{array}{l} \sim CHCl-CH_2-CH-CH_2-CHCl \sim \\[4pt] \underset{\underset{\displaystyle O}{|}}{} \\[4pt] \sim CHCl-CH_2-CCl-CH_2-CHCl \sim \end{array}$$

Structuring in the reaction of a macroradical with a polyene system:

$$\sim CHCl-CH{=}CH-CH{=}CH-CH{=}CH \sim + R^{\cdot} \to CHCl-CHR-\overset{\cdot}{C}H-CH{=}CH-CH{=}CH$$

$$\left. \begin{array}{l} \sim CHCl-CHR-\overset{\cdot}{C}H-CH{=}CH-CH{=}CH \sim \\[4pt] \underset{\underset{\displaystyle O-\overset{\cdot}{O}}{|}}{} \\[4pt] \sim CHCl-CH_2-\overset{\cdot}{C}Cl-CH_2-CHCl-CH_2 \sim \end{array} \right\} \to \begin{array}{l} \sim CHCl-CHR-CH-CH{=}CH-CH{=}CH \sim \\[4pt] \underset{\underset{\displaystyle O}{|}}{} \\[4pt] \underset{\underset{\displaystyle O}{|}}{} \\[4pt] \sim CHCl-CH_2-CCl-CH_2-CHCl-CH_2 \sim \end{array}$$

The shortcoming of the scheme under consideration is the fact that it does not consider the influence of terminal unsaturated groups on the rate of polymer decomposition, which, as has already been mentioned, is noted by many researchers.

The following equation is proposed to express the quantitative dependence of the rate of dehydrochlorination of polyvinyl chloride on the molecular weight:

$$v = a + k\frac{1}{M}$$

in which v is the rate of dehydrochlorination of the polymer in micromoles per gram, M is the molecular weight of the polymer, a and k are constants for the conditions of decomposition adopted.

The treatment of the experimental data showing the interrelationships between the rate of polymer decomposition and the concentrations of the terminal group has made it possible to determine the values of the constants a and k: in nitrogen medium a = 0, k = 7 × 10⁵, in air medium a = 6, k = 6 × 10⁵, in oxygen medium a = 18, k = 3 × 10⁵.

The variation of the values of the quantities a and k as a function of the medium shows that the rate of dehydrochlorination is substantially increased when the oxygen concentration in the system is increased, and that the sharp variation of the molecular weight of the polymer influences the rate of thermal decomposition to a greater degree than the rate of thermooxidative decomposition. These data have served as the basis for the following conclusions:

a) Thermal decomposition begins with the unsaturated terminal groups and is accompanied by the formation of conjugated systems; moreover, radical chain reactions, leading to destruction of the polymer chains and to recombination of the macroradicals, are possible.

b) Thermooxidative decomposition is a chain radical process with degenerate branching, in which initiation on account of the unsaturated terminal groups plays a significantly smaller role than in decomposition in the absence of oxygen [21].

In considering the influence of the chemical structure of the polyvinyl chloride macromolecules on its stability, one must consider the extremely low stability of the polymer, which possesses a basic chain constructed according to the "head-to-tail" principle. Such a polymer with a 1,2-structure is capable of splitting out hydrogen chloride and forming an insoluble product at low temperatures, and even upon repeated reprecipitation [22]. The formation of 1,2-structures during the process of vinyl chloride polymerization requires greater energy expenditures than the formation of 1,3-structures, since the formation of the former requires that the forces of mutual repulsion in the approaching monomer molecules with polar substituents be overcome. The relative content of 1,2-structures in the polymer will be greater, the higher the temperature of vinyl chloride polymerization [23].

Up to the present, the question of the influence of hydrogen chloride on the rate of decomposition of polyvinyl chloride remains debatable. Some researchers believe that hydrogen chloride catalytically accelerates decomposition both in the presence of oxygen and in inert medium [24, 25]; others deny the autocatalytic character of this process [3]; still others believe that hydrogen chloride per se does not influence the rate of decomposition, but, giving rise to corrosion of the apparatus, pro- motes the formation of iron salts, which are catalysts of oxidation-re- duction processes, accelerating the decomposition of the polymer [8,11].

Some peculiarities of the thermal and thermooxidative decomposition of polyvinyl chloride depend on the conditions of its production. Thus, it is known that samples of the polymer produced by initiating the poly- merization of vinyl chloride with ultraviolet irradiation possess higher stability in comparison with samples produced in polymerization under the action of chemical agents [26, 27]. Reversibility of the process of dehydrochlorination in the decomposition of samples of polyvinyl chlo- ride produced by the latex method is noted, while in the process of de- composition of suspension polymer, the phenomenon of reversibility is not observed [21]. It has been shown that the rate of dehydrochlorination of the latex polymer is significantly higher than that of the suspension polymer under the same conditions [21]. It has been established that the polymerization of vinyl chloride in the presence of oxygen leads to the formation of unstable peroxide groups, which can initiate decomposition of the polymer [28, 29]. It is noted that an extremely substantial in- fluence on the stability of polyvinyl chloride is exerted by the purity of the monomer, as well as the presence of impurities of metals of variable valence [28].

The studies discussed pertain to the decomposition of polyvinyl chlo- ride at temperatures not exceeding 250°C; pyrolysis at higher temper- atures has been less completely studied.

The decomposition of samples of polyvinyl chloride produced by polymerization with initiation by γ-irradiation, benzoyl peroxide, and azoisobutyrodinitrile, has been investigated by the method of IR-spec- trometry. Such samples behave practically the same under vacuum at 300-400°C. Up to 300°C, decomposition is accompanied by the liber- ation of hydrogen chloride and a small amount of benzene; increasing the temperature to 400°C leads to a substantial change in the composition of the volatile portion, to an increase in its benzene content, and to the ap-

TABLE 12

λ, mμ	Amount of HCl split out in 24 hr, mg per g of PVC	λ, mμ	Amount of HCl split out in 24 hr, mg per g of PVC
340 up to visible	0.00000	280 up to visible	0.00016
300 up to visible	0.00010	235 up to visible	0.00110

pearance of other hydrocarbons. Pyrolysis is accompanied by an increase in the branching of the carbon chains in the solid residue. The presence of aromatic compounds in the decomposition products results from recombination of the free radicals formed during the process of decomposition, together with the polyene structures. Aromatization of the decomposition products proves energetically profitable; in addition, the weakening of the carbon-carbon bond in the α-position to the aromatic nucleus promotes splitting out of the benzene rings and further cyclization of the remaining polyene chains [26].

A more detailed analysis of the composition of the volatile portion by the method of mass spectrometry showed that the products liberated from polyvinyl chloride at 400°C under vacuum in 30 min contain not only hydrogen chloride, but also 26 different aliphatic and aromatic compounds: saturated and unsaturated hydrocarbons, dichloroethane, alkyl- and alkylenebenzenes are detected; ethylene, propylene, ethane, pentane, hexane, benzene, and toluene predominate quantitatively [30]. It has been found by the methods of chromatography and IR- and UV-spectrometry that when the pyrolysis temperature is raised to 450-500°C, substances with three to five condensed aromatic nuclei appear among the volatile decomposition products of polyvinyl chloride [31].

In considering the processes of decomposition of high-molecular compounds under the influence of radiations, it is the custom to distinguish photodecomposition and radiation decomposition. Both types of decomposition are related to phenomena of electronic excitation and to the formation of free radicals or radical ions, which initiate chain reactions and cause the breakdown of the macromolecules.

The energy of radioactive radiations (α and β particles, neutrons, and γ-radiation) is many orders of magnitude higher than the energy of optical photons. Optical photons of the visible and ultraviolet regions of the spectrum possess energies of approximately the same order of magnitude as chemical bonds; they are absorbed in the surface layers of the substance, thanks to which photochemical reactions are inhomogeneous. Radioactive radiations possess high penetrating ability; hence radiochemical reactions in the irradiation medium proceed rather uniformly over the entire volume of the substance [32].

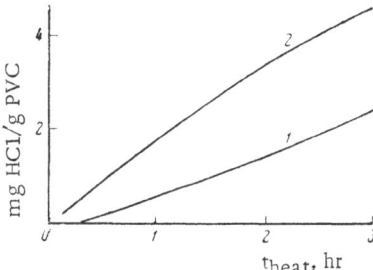

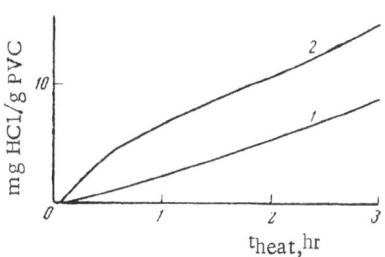

Fig. 90. Influence of UV-irradiation on the rate of dehydrochlorination of PVC in the case of heating in nitrogen. 1) Before irradiation; 2) after irradiation.

Fig. 91. Influence of UV-irradiation on the rate of dehydrochlorination of PVC in air. 1) Before irradiation; 2) after irradiation.

According to the basic premise of photochemistry, radiation corresponding to a definite wavelength range will exert some action on matter only in the case when the latter is capable of absorbing in the given region of the spectrum. The energy absorbed by the substance can be consumed for the initiation of photochemical reactions (direct photolysis or sensitized processes) or converted to radiation (fluorescence, phosphorescence) and thermal energy [33-35]. Extremely often polymer compounds possess the ability to absorb in the same regions of the spectrum as the monomer substances corresponding to them. For the C – C bond of vinyl chloride, the absorption maximum lies in the region of $\sim 300\,m\mu$ [36, 37]. In polyvinyl chloride the absorption in the ultraviolet region of the spectrum is apparently related to the presence of unsaturated structures in the polymer molecules [14].

The mechanism of the photodecomposition of polyvinyl chloride has not been strictly proven. It is customarily considered that it is analogous to the mechanism of thermal decomposition [12, 14]. From works devoted to the study of the photodecomposition of polyvinyl chloride, it follows that mainly the same phenomena occur under the action of light in the ultraviolet region of the spectrum as when the polymer is heated. The basic routes of decomposition are dehydrochlorination, oxidation, destruction, and structuring; the rate of photodecomposition in the presence of oxygen is greater than in neutral media or under vacuum; in the presence of oxygen, destruction reactions predominate, as a result of which the molecular weight of the polymer decreases; in inert media or under vacuum, structuring processes predominate, resulting in the formation of a three-dimensional insoluble polymer [3, 12, 27, 36-46]. After irradiation, the polymer becomes less thermally stable (Figs. 90 and 91). This aftereffect phenomenon can be explained by the formation of free radicals capable of initiating thermal and thermooxidative decomposition during the process of irradiation. Table 12 presents the values of the rate of dehydrochlorination of polyvinyl chloride in the case

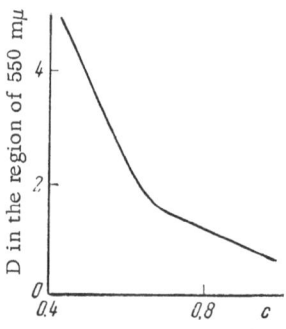

Fig. 92. Variation of the optical density (D) of PVC after β-irradiation as a function of the relative concentration (c) of free radicals with half-life 1630 hr.

of irradiation by ultraviolet light of various wavelengths [42].

In recent years, in connection with studies on the modification of the properties of high-molecular compounds, investigations of the influence of various types of radiation on the stability of polymers, including polyvinyl chloride, have begun [47-51].

Samples of polyvinyl chloride, subjected to β-irradiation in doses from 0.5 to 3 Mrad, followed by heating, illumination, and the action of oxygen, ozone, and sulfur dioxide, have been investigated by the methods of EPR and optical spectrometry in the wavelength region from 200 to 700 mμ. It was found that in this case three types of stable radicals are formed, the half-lives of which are equal to 1630, 63, and 4.5 hr. A comparison of the rates of destruction of the free radicals with the rate of variation of the depth of the intensity and color of the sample according to absorption in the wavelength region around 550 mμ showed that the color does not depend on the concentration of the radicals with half-lives 63 and 4.5 hr. The absorption in the spectral region ~550 mμ proved to be inversely proportional to the relative concentration of the long-lived radicals (Fig. 92). The break on the curve is interpreted as an indication of a change in the type of color substance formed as a result of the destruction of the free radicals, after they have already been about one-third destroyed. On the basis of these data, we can conclude that the appearance of color observed by researchers in the irradiation of polyvinyl chloride by high-energy sources is related to the formation of polyene structures during dehydrochlorination of the polymer [50].

For samples of polyvinyl chloride irradiated by Co^{60} in a dose of 8 Mrad, the rate of variation of the free-radical concentration and the rate of dehydrochlorination at 80°C have been investigated. It has been found that three types of free radicals, the half-lives of which are equal to 6500, 345, and 28 min, are formed in the γ-irradiation of polyvinyl chloride. The results of a simultaneous measurement of the free-radical concentration in the sample and the rate of dehydrochlorination at 80°C are cited in Fig. 93. A comparison of the intensity of the color of the sample, determined by means of a spectrometer, with the variation of the free-radical concentration showed that when the irradiated polymer is exposed at 80°C for 900-3000 min, only long-lived radicals are present in it, the concentration of which is inversely proportional to the color intensity [51].

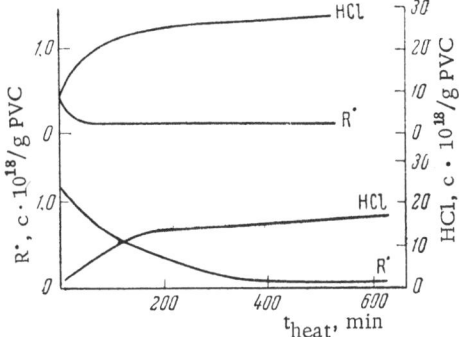

Fig. 93. Dependence of the concentration (c) of
free radicals R˙ and the amount of HCl split out
on the time of heating at 80°C for PVC irradiated
by Co[60].

In one of the works devoted to the study of the thermal decomposition
of polyvinyl chloride, the role of polyene structures in the formation of
the color of the polymer is doubted. It is proposed that the color that appears when polyvinyl chloride is heated cannot be due to the formation of
any chromophore groups, including polyene chains, since their concentration at the initial stage of the decomposition is extremely small. The
opinion is expressed that the original color is due to the formation of
stable free radicals. Data on the decolorization of destructed polyvinyl
chloride by solvents and radical-type stabilizers, as well as on the inhibition of the polymerization of ethyl acrylate by colored polyvinyl chloride, are cited in favor of such a hypothesis [52]. However, the assumption of the appearance of color on account of the formation of free radicals cannot be considered correct. The low average value of the unsaturation at the initial stage of decomposition, determined according to
the amount of hydrogen chloride split out, cannot serve as evidence of in-
sufficiency of conjugation, since it does not characterize the numerical
distribution of double bonds among the remaining polyene chains; in all
probability, there is a sufficient number of macromolecules, in which
the number of double conjugated bonds reaches or exceeds the amount
needed for the appearance of color, even at the initial stage of the decomposition in polyvinyl chloride. The decolorization of the destruc--
tured polymer by solvents or radical-type stabilizers, as well as its inhibition of the polymerization of ethyl acrylate are easily explained by
the high reactivity of conjugated systems in free-radical processes, as
well as by the ease of their entry into the formation of π-complexes
[53, 54].

As has already been noted, the stability of vinyl chloride copolymers has been studied less completely than the stability of the homopolymer. We cannot draw any unambiguous conclusion on the greater or

lesser stability of vinyl chloride copolymers in comparison with the homopolymer on the basis of the data available in the literature. Copolymers of vinyl chloride with vinylidine chloride [21, 55, 56] and with vinyl acetate [56] have been investigated in the greatest detail; there are data on the stability of vinyl chloride copolymers with ethyl acrylate [57, 58] and with vinylisobutyl ether [56].

Copolymers of vinyl chloride with vinylidine chloride, produced by the suspension method, are more stable than the latex copolymers. The rates of thermal decomposition of polyvinyl chloride and the copolymer of vinyl chloride with vinylidine chloride, produced by the latex method, are practically the same; for the latex copolymer, just as for the latex homopolymer, reversibility of the process of dehydrochlorination is observed [21]. In an investigation of the stability of copolymers of vinyl chloride with vinyl acetate, vinylidine chloride, and with vinylisobutyl ether in nucleophilic substitution reactions, it was found that the copolymer with vinyl acetate is the least stable to the action of alcoholic alkali; the copolymers with vinylidine chloride and vinylisobutyl ether proved more stable [56]. The stability of the copolymer of vinyl chloride with methyl acrylate is substantially increased when the degree of homogeneity of the copolymer with respect to composition is increased, and when monomers with a smaller content of impurities are used, as well as when the copolymerization is conducted in the presence of chain carriers [57, 58].

There are indications that the introduction of small amounts of a second monomer during the process of polymerization leads to a disturbance of the regular structure of the basic chain of the macromolecule and to increased stability for high-molecular compounds that decompose with high yields of the monomer [59]. However, the information on the increase in the thermal stability of polyvinyl chloride by copolymerization with small amounts of glycidyl methacrylate [21] cannot serve as a basis for a general evaluation of the effectiveness of this method of stabilizing polyvinyl chloride.

Analyzing the results of studies of the decomposition of polyvinyl chloride, we can mention the following basic factors, determining the stability of the polymer: structure of the macromolecules, physical state of the polymer, reactivity of the medium, nature and intensity of the energy influence, presence of impurities capable of accelerating or decelerating decomposition. A substantial influence on the stability of vinyl chloride copolymers is exerted not only by the enumerated factors, but also by the nature of the second monomer, ratio of the monomers, and degree of homogeneity of the copolymer with respect to composition. Each of the factors mentioned can influence the rate of initiation of decomposition and the length of the kinetic chains to a greater or lesser degree.

We should take the structure of the polymer to mean the chemical structure of the macromolecules, as well as their steric configuration and mutual arrangement, including the degree of crystallinity. The elements of the chemical structure determining the high degree of stability of polyvinyl chloride are: absence of functional groups more capable of destructive processes than the alternating groups $-CH_2-CHCl-$ in the macromolecule; minimum degree of branching of the basic chain of the polymer; absence of 1,2-halogen structures; and minimum concentration of double bonds. The production of polyvinyl chloride corresponding to the enumerated conditions in the structure of the macromolecules is theoretically possible when the polymerization is conducted in the absence of oxygen, at as low a temperature as possible, using a monomer containing no impurities, in the presence of chain carriers, as well as under conditions guaranteeing the absence of free radicals in the polymer at the end of the polymerization process.

On the basis of the general concepts of the energy state of amorphous and crystalline structures, we might assume that the stability of a polymer is directly dependent on the degree of its crystallinity. However, we must consider the fact that the high degree of lability of the macromolecules in the crystalline state at increased temperatures can promote a more intensive development of the decomposition processes [60].

The dependence of the stability of polyvinyl chloride on its physical state is related to the possibility of the development of decomposition processes in the medium, where a mutual displacement of the macromolecules or individual parts of them either can or cannot occur.

The influence of the chemical reactivity of the medium on the decomposition is extremely complex; however, the increased rate of decomposition in the presence of oxygen can have an effect on the end result of the individual, sometimes oppositely directed processes.

We should scarcely dwell on the role of the nature and intensity of the energy influence in the process of decomposition. We should mention especially that, depending on the nature and intensity of the influence, the same impurities, contained in the polymer, can either accelerate or decelerate decomposition. Of the impurities that reduce the stability of polyvinyl chloride, compounds containing metals of variable valence, including iron salts, are especially undesirable. When present in the polymer even in negligible amounts, such salts extremely actively catalyze oxidative decomposition.

The experimental facts currently available permit us to assume that the decomposition of polyvinyl chloride under an energy influence proceeds in an interconnected manner, according to a mechanism including ionic-molecular and radical reactions. The ratio of the rates of these reactions depends on the chemical structure of the polymer, reactivity of the medium, and degree of decomposition of the polymer.

In purely thermal decomposition, ionic-molecular reactions predominate at the initial stage, while as the decomposition progresses, radical processes also begin to develop. Thermooxidative decomposition also begins with the splitting out of hydrogen chloride on account of polarization of the chlorine-carbon bonds. However, the radical processes in this case already proceed at the early stage of decomposition.

The presence of polarized bonds in the macromolecules leads to an intensification of decomposition according to an ionic-molecular mechanism, while the presence of free radicals serves as a cause for intensifying radical processes.

The presence of isolated or conjugated unsaturated bonds, situated at the ends and in the chain of the macromolecule, can also give rise to ionic-molecular and radical decomposition reactions. The role of polyene structures, formed during dehydrochlorination, can be positive or negative, depending on the conditions of the energy influence and the reactivity of the medium. At low temperatures in an inert medium, the formation of polyene structures lead to self-stabilization; at increased temperatures or in the presence of oxygen, the formation of polyene structures can be one of the causes of further progression of decomposition.

The above is illustrated by the following scheme.

Thermal Decomposition. The dehydrochlorination chain begins according to an ionic-molecular mechanism, when splitting out of a hydrogen chloride molecule leads to transition to a more profitable energy state on account of the formation of conjugated double bonds:

This process continues until polarization as a result of the electronegativity of the chlorine atom is compensated for by the energy of conjugation of the system:

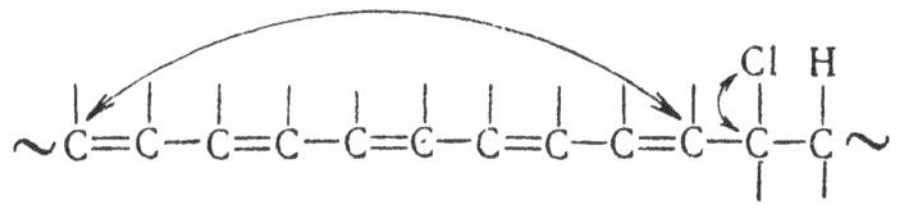

Such self-stabilization by polyene structures is possible only at low temperatures, insufficient for their transition to the triplet state. If the temperature is sufficiently high, then the transition of the polyene structures to the triplet state promotes a progression of dehydrochlorination according to an ionic-molecular mechanism. In addition, at increased temperatures, before a length of the polyene chain sufficient for self-stabilization of the decomposing macromolecule is reached, initiation of decomposition according to a radical mechanism becomes possible:

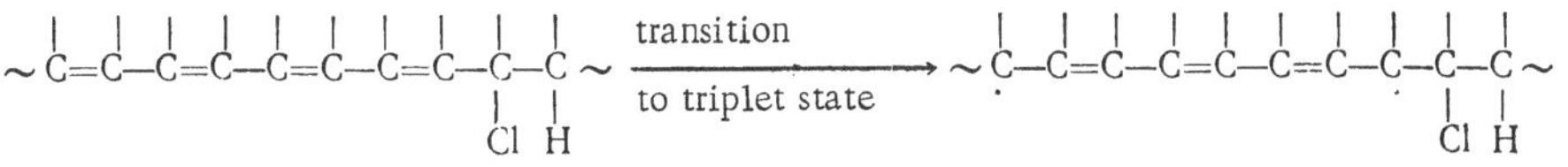

further progression of decomposition on account of the polarization effect:

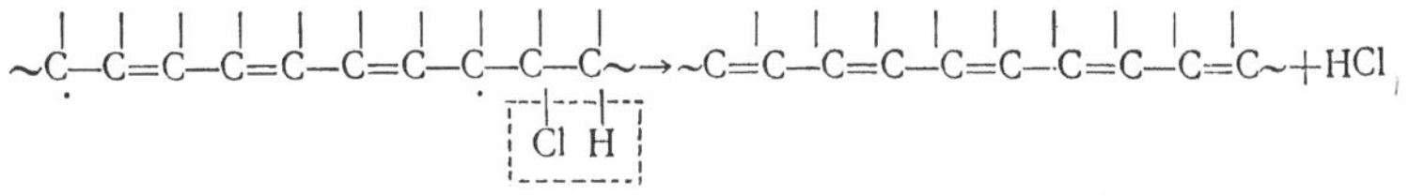

the beginning of decomposition according to a radical mechanism:

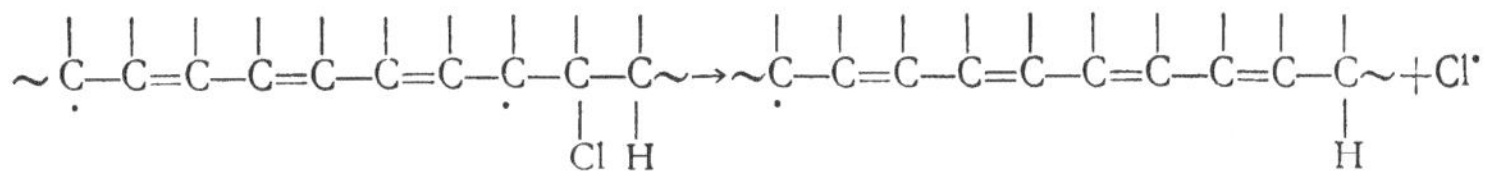

Initation of decomposition according to a free-radical mechanism leads to a development of a whole series of varied reactions, considered above.

The beginning of decomposition according to an ionic-molecular mechanism, with further development of interlinked ionic-molecular and radical processes, is due to the presence in the polymer chain of any functional groups that facilitate the polarization of the chlorine-carbon bond, including double bonds, oxygen-containing functional groups, as well as free radicals formed during the decomposition process, or as a result of the appearance of π-complexes in the interaction of excited polyene structures with the free radicals present in the system [53, 54].

It follows from the above that thermal decomposition according to an interlinked ionic-molecular and radical mechanism will be more intensive, the more functional groups that disturb the ideal structure of polyvinyl chloride, present in the macromolecules of the polymer.

Thermooxidative Decomposition. The initiation of thermooxidative decomposition is related not only to the presence of free radicals, which are radicals of the initiator or products of polymer oxidation. It is also produced by the splitting out of hydrogen chloride, thanks to the effect of polarization of the chlorine-carbon bonds, due to deviations of the structures of the chain from an ideal structure with alternating $-CH_2-$ and $-CHCl-$ groups.

In the presence of oxygen, the interlinked ionic-molecular and radical processes of polymer decomposition require smaller energy expenditures. This is due to the fact that in the presence of oxygen, the oxidation processes increase the free-radical concentration, which facilitates transitions of the polyene structures to the triplet state, the formation of π-complexes, and initiation of radical decomposition by the splitting out of atomic chlorine. In addition, in the oxidation of the polymer, the concentration of chlorine-carbon bonds polarized by neighboring oxygen-containing groups increases.

As thermooxidative decomposition of polyvinyl chloride progresses, the reactions considered above can occur [12, 13].

The proposed scheme makes it possible to more profoundly disclose the mechanism of the decomposition of polyvinyl chloride, since it considers the interrelationship between the ionic-molecular and radical reactions that occur during thermal and thermooxidative decomposition of the polymer.

II. GENERAL PRINCIPLES OF THE STABILIZATION
OF POLYVINYL CHLORIDE

Three methods can be used in principle to slow down the thermal and thermooxidative decomposition of polyvinyl chloride and vinyl chloride copolymers: 1) suppression of chain reactions developing during the process of thermal and thermooxidative decomposition; 2) the creation of conditions under which the substances formed during the process of breakdown prevent more profound decomposition of the polymer, and 3) the creation of conditions under which decomposition proceeds reversibly. In practice, the first method is used in the overwhelming majority of cases in the stabilization of halogen-containing high-molecular compounds, while the substances used as stabilizers, as a rule, are decelerators of chain decomposition reactions.

During the process of thermal and thermooxidative decomposition of polyvinyl chloride, as has already been indicated, both ionic-molecular

and radical reactions are possible. It is customarily believed that acceptors of hydrogen chloride – metal salts of organic and inorganic acids –
are typical decelerators of ionic reactions, while antioxidants and heteroorganic compounds, in particular, dialkylstannane derivatives, are decelerators of free-radical chain reactions [3, 12, 14, 38, 42, 61-80].

In spite of the fact that in practice, rare compositions based on polyvinyl chloride or vinyl chloride copolymers do not include metallic salts
of organic or inorganic acids, the problem of the mechanism of the stabilizing action of these compounds still remains not entirely clear. On the
basis of the available experimental data, however, we can assert that the
role of metallic salts does not reduce the bonding of hydrogen chloride
alone.

The effectiveness of the reaction of accepting of hydrogen chloride
is not the only factor determining the activity of the stabilizer, in particular, its ability to prevent the formation of stable free radicals and
polyene structures, which give rise to the appearance of color. All the
stabilizers – salts providing high color stability – are extremely effective acceptors of HCl; however, far from all strong hydrogen chloride
acceptors prevent the decomposition and appearance of color when the
polymer is heated [38]. It is quite possible that the effect of stabilization in the use of salts of organic acids is achieved thanks to the replacement of labile chlorine atoms in the polymer macromolecules by stable,
carboxylate groups, as well as by the reduction of the length of the kinetic
decomposition chain when the regular alternation of $-CH_2-$ and $-CHCl-$
groups in the chains of the macromolecules is disturbed. Evidence of
this can be found in the experimental facts obtained by two different
methods of analysis – IR-spectrometry and the use of labelled carbon
atoms – and showing that when polyvinyl chloride is heated in the presence of salts of barium, cadmium, and zinc, the acid groups prove to be
chemically incorporated into the composition of the macromolecular
chains [69].

There are data on the basis of which we can conclude that the metals
contained in the salts participate in the decomposition reactions of polyvinyl chloride according to a radical mechanism. Metals salts of stearic
acid, depending on the position of the metal in the period system, temperature, and concentration, can accelerate or decelerate dehydrochlorination. The greatest influence on the rate of dehydrochlorination is exerted by salts of those metals in which the d-layer of the electron shell
following the outer shell is filled, namely salts of lead, cadmium, and
zinc. A relatively weak influence on the rate of dehydrochlorination is
exerted by the stearates of calcium and barium, which contain no d-electrons in the shell following the outer shell [73]. The observed phenomenon agrees with the representations of the catalytic action of metals of

variable valence in oxidation-reduction reactions, in particular, in processes of oxidative decomposition of high-molecular compounds [81, 82].

Lead, cadmium, and zinc are not typical metals of variable valence; however, they do have a d-layer filled with electrons in the shell following the outer shell; in those cases when the d-electrons take part in the formation of a bond with the activated molecule, they can act like transition metals [83]. This can explain the relatively weak influence of salts of calcium and barium on the rate of dehydrochlorination of polyvinyl chloride [73].

One of the facts leading to the assumption of participation of metallic salts in reactions of polyvinyl chloride decomposition according to a free-radical mechanism is the influence of the chlorides of certain heavy metals on the decomposition rate. For the same metal, it can be positive or negative, depending on the temperature and concentration [68]. The effectiveness of the stabilizing action is sometimes explained by the ability of the chlorides to react with organic acids, i.e., by the reversibility of the reaction of the stabilizer with hydrogen chloride [67].

Of great interest are the results of investigations of organotin compounds as stabilizers of polyvinyl chloride, as well as investigations of substances belonging to typical inhibitors of radical chain reactions — organic inhibitors with a labile hydrogen atom in the molecule.

The high effectiveness of the stabilizing action of organotin compounds, noted in many investigations and patents, can be explained by their ability to suppress free-radical chain processes. Dibutyltin dilaurate, widely used as a stabilizer of polyvinyl chloride, proves to be an effective inhibitor of the radical decomposition of tertiary butyl hydroperoxide, initiated by cobalt octoate [12].

A free-radical mechanism of the stabilizing action of dialkylstannane derivatives in the process of photodecomposition of polyvinyl chloride has been proposed in an interpretation of the results obtained in an investigation of the activity of dibutyltin diacetate, labelled with carbon C^{14} in the butyl group. After the irradiation, the butyl group is detected in polyvinyl chloride. Apparently the formation of cross-links and oxidation of the macromolecules during UV irradiation of the polymer is prevented when the butyl group is added to the macroradicals. The possible mechanism of the action of dibutyltin diacetate is described by the scheme [42]:

$$R^{\cdot} + (C_4H_9)_2 \, Sn \, (OCOCH_3)_2 \rightarrow RC_4H_9 + C_4H_9Sn \, (OCOCH_3)_2$$

In an investigation of the stabilizing action of alkoxy, mercapto, and acid derivatives of dibutyltin, it was shown that they are decelerators of dehydrochlorination when polyvinyl chloride is heated. The activity of acid and alkoxyl derivatives is approximately the same and is practically independent of the number of carbon atoms in the acid or alkoxyl groups; the effectiveness of the action is considerably intensified when the alkoxyl group is replaced by the mercapto group. The investigated dibutyltin derivatives are not typical HCl acceptors, in spite of the fact that they can react with hydrogen chloride and form chlorides and dichlorides of dialkylstannane. One of the peculiarities of the stabilizing action of these compounds has been noted: in their presence the decomposition of polyvinyl chloride proceeds with the formation of substances containing covalently bonded chlorine. If the polymer is heated in the presence of dibutyltin dichloride and an organotin stabilizer, then as the duration of the heating increases, the concentration of ionic chlorine in the composition decreases, no corresponding increase in the amount of ionic chlorine in the volatile products being observed. It is not entirely clear whether the observed phenomenon is related to the peculiarity of the inhibiting action of organotin compounds or to the addition of hydrogen chloride to the unsaturated bonds of the polymer under the catalytic action of the tin compounds [76, 77].

An investigation of the influence of definite types of inhibitors of free-radical chain reactions – phenols, amines, mercaptans, hydroxyketones – the action of which is not related to the accepting of hydrogen chloride, on the rate of thermal and thermooxidative decomposition of polyvinyl chloride showed that these compounds with a labile hydrogen atom in the molecule suppress dehydrochlorination when the polymer is heated. The decelerating action depends on the chemical nature of the inhibitor, its concentration in the composition, temperature, the presence of oxygen in the system, as well as on the intrinsic stability of the polymer. The influence of the same compound on individual routes of decomposition of polyvinyl chloride can differ: among phenols, hydroxyketones, and certain other aromatic compounds, the ability to decelerate dehydrochlorination is not always combined with an ability to prevent oxidation or structuring. Phenols, hydroxyketones, mercaptans, in particular, 2,4-dihydroxyacetophenone, hydroquinone, 1-ethyl-2,4-dihydroxybenzene, and dodecyl mercaptan, decelerate both thermal and thermooxidative decomposition of polyvinyl chloride. This is explained by the ability of the investigated compounds to interact with radicals of the type $R^{\cdot}$, $RO^{\cdot}$, $ROO^{\cdot}$, and with atomic chlorine under the conditions of decomposition [75, 78, 84-86].

As has already been indicated (see Chapter I), an extremely important shortcoming of inhibitors that contain readily labile hydrogen atoms and can form stable radicals is the fact that the possibility of transfer of the decomposition reaction chain is not excluded in their presence. Under definite conditions, in particular, when the temper-

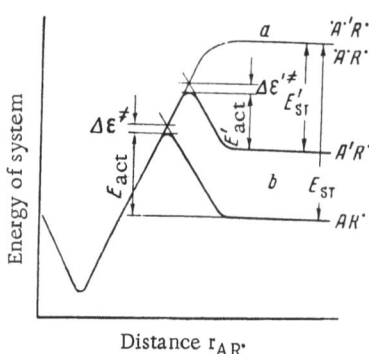

Fig. 94. Energy diagram of the interaction of radicals with substances excited in the triplet state. A and A')
Substances in the singlet state; ˙A˙ and
˙A˙') substances in the triplet state; energy of conjugation in the singlet state
for A' greater than for A.

ature is raised to 200-250°C, In radicals can initiate decomposition processes.

In view of this, an investigation of
substances in the presence of which
transfer of the reaction chain of polymer decomposition is not observed as
stabilizers of polyvinyl chloride is of
great theoretical and practical interest. As has been discovered, certain
polymers with a system of conjugated
bonds can serve as stabilizers of this
type [87-89]. The data of a number of
theoretical studies devoted to elucidating the influence of the structure of organic compounds on their reactivity in
radical processes have served as a
basis for testing these substances as
decelerators of the decomposition of
high-molecular compounds.

It is known that many substances with an aromatic conjugated chain
(acenes, phenylenes, pyrenes) possess high reactivity with respect to
radicals, representing a quantity of the same order of magnitude as the
reactivity of typical vinyl monomers [53]. The cause of such a high affinity for radicals of formally valence-saturated compounds lies in the
following. As the energy of conjugation in the singlest state increases,
the distance between the singlet and triplet levels decreases [90, 91],
which in Fig. 94 corresponds to a decrease in the quantity E_{ST} with displacement of the repulsion curve upward parallel to itself. We might as-
sume that the energy of conjugation in the transition state ($\Delta\varepsilon^{\neq}$) increases
with increasing conjugation energy in the singlet state [53]. A decrease
in E_{ST} and increase in $\Delta\varepsilon^{\neq}$ lead to a decrease in the activation energy of
the addition of radicals to the substances with a system of conjugated
bonds. For such substances, the probability of formation of a transition
complex increases in the presence of free radicals; moreover, the activation energy (E_{act}) will be smaller the smaller the excitation energy
in the triplet state. In other words, in the system decomposing poly-
mer-stabilizing substance with a developed system of conjugation, active biradicals can form, which terminate the decomposition chain, but
are not capable of initiating it.

Polymer substances with a developed conjugation chain should be
considered as a set of polymer homologs, distinguished by their values
of E_{ST} and E_{act}. We can consider it proven that the polymer substances

TABLE 13. Dependence of the Effectiveness of the Stabilizing Action of Polyvinylacetylene and the Product of Total Dehydrochlorination of Polyvinyl Chloride on the Concentration and Temperature

Additive to polyvinyl chloride Name	Concentration, %	Temperature at which the rate of dehydrochlorination was determined, °C	Duration of induction period up to beginning of dehydrochlorination, min	Average integral rate of dehydrochlorination in 180 min, mg HCl per g PVC	Ratio of the rates of dehydrochlorination of stabilized and unstabilized PVC, %
PVC without	–	175	7	3.91	–
additives	–	185	6	10	–
	–	195	5	20.7	–
Polyphenyl-	1	175	7	1.34	34
acetylene	1	185	5	6.3	63
	5	185	6	6.2	62
	10	185	7	6.17	62
	1	195	3	19	92
	5	195	3	20	96
	10	195	3	20.8	100
Dehydro-	1	175	9	3.07	78
chlorination	1	185	8	7.8	78
products of	5	185	6	6.4	64
PVC with the	10	185	7	9	90
composition	1	195	6	27	130
$(-CH=)_n$	5	195	5	20.7	100
	10	195	6	31.2	150

with a system of conjugated chains thus far synthesized contain stable paramagnetic particles. These paramagnetic fractions of higher polymer homologs apparently represent stable biradicals, forming π-complexes with the lower-molecular diamagnetic homologs, which make up the bulk of the substance. The formation of such complexes should decrease the energy of excitation in the triplet state.

The above permitted us to assume that in the case of polymers with conjugated bonds, we can count on a substantial increase in $\Delta\varepsilon^{\neq}$ and a decrease in the quantities E_{ST} and E_{act} when they interact with free radicals formed during the process of decomposition of polymers.

Experimental verification of the influence of polymers with alicyclic and aromatic systems of conjugated bonds on the rate of decomposition of polyvinyl chloride confirmed the correctness of the concepts presented above.

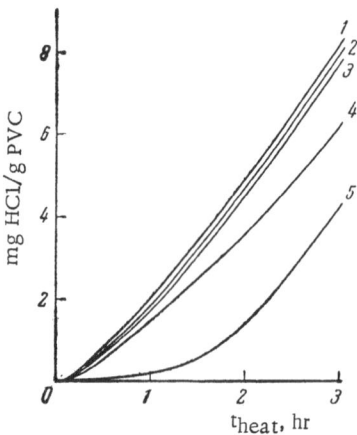

Fig. 95. Dependence of the dehydrochlorination of PVC on the time of heating at 175°C in a stream of air without
additives (1) and with additions of 1%
lead silicate (2), dibutyltin maleate (3),
diphenylolpropane (4), and polyphenylacetylene (5).

Figure 95 shows the kinetic curves of the splitting out of hydrogen chloride from polyvinyl chloride at 175°C in a stream of air in the presence of lead silicate, dibutyltin maleate, diphenylolpropane, and polyphenylacetylene. These curves show the high effectiveness of the inhibiting action of polyphenylacetylene. Table 13 presents data characterizing the influence of polyphenylacetylene and the complete dehydrochlorination product of polyvinyl chloride, corresponding in composition to the formula $(-CH=CH-CH= =CH-CH=CH-)_n$, on the rate of decomposition of polyvinyl chloride at various temperatures; these data give evidence of a reduction of the effectiveness of the stabilizing action of polymers with an alicyclic conjugated chain at increased temperature.

Polymers with an aromatic conjugated chain differ sharply in their stabilizing action from polymers with an alicyclic conjugated chain. The curves of Figs. 96 and 97 show that the stabilizing action of polymer aromatic compounds does not decrease with increasing temperature, but, on the contrary, is intensified. The observed difference in the action entirely corresponds to the data cited above on the excitation energies in the triplet state for polyphenylenes and polyvinylenes: since E_{ST} is greater for polyphenylenes than for polyvinylenes, the inhibiting action of the former is manifested at higher temperatures.

The increase in the activity of polymer aromatic compounds with increasing temperature can serve as a confirmation of the correctness of the assumption that such substances can be high-temperature stabilizers in the case of relatively small conjugated chains. In this case the values of E_{ST} and E_{act} are still rather great; hence higher temperatures are needed for the formation of biradicals and the manifestation of their inhibiting action. When high-molecular compounds of fractions of polymer homologs with a broad range of molecular weights are used as stabilizers, the temperature limits of effective action will apparently depend on the degree of polymerization of the stabilizer or, more accurately, on the average number of conjugated π-electrons in its molecule. The

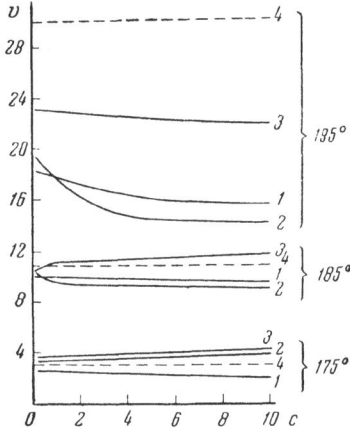

Fig. 96. Dependence of the stabilizing activity of polyphenylene A (1), polyphenylene B (2), and polyphenylene C (3) on their concentration in PVC and the temperature. v) Amount of HCl (mg) split out from PVC (1 g) in 180 min in a stream of air; c) parts by weight of stabilizers (1-3) per 100 parts by weight PVC; 4) values of the quantity v for unstabilized PVC.

temperature optimum of the stabilizing action should lie lower, the higher the molecular weight of the fraction.

In considering the index of the effectiveness of the stabilizing action of polymers with conjugated bonds, it is still essential to consider the possibility of chain transfer through the inhibitor. As is shown in Fig. 98, the stabilizing action of polyphenylacetylene is sharply reduced after its thermal processing. Moreover, an increase in the magnetic susceptibility and EPR signal is observed. This phenomenon is apparently related to "opening" of the double bonds of the polyene structures and to the formation of radicals containing unpaired electrons, unconnected with the system of conjugation. The activity of such radicals significantly exceeds that of the biradicals arising in the case of triplet excitation of conjugated systems. Although the former cannot only terminate chain decomposition, but also initiate it, the latter, being relatively inactive, are capable mainly of reacting only with radicals that carry the decomposition chain. The summary effect will depend on the degree to which one state or another is realized for the given system under the conditions used.

The use of polymers with conjugated bonds as stabilizers of high-molecular compounds opens up new prospects, since it makes inhibition of the development of the chain during decomposition possible within a broad range of temperatures, using fractions of polymer homologs of various structures and average molecular weights, differing in values of E_{ST} and E_{act}.

Such stabilizers can be of interest in the solution of the problem of raising the temperature limit of the use of the known organic and hetero-organic polymers.

We should especially consider the question of the influence exerted by thermal stabilizers on the reprocessing process. As is well known, the reprocessing of polyvinyl chloride and vinyl chloride copolymers is hindered by the fact that in these polymers the temperature of transition to the viscous-fluid state, as a rule, exceeds the temperature of the begin-

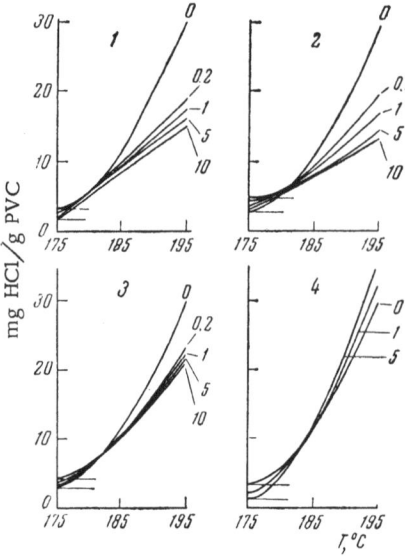

Fig. 97. Dependence of the stabilizing activity of polyphenylene A (1), polyphenylene B (2), polyphenylene C (3), and lead stearate (4) on their concentration in PVC (numbers on the curves) and the temperature. Parts by weight taken per 100 parts by weight PVC.

ning of decomposition. As has already been mentioned, the basic routes of thermooxidative decomposition of chlorine-containing high-molecular compounds are dehydrochlorination, oxidation, destruction, and structuring. Since a relatively small increase in the temperature leads to an extremely substantial increase in the rates of the enumerated routes of decomposition, the question arises of the minimum temperature at which the compositions can be reprocessed and an object with the necessary physicomechanical properties can be obtained.

The process of polyvinyl chloride reprocessing is very sensitive to impurities capable of inhibiting or initiating chain oxidation processes. The presence of such substances which disturb the equilibrium between the processes of structuring and destruction   influences the process of chemical flow. The introduction of stabilizer-antioxidants, which decelerate chain oxidation reactions, leads, with all other conditions equal, to a displacement of the equilibrium in the direction of destruction, to the development of chemical flow, and to a facilitation of reprocessing. On the contrary, the presence of substances that initiate oxidative processes, in particular, metals of variable valence, leads to a reduction of the chemical fluidity and to the necessity of introducing stabilizers into the composition, to suppress the harmful influence of such impurities.

The introduction of small amounts of inhibitors of thermal and thermooxidative decomposition of polyvinyl chloride suppresses structuring upon heating of the polymer and under conditions when it is not subjected to the action of mechanical forces; this has been demonstrated on compositions containing dinaphthylmethane, hydroquinone, and diphenanthrylmethane [92].

In a study of the interrelationship between the ability of stabilizers to slow down dehydrochlorination and prevent the formation of crosslinks during heating of polyvinyl chloride, it was established that stabilizers – acceptors of hydrogen chloride, which do not inhibit dehydro-

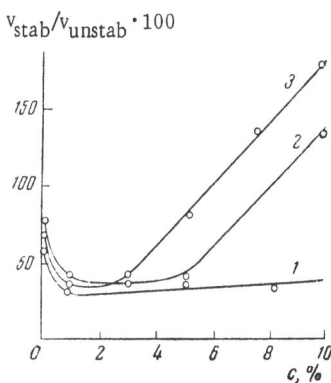

$v_{stab}/v_{unstab} \cdot 100$

Fig. 98. Rate of dehydrochlorination of PVC at 175°C in a stream of air in the presence of: (1) polyphenyl-acetylene up to the temperature of treatment; (2) polyphenylacetylene after exposure for 6 hr at 300°C; (3) polyphenylacetylene after exposure for 6 hr at 400°C. v_{stab}/v_{unstab} is the ratio of the rates of dehydrochlorination of the stabilized and unstabilized polymer; c is the concentration of polyphenylacetylene.

chlorination (barium, calcium, and cadmium stearates) do not prevent the loss of fluidity. Lead stearate inhibits dehydrochlorination and slows down the loss of fluidity; derivatives of phenols, aromatic hydroxyketones, and other analogs of these compounds, which exert essentially no influence on the rate of dehydrochlorination, increase the time during which the ability of polyvinyl chloride to flow is preserved [84].

Considering the possible means of suppressing processes of decomposition of polymers and copolymers of vinyl chloride under the action of radiant energy, we must proceed from the following theoretically possible lines. One of them is the selection of substances that absorb in the region of the same wavelengths as the functional groups contained in the polymer. In this case the necessary condition is high intrinsic stability of such "light-filtering" agents. Only when this condition is observed will the UV-absorbers absorb photons and, without being decomposed, transform them to thermal energy. Otherwise their conversion products can initiate photosensitized decomposition of the polymer. Complex esters of aromatic acids and phenols, as well as derivatives of benzophenone, are used as light-filtering agents in the stabilization of polyvinyl chloride. A second means of light stabilization of polyvinyl chloride consists of introducing compounds capable of slowing down free-radical chain reactions during photodecomposition or capable of bonding the hydrogen chloride split out from the polymer, into the composition. Such substances include metallic salts of organic and inorganic acids, organotin compounds, epoxy compounds, as well as substances with a labile hydrogen atom in the molecule — phenols and amines. Just as in thermal stabilization, their action is related to the accepting of hydrogen chloride, blocking of free radicals carrying the reaction of photodecomposition, or to termination of this chain as a result of the formation of stable radicals.

Apparently many of the aromatic sensitizers can combine light-filtering action with an ability to suppress chain photodecomposition. As an example we might cite the data of [75, 78], in which it was shown that the ability to prevent the aftereffect during ultraviolet irradiation of poly-

Absorption coefficients

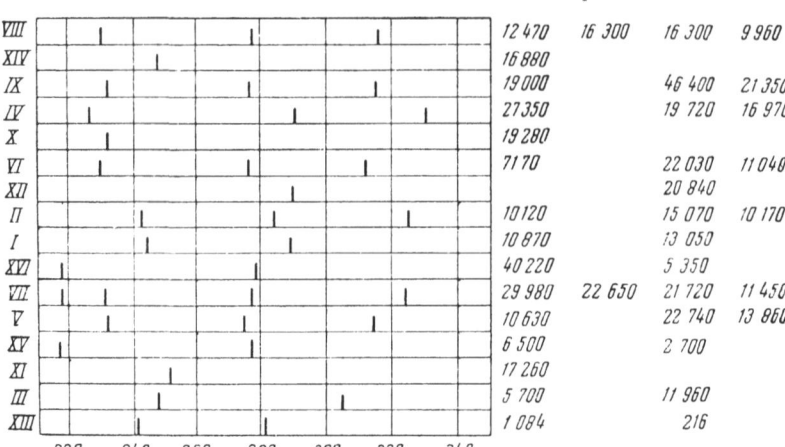

Fig. 99. Position of the absorption maxima in the ultraviolet region of the spectrum and values of the absorption coefficients of certain compounds. (Indices the same as in Table 14.)

vinyl chloride, possessed by 2-hydroxy-substituted and unsubstituted benzophenones and acetophenones, alkyl- and alkyleneresorcinols, and certain analogs of these compounds, as well as by derivatives of anilino-sym-triazine, is related to light-filtering action and to an ability to suppress photochemical reactions.

Figures 99 and 100 show the positions of the maxima and values of the absorption coefficients of these compounds; Tables 14 and 15 present data on their influence on the rate of thermooxidative decomposition of polyvinyl chloride and on the aftereffect in UV-irradiation of the compositions.

One of the most important factors influencing the effectiveness of the stabilization of polyvinyl chloride is synergism. The significance of this phenomenon lies not only in the fact that synergic mixtures give a greater effect than might be expected on the basis of additivity of the action of the stabilizers contained in the mixture. Of incomparably greater significance is the fact that when synergic mixtures are used, effects unachievable in the case of individual use of the most active of the components contained in the mixture can be obtained. Mixtures of substances belonging to the same and to different types of stabilizers can give a synergic effect.

The data cited in the survey [93], where the synergic action of hydrogen chloride acceptors – salts of barium and cadmium, calcium and zinc – is noted, as well as the data of [71, 73], which characterize the synergic action of mixtures of the stearates of lead, barium, calcium, cadmium, and zinc, can be cited as examples of mixtures of monotypic stabilizers.

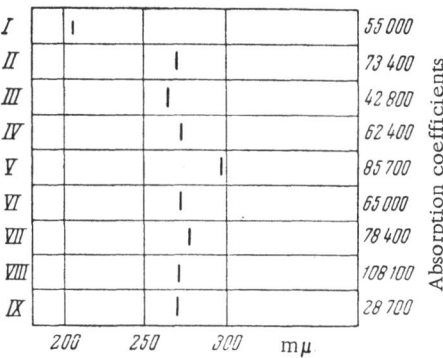

Fig. 100. Positions of the absorption maxima in the ultra-violet region of the spectrum and values of the absorption coefficients of derivatives of symmetrical triazine. (Indices of the compounds the same as in Table 15.)

The manifestation of synergic action among mixtures of substances belonging to different types of polyvinyl chloride stabilizers is mentioned not only in abundant patent literature, but also in many investigations. Data are cited on the high activity of mixtures of epoxy compounds with salts of cadmium, zinc, and organotin compounds [93-97].

A synergic action of a mixture of epoxy resins with cadmium stearate has been demonstrated in the reaction of radical decomposition of hydroperoxide. This mixture exhibits activity in the inhibition reaction, exceeding not only that expected from a calculation on the basis of additivity, but also the activity of the most energetic inhibitor of peroxide decomposition – cadmium stearate (Fig. 101) [12]. Synergic action in the stabilization of polyvinyl chloride by mixtures of epoxy compounds with lead silicate, mixtures of the autocondensation products of cyclohexanone with lead silicate, as well as mixtures of the autocondensation products of cyclohexanone with organotin compounds and with calcium stearate, is described in [74, 77, 98].

The synergic action of three types of mixtures of polyvinyl chloride stabilizers: mixtures of two decelerators of thermal dehydrochlorination that do not bond hydrogen chloride, mixtures of two stabilizers that are hydrogen chloride acceptors, and mixtures of hydrogen chloride acceptors with decelerators of dehydrochlorination, possessing no acceptor properties, was investigated in [54, 73, 86].

Mixtures of decelerators of dehydrochlorination, the action of which is not related to the accepting of hydrogen chloride (2,4-dihydroxyacetophenone, 1-ethyl-2,4-dihydroxybenzene, hydroquinone, dodecyl mercaptan) give a synergic effect in both thermal and thermooxidative decomposition of polyvinyl chloride. The synergic action of the indicated compounds can be explained by the differing activity of the stabilizers

TABLE 14. Influence of Aromatic Hydroxyketones, Phenols, and Certain Analogs of These Compounds on the Rate of Thermal Dehydrochlorination of PVC and Aftereffect in Ultraviolet Irradiation of the Compositions

Index of additive	Additive	Ratio of rates of dehydro-chlorination of stabilized (v_3 and v_4) and un-stabilized (v_1 and v_2) PVC in 180 min at 175°C, %	
		v_3/v_1 before UV-irradiation	v_4/v_2 after UV-irradiation
	PVC without additives	100	100
I	2,4-Dihydroxybenzophenone	98	95
II	2-Hydroxy-4-methoxybenzophenone	90	85
III	2,4,6-Trihydroxybenzophenone	50	60
IV	2,4-Dihydroxy-3'-nitrobenzophenone	110	95
V	2,2'-Dihydroxy-4,4'-dimethoxybenzophenone	90	93
VI	2,4-Dihydroxyacetophenone	100	93
VII	2,4-Dihydroxy-5-ethylacetophenone	85	85
VIII	2,2',4,4'-Tetrahydroxyadipophenone	116	90
IX	2,2',4,4'-Tetrahydroxysebacephenone	123	102
X	Di-(2,4-dihydroxybenzoyl)-p-xylylene	100	90
XI	Benzophenone	100	105
XII	4-Phenylbenzophenone	100	95
XIII	Acetophenone	95	120
XIV	Diphenyl	95	75
XV	Ethylresorcinol	75	78
XVI	Di-(2,4-dihydroxyphenol)decane	55	55

Note. Concentration of additives 0.00025 mole per 10 g PVC; the average integral rates of splitting out of HCl at 1705°C in a stream of air for the unstabilized polymer are: before UV-irradiation (v_1)= 2.70 mg HCl per g PVC; after UV-irradiation (v_2)= 6.00 mg HCl per g PVC.

themselves and their conversion products in reactions with free radicals of the type $R^{\cdot}$, $RO^{\cdot}$, $ROO^{\cdot}$, and with atomic chlorine.

In an investigation of the synergic action of two stabilizers that act as hydrogen chloride acceptors, the basic attention was paid to revealing the interrelationship between the synergic action and the effect of deceleration of dehydrochlorination, as well as the interrelationship between the synergic action and the acceptor capacity of the mixtures. When mixtures of the stearates of lead, barium, calcium, cadmium, and zinc are introduced into polyvinyl chloride, an effect of deceleration of dehydrochlorination is not detected in those cases when it did not occur after individual introduction of the salts. If an effect of deceleration of

TABLE 15. Influence of Triazine Derivatives on the Rate of Thermal
Dehydrochlorination of PVC and the Aftereffect in Ultraviolet
Irradiation of the Composition

Index of additive	Additive	Ratio of rates of dehydro-chlorination of stabilized and unstabilized PVC in 180 min at 175°C, %	
		v_3/v_1 before UV-irradiation	v_4/v_2 after UV-irradiation
	PVC without additives	100	100
I	2,4,6-Triaminotriazine	125	117
II	2,4,6-Trianilino-sym-triazine	121	141
III	2,4,6-Tri-(o-methylanilino)-sym-triazine	169	118
IV	2,4,6-Tri-(p-methylanilino)-sym-triazine	150	126
V	2,4,6-Tri-(p-sulfamidoanilino)-sym-triazine	132	109
VI	2,4,6-Tri-(o-chloroanilino)-sym-triazine	92	117
VII	2,4,6-Tri-(p-chloroanilino)-sym-triazine	92	118
VIII	2,4,6-Tri-(o,p-dichloroanilino)-sym-triazine	100	119
IX	2-Chloro-4,6-di-(o,p-trichloroanilino)-sym-triazine	84	92

Note. Concentration of additives 0.00025 mole per 10 g of the polymer. Average integral rates of splitting out of HCl for the unstabilized polymer at 175°C in a stream of air are: before UV-irradiation (v_1) = 2.43 mg HCl per g PVC; after UV-irradiation (v_2) = 4.90 mg HCl per g of PVC.

the splitting out of HCl was observed in the individual use of the stearates, then the joint use of the salts did not lead to its intensification. The acceptor capacity of the mixtures proves to be considerably greater than the additive values for most of the investigated pairs of stearates (Fig. 102).

In revealing the characteristics of the synergic action of mixtures of two stabilizers, one of which is a hydrogen chloride acceptor, while the second is a decelerator of decomposition, which does not bond HCl, the fact that stabilizer-acceptors, in particular, metallic salts of organic acids, can decelerate or accelerate decomposition of the polymer was taken into consideration. In this case, as has already been mentioned, the greatest influence on the rate of decomposition is exerted by salts of

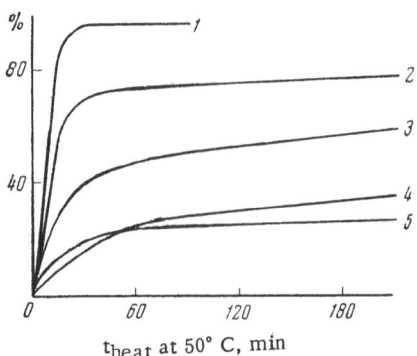

Fig. 101. Influence of stabilizers of PVC on the rate of decomposition of tert-butyl-hydroperoxide in the presence of cobalt octoate. 1) Without stabilizer; 2) with epoxy resin "Epone 834"; 3) with cadmium soap; 4) with mixture of epoxy resin and cadmium soap; 5) with dibutyltin dilaurate. Along the vertical axis — amount of tert-butylhydroxy-peroxide decomposed.

those metals in which the d-layer of the electron shell is filled, namely salts of lead, cadmium, and zinc. If free radicals are generated in the presence of such salts as a result of oxidation-reduction processes, then the variation of the concentration in the polymer-stabilizer system will be one of the basic factors in the influence of the stabilizer on the rate of polymer decomposition. Moreover, the positive or negative character of the influence of the additive depends on the reactivity of the radicals formed. Thus, the mutual influence of metallic salts and typical inhibitors of radical chain reactions, on the strength of which the stabilizing action of their mixtures can be intensified or weakened, should be sharply pronounced in salts of metals with a filled d-layer and weakly pronounced in salts of metals with an unfilled d-layer. Table 16 characterizes the stabilizing action of mixtures of salts of lead and calcium, which accept hydrogen chloride, with hydroquinone and hydroxyketone. These data confirm the correctness of the concepts presented above.

In the investigation of the activity of certain decelerators of dehydrochlorination of polyvinyl chloride in the temperature range from 150 to 200°C, it was established that definite temperature intervals exist for each stabilizer, within which its introduction into the polymer guarantees deceleration of decomposition.

The existence of upper and lower temperature limits for stabilizers of various types is confirmed by the data of Table 17. The temperature limits of the optima are evidently determined by the values of the total activation energy of the reactions in the polymer-stabilizer system, as well as the selective action of the stabilizer and its conversion products with respect to the decomposition products of the polymer. On the basis of the principle of the action of inhibitors of chain reactions initiated by radiations, we might assume that limits of optimum irradiation intensities exist for them, just as for thermal stabilizers. When the intensity of the influence is lower than the optimum value, the additives remain inactive; when the intensity of the influence is higher than the

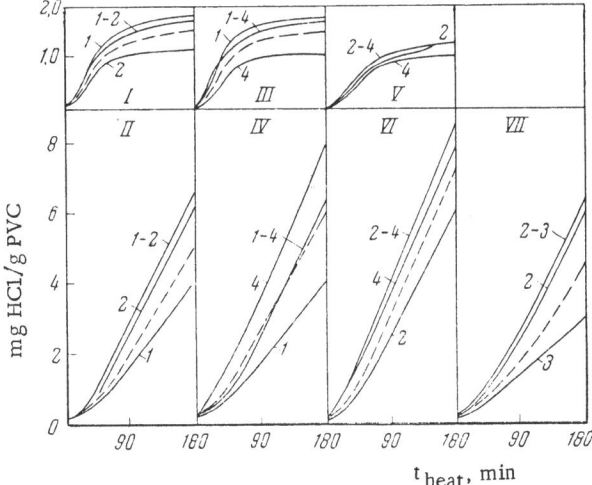

Fig. 102. Rate of bonding of hydrogen chloride at 175°C in
a stream of air by mixtures of stearates. I) Lead (1) and
barium (2), 1% each; II) the same, 5% each; III) lead and cadmium (4), 1% each; IV) the same, 5% each; V) barium and
cadmium, 1% each; VI) the same, 5% each; VII) barium and
calcium (3), 5% each. Dotted lines show the rates of bonding
of HCl in the case of additive action of the stearates.

optimum value, the reactions of conversion of the stabilizer itself can
"outstrip" the decomposition of the polymer; if the products formed as a
result of conversions of the stabilizer are active in reactions with the
initial polymer, then they will accelerate its decomposition.

According to the data on the mechanism of the decomposition of polyvinyl chloride and vinyl chloride copolymers, as well as on factors influencing the rate of decomposition under the action of heat, light, ionizing radiations, oxygen, ozone, microorganisms, various chemical reagents, and other factors, stabilizers should guarantee: deceleration
of thermal and thermooxidative, photolytic, radiochemical, and sensitized free radical chain reactions of polymer decomposition; absorption
of radiation in the region of 200-400 mμ; suppression of the negative action of the polymer decomposition products (hydrogen chloride, macroradicals, readily oxidized unsaturated structures); suppression of the
harmful influence of impurities that accelerate decomposition, in particular, metals of variable valence, in addition to providing antiseptic
protection under the action of microorganisms on the material.

Consequently, the stabilizing system should contain a hydrogen chloride acceptor, an antioxidant, a UV-absorber, and a substance suppressing the action of metals of variable valence. Since one substance, as a
rule, is not characterized by all these functions, but only some of them,

TABLE 16.   Influence of 2,4-Dihydroxyacetophenone and Hydroquinone on the Temperature of the Beginning of Decomposition t_d and the Duration of the Induction Period before the Beginning of Liberation of HCl (τ) in the Heating of PVC Stabilized by the Stearates of Lead and Calcium

| Additive | | | t_d, °C | τ at 175°C, min | |
| Name | Amount per 10 g of PVC | | | In a stream of air | In a stream of nitrogen |
	Moles	g			
PVC without additives	—	—	155	6	10
2,4-Dihydroxyaceto-phenone	0.00025	0.038	148	9	9
Hydroquinone	0.00025	0.027	153	10	12
Lead stearate (d-layer filled)	0.00025	0.200	190	37	47
Lead stearate	0.00025	0.200	240	40	51
2,4-Dihydroxyaceto-phenone	0.00025	0.038			
Lead stearate	0.00025	0.200	220	47	55
Hydroquinone	0.00025	0.027			
Calcium stearate (d-layer unfilled)	0.00025	0.150	181	17	18
Calcium stearate	0.00025	0.150	186	21	22
2,4-Dihydroxyaceto-phenone	0.00025	0.038			
Calcium stearate	0.00025	0.150	186	23	24
Hydroquinone	0.00025	0.027			

in practice stabilizing mixtures are made up of two to three components. It must be kept in mind that a substance used as a stabilizer should satisfy a number of requirements not related to the effectiveness of its action: it should be compatible with the polymer, should cause no deterioration of the physical, physicomechanical, and chemical properties and external appearance of the object, should be nontoxic, etc.

III. METHODS OF EVALUATING THE EFFECTIVENESS OF THE ACTION OF POLYVINYL CHLORIDE STABILIZERS

Methods of evaluating the effectivness of the action of stabilizers of polyvinyl chloride and vinyl chloride copolymers have been rather completely discussed in the monograph [65], in the surveys [41, 61, 99-109], and certain other works.

To determine the activity of thermal stabilizers, one investigates the variations of the properties of unreprocessed powdered compositions when stabilizers are introduced into them, the variation of the properties

TABLE 17. Influence of Certain Substances on the Rate
of Dehydrochlorination of PVC in the Temperature Range from 150 to 190°C

Additive		Ratio of average integral rates of liberation of HCl from stabilized and unstabilized PVC in 180 min in a stream of air (in %) at a temperature, °C			
Name	Number of parts by weight per 100 parts by weight PVC	150	175	185	195
PVC without additives	—	100	100	100	100
Lead stearate	1	—	97	100	133
	5	—	80	100	140
Di-n-hexyl ester of iso-propoxyphosphinic acid	1	—	94	73	67
	2.5	—	94	74	57
	5	—	91	75	47
Di-n-octyl ester of iso-propoxyphosphinic acid	1	—	91	82	67
	2.5	—	87	82	60
	5	—	87	74	50
Diphenyl ester of iso-propoxyphosphinic acid	1	—	94	—	83
	5	—	91	—	54
Dodecyl mercaptan	0.5	—	86	—	65
	1	—	80	—	54
Phenyl-β-naphthylamine	0.05	70	120	—	230
Hexaphenyldiplumbane	0.25	74	350	—	—

of the compositions during their reprocessing, as well as the influence
of stabilizers on the properties of the finished objects. The evaluation
of the activity of thermal stabilizers is based on a qualitative or quantitative comparison of the rates of dehydrochlorination, oxidation, destruction, and structuring of the unstabilized and stabilized polymer at
temperatures from 150 to 190°C. In this case the experiment is based
on chemical, physicochemical, physicomechanical, or optical methods
of analysis.

Most often some quantity characterizing the influence of the thermal
stabilizer on the rate of dehydrochlorination is used as the criterion of
its activity. For hydrogen chloride acceptors, influence on the true rate
of dehydrochlorination and on the rate of liberation of hydrogen chloride
from the composition in the free state are distinguished. The influence
of the stabilizer on the temperature of the beginning of decomposition of
the polymer (t_d in °C), on the duration of the induction period before the
beginning of the liberation of hydrogen chloride, which is customarily
called thermal stability (T in min), as well as on the rate of evolution
of HCl at the end of the induction period (v in % of HCl liberated or in mg
HCl per unit weight of the polymer) is usually determined. In determining the quantities t_d and T, the hydrogen chloride liberated in the

free state is detected qualitatively according to a color change of in-
dicator paper or according to the appearance of opalescence when it is
trapped by an aqueous solution of silver nitrate [101].

In determining the value of v, hydrogen chloride is absorbed by water
and then determined quantitatively by volumetric titration with a solution
of silver nitrate [101] or by potentiometric [38, 73, 110], conductometric
[98, 111], and nephelometric [21] methods. Since in all cases of qualitative or quantitative determination of hydrogen chloride liberated in the
free state, the fraction of it bonded by the stabilizer-acceptor is not considered, the enumerated methods make it possible to evaluate only the
apparent values of t_d, T, and v. An attempt to determine the true rate
of dehydrochlorination in the presence of stabilizer-acceptors was made
in [112]. The author investigated the variation of the electroconductivity
of the sample during its heating and arrived at the conclusion that the
rate of splitting out of hydrogen chloride from polyvinyl chloride in the
presence of lead and zinc salts depends on the temperature, but does not
depend on the nature and concentration of the stabilizer. A more ac-
curate method of evaluating the influence of hydrogen chloride acceptors
on the rate of dehydrochlorination was proposed in [38]. It was proposed
that the fraction of hydrogen chloride bonded by the stabilizer be determined by dissolving the samples in cyclohexanone after heat treatment,
then extraction of the chlorides with water, and determination of the
amount of ionic chlorine in the extract by potentiometric titration with a
solution of silver nitrate with a mercury-silver electrode. It is indicated
that in the investigation of organotin compounds, the extraction of the
chlorides should be performed with an aqueous solution of ammonia. In
[73] this method was somewhat modified: it was recommended that the
samples be dissolved after heat treatment in a mixture of cyclohexanone
and dioxane, which makes it possible to determine the chloride ion po-
tentiometrically in the mixture of solvents without extraction of the chloride. For inhibitors of dehydrochlorination that are not hydrogen chloride acceptors, the determination of the quantities t_d, T, and v makes it
possible to evaluate their influence on the true rate of dehydrochlorination.

Methods of evaluating the activity of thermal stabilizers according to
their influence on the rate of dehydrochlorination also include the determination of the electrical insulation properties of the finished articles.
Methods have been described for determining the thermal stability of objects according to the variation of the value of the breakdown voltage
during heating of insulation hoses at temperatures from 150 to 230°C
[113], as well as according to the variation of the electroconductivity
of insulation plastic as a function of the duration of exposure of the sample at temperatures from 100 to 175°C [114].

An idea of the influence of stabilizers on the rate of thermal dehydrochlorination of the polymer is also given by evaluating the degree of color change upon heating of the sample. Some authors believe that an evaluation of the activity of an additive according to color stability can be only approximate, since the color change is a result of several, sometimes directly opposed processes [38]. In spite of this, the method is finding wide use in the testing both of powders and of finished objects. In the testing of powder, the stabilized or unstabilized polymer is exposed for a definite time at a set temperature, then pulverized and screened; the color of the sample is compared with a standard visually or on a reflection colorimeter [52, 101]. The color change of the composition during reprocessing and during heating of the finished object is also observed visually, or more accurately, by means of instruments [101, 115].

Studies have been published in which an attempt was made to give a quantitative estimate of the activity of thermal stabilizers. For a comparative quantitative estimate of the activity of a number of stabilizers − hydrogen chloride acceptors − it has been proposed that the duration of the induction period (τ) be expressed by the equation

$$\tau = k\sigma^{\frac{1}{n}}$$

where σ represents the percent by weight of the stabilizer; k and n are quantities that are constant for a given series of experiments [116]. In [84-86, 89], the activities of stabilizers were evaluated according to the ratio of the rates of dehydrochlorination of stabilized and unstabilized polyvinyl chloride in percent or in fractions of a unit; the rate of dehydrochlorination of the unstabilized polymer was taken as 100% (unit).

The values of the rates of oxidative processes in the evaluation of the thermal stability of compositions based on chlorine-containing polymers are not as widely used as the values of the rates of the dehydrochlorination reaction. The methods of determining the intensity of oxidative processes have been most fully developed as applied to rubbers, including halogen-containing rubbers [124].

However, such indices as the rate of absorption of oxygen, rate of consumption of the antioxidant, duration of the induction period before the beginning of oxidation of the polymer, and duration of oxidation at a constant rate are not often used in the testing of polyvinyl chloride. As an example of the use of the indicated methods in the testing of the oxidizability of polyvinyl chloride, we might cite [66]. The oxidizability of polyvinyl chloride is judged considerably more often according to the influence of oxygen on the rate of the processes of dehydrochlorination [9, 107], destruction [14, 41], and structuring [14, 41]. In [89], the following index is proposed for a quantitative evaluation of oxidizability:

$$I = \frac{V_a - V_n}{V_n}$$

where V_a and V_n are the rates of dehydrochlorination of the polymer in air and nitrogen medium.

Powders and objects made from polyvinyl chloride are evaluated extremely often according to their thermal stability, i.e., according to the rate of the processes of destruction and structuring. The following are used as indices characterizing the rates of these processes: variation of the absolute, relative, and characteristic viscosity of solutions of the polymer [99, 118, 119], rate of formation of insoluble three-dimensional structures [120], variation of the thermomechanical properties of the polymer [39, 121], variation of the properties of the composition during its reprocessing, including variation of the duration of rolling until the mass begins to stick to the rollers [122-125], variation of the degree of branching and molecular weight distribution [119, 126-128].

Optical methods of analysis, in particular, the method of IR-spectrometry, have received widespread use in the investigation of the thermal and thermooxidative decomposition of polyvinyl chloride. Investigation of the IR-spectra of polyvinyl chloride has encountered difficulties on account of the absence of a suitable solvent, giving no background. The method of taking the IR-spectra of polyvinyl chloride in thin discs [40] is used in the study of the mechanism of the decomposition of polyvinyl chloride and the action of stabilizers, as well as in evaluating the stability of compositions [30, 41, 69, 129-137]. Recently, the use of the methods of EPR [50, 51, 138, 139] and deuteration [136, 140, 141] has begun to find use in the investigation of the properties and character of the transformations of chlorine-containing high-molecular compounds.

Independent of the indices according to which the rate of thermal or thermooxidative decomposition of the polymer is determined, the most improved methods of determining the stability of the composition during reprocessing are those that consider the influence of mechanical forces, i.e., all methods including observation of the change in properties of the composition directly in the reprocessing process. Most complex is the evaluation of the stability of nonplasticized compositions, designed for reprocessing by methods of extrusion and casting under pressure. As is noted in the literature, a reliable characterization of stability under dynamic conditions is given only by direct testing of the compositions on experimental extrusion or casting machines. In [142], it is recommended that the thermal stability of polyvinyl chloride be tested under continuous circulation of a definite amount of material through the extruder until the polymer begins to decompose. A survey of methods of reprocessing polyvinyl chloride and certain requirements set for thermal stabilizers is given in [143-146].

The aging of objects based on polyvinyl chloride and vinyl chloride copolymers can occur under the action of various agents, depending on the conditions of their operation: heat, light, oxygen, ozone, microorganisms, etc. Thus, the factors producing aging of electrical insulation materials are heat and chemical reagents, in particular, oxygen, ozone, and metals with which the object is in contact; in the operation of packaging and construction materials, they are acted upon by light, oxygen, and ozone; aging under the action of the most varied chemical agents occurs during the operation of chemical apparatus made of materials based on polyvinyl chloride.

The rate of aging depends on the nature and intensity of the influence of the agent producing it, on the physical state of the material, determined by the operating temperature and amount of plasticizer in the composition, on the presence of inhibitors or promoters of decomposition in the material, as well as on the conditions of reprocessing the composition into the object. In practice, it is difficult to find cases under the conditions of operation when aging is produced by some one factor. As a rule, the change in the physical, chemical, and physicomechanical properties is a result of the simultaneous influence of several factors.

In spite of the fact that in all cases of the aging of materials based on polyvinyl chloride and vinyl chloride copolymers, dehydrochlorination, oxidation, destruction, and structuring processes occur, the concrete ratio of the rates of these four routes can be varied. The general direction of aging, in accord with this, can be expressed in a deterioration of the external appearance, strength and physicomechanical indices of the material, or in a reduction of its stability to the action of various chemical and physical agents.

Since, as was already indicated, many chemical reactions that take place in the aging of polyvinyl chloride are radical in character, the rate of aging will be determined, with all other conditions equal, by the free-radical concentration in the material and by their activity. The free-radical concentration depends on the conditions of production of the polymer, as well as on the composition and method of reprocessing the mixture. The reactions of free radicals with undecomposed polymer molecules will accelerate aging, while their recombination with one another forming inactive products will decelerate it. The end result is determined by the ratio of the rates of the reactions leading to the accumulation or destruction of active groups in the material, as well as by the variation of the composition of the radicals.

In addition to the activity and concentration of free radicals, the rate of aging depends on the physical state of the material, which guarantees or, on the other hand, excludes the possibility of the approach of reactive groups: in raw and cured rubbers, the rate of aging in the vitreous state is substantially lower than in the highly elastic state; in polyvinyl chloride, plasticized compositions frequently prove more stable to aging than rigid compositions. During the process of aging, a vital role is played by the size and state of the surface of the object or of the polymer particles. In particular, the ratio of the fractions of reflected and absorbed energy of the incident radiation, as well as the rates of the chemical reactions related to the phenomena of sorption or diffusion of reagents, depends on the size and state of the surface.

In evaluating the effectiveness of antiaging agents, their preliminary testing should be conducted under conditions differing from the operating conditions. The methods of preliminary testing of antiaging agents are usually analogous to the methods used in the study of the mechanism of the aging of polymers or the mechanism of the stabilizing action of antiaging agents. They are based on a comparison of the rates of change of the properties of the powders during aging in the absence and in the presence of the stabilizers. The final judgment of the suitability of one substance or another for use as an antiaging agent can be given only by a study of the aging of the finished object. The methods of investigating the stability of materials and finished objects to aging can also be accelerated, and designed for long-term changes of properties under natural conditions.

The basic method of accelerating aging is the use of the same agents that act on the material under the conditions of operation, but at considerably higher intensities of the influence. An example is weatherometric testing of objects, in which light sources are used, approximately corresponding in wavelength to irradiation under natural conditions, but substantially exceeding it in intensity.

However, here we should keep in mind the following. As has already been indicated, the rate of aging, with all other conditions equal, depends on the concentration and reactivity of the active centers that initiate decomposition of the polymer. The reactivity of such functional groups or free radicals is determined by the activation energies in the interaction with the polymer. Consequently, the concentration of reactive groups that are capable of initiating the decomposition chain and guaranteeing its continuance depends on the intensity of the influence. From this sometimes the observed irreproducibility not only of the quantitative, but also of the qualitative indices in the testing of materials based on polyvinyl chloride under conditions of various intensities of the influence of the same agent becomes comprehensible. It is evident that

mathematical conversion of the rates of aging of the polymers and activity of additives under various conditions, upon which the consideration of the amount of energy that the material can receive is based, can
give correct results only in the case when changing the intensity of the
influence does not entail a change in the concentration of active centers.

In preliminary primary testing of antiaging agents and in the final
tests of the stability of materials or finished objects to aging, the basis
of the experiment, as in the determination of the activity of thermal
stabilizers, can be chemical, physicochemical, physicomechanical, or
optical methods of analysis.

Chemical methods of analysis in most cases reduce to a qualitative
or quantitative determination of chlorine or hydrogen ions formed as a
result of dehydrochlorination of the polymer. Of the physicochemical
methods of investigating the stability of polyvinyl chloride, the EPR
method has received widespread use. The determination of the rate of
change of the electroconductivity of the material has been recommended
for the testing of electrical insulation articles [147]. In the investigation of the photodecomposition of polyvinyl chloride, the method of measuring the rate of diffusion of hydrogen chloride through a membrane of the
polymer has been used [43]. Extremely often, especially in weatherometric and bench tests, an examination of the strength and physicomechanical indices of the material is used; in this case the breaking
force and elongation are most often determined. Data have been published on the results of weatherometric tests conducted with a determination of the work expended for cutting the material [46]; a method of
evaluating the weather stability of polyvinyl chloride according to load-
elongation diagrams has been described [148, 149]. In [50], the impact
strength was selected as the index characterizing the rate of aging of material based on polyvinyl chloride. Data are available on the use of the
thermomechanical method in the investigation of the aging of polyvinyl
chloride.

Optical methods, based on observation of the change in the IR and
UV spectra of the polymers under the action of agents producing aging,
have received widespread use in the investigation of aging processes. As
a rule, experimental data obtained by several different methods of analysis are compared in studies devoted to the problem of the aging of polyvinyl chloride.

IV. STABILIZERS OF POLYVINYL CHLORIDE
AND VINYL CHLORIDE COPOLYMERS

The substances used as light and thermal stabilizers of polyvinyl
chloride and vinyl chloride copolymers have been described in mono-

graphs [65, 151], surveys [61-63, 80, 93, 94, 143-146, 148, 152-164], and in extensive patent literature. The stabilizers are classified according to various features: character of the stabilizing action, presence of certain functional groups in the molecule of the substance, or affiliation to one type of chemical compound or another.

Classification according to affiliation to one type of chemical compound or another is the most expedient.

1. Metallic Salts of Inorganic and Organic Acids

Of the metallic salts of inorganic acids capable of accepting hydrogen chloride, salts of lead, sodium, and magnesium have found wide use in industry. The most effective of them are the silicate, basic sulfate, basic carbonate, and basic phosphate of lead [165, 166, 176, 177]. Salts of inorganic acids exhibit no lubricating action and hence can be used only in mixtures with lubricants. To intensify the stabilizing action of salts of inorganic acids and oxides of certain metals, it is recommended that they be mixed with salts of organic acids [167-169], metal chlorides [170], as well as used in the form of fine powders, produced by sublimation [171].

We should especially take up salts of phosphoric acids. It is known that sodium pyrophosphate, hypophosphate, and metaphosphate, as well as polyphosphates ($Na_9P_5O_{17}$, $Na_5P_3O_{10}$, $Na_6P_4O_{13}$, or $Na_{12}P_{10}O_{18}$) are splendid stabilizers, which prevent decomposition of peroxide compounds. It is extremely possible that the stabilizing action of such salts is related to their ability to suppress the harmful catalytic action of polyvalent metals. Thus, for example, the hexametaphosphate bonds cations of divalent metals extremely strongly by a partial exchange decomposition, forming mixed salts $Na_4MeP_6O_{18}$ or $Na_2Me_2P_6O_{18}$ [172]. A high stabilizing effect is given by mixtures of phosphates with phenols; in this case the phosphoric acid liberated in the reaction of the stabilizer with hydrogen chloride leads to a substantial increase in the activity of the phenols in chain oxidation reactions [178].

Metallic salts of organic acids, both aliphatic and aromatic, are among the most widespread type of stabilizers. In addition to their stabilizing action, many of them possess properties of lubricants [173]. In the survey article [174], devoted to a generalization of the data on the use of soaps for the stabilization of polyvinyl chloride, it is noted that in recent years salts of inorganic acids have been gradually displaced by soaps, the most widespread among which are lead stearate and barium and cadmium laurates. Salts of lead, tin, barium, calcium, cadmium, strontium, sodium, and lithium of such acids as formic, oxalic, maleic, caprylic, undecylenic, lauric, stearic, ricinoleic, etc., have been proposed and are being used as stabilizers of polyvinyl chloride and vinyl chloride copolymers [175, 179-184]. Metallic salts of aromatic acids – substituted benzoic [185], phthalic, benzoic, gallic, salicylic, β-phenylpropionic, and β-phenyl-α- aminopropionic [186] – have been recommended.

It is frequently recommended that nitrogen-, oxygen-, sulfur-containing and other functional groups, which intensify the stabilizing action, be included in the acid radical of the salt. The following have been proposed as such polyfunctional stabilizers with a complex acid radical: cadmium salts of vinylphosphinic acids with the general formula [187]:

$$\left[\begin{array}{c} CH_2=CH \\ RO \end{array} \!\! P \!\! \begin{array}{c} O \\ O^- \end{array} \right]_2 Cd$$

barium salts of esters of thiophosphoric acid [188]:

$$\left[\begin{array}{c} RO \\ R'O \end{array} \!\! P \!\! \begin{array}{c} S \\ S^- \end{array} \right]_2 Ba$$

lithium, sodium, potassium, barium, and lead salts of phenylphosphinic acid [189]:

$$\left[\begin{array}{c} C_6H_5 \\ O \end{array} \!\! P \!\! \begin{array}{c} O^- \\ O^- \end{array} \right]_n Me$$

salts of thiocarbamic acid [190]:

$$\left[\begin{array}{c} R_1 \\ R_2 \end{array} \!\! N\!\!-\!\!C \!\! \begin{array}{c} S \\ S^- \end{array} \right]_n Me, \quad \left[\begin{array}{c} R_3 \\ R_4 \end{array} \!\! N\!\!-\!\!C \!\! \begin{array}{c} S \\ S^- \end{array} \right]_n Me$$

salts of aliphatic thioacids [191], epoxystearic [192], epoxysuccinic [193], xanthic [194], and other acids [195-198].

Salts of dienophilic acids, capable of reacting with polyene structures according to a Diels-Alder reaction, are used to obtain colorless objects based on polyvinyl chloride and vinyl chloride copolymers. Basic lead maleate, as well as alkaline earth metal salts of α-furylacrylic and sorbic acids have been proposed to guarantee good color of polyvinyl chloride objects [199].

As a rule, not individual salts, but mixtures, which permit an intensification of the effectiveness of the stabilizing action, are used in practice. Such mixtures can contain monotypic stabilizers, for example, mixed or coprecipitated salts of barium, cadmium [200-203], as well as additives of other types of stabilizers that intensify the action of the salts. Oxides of lead, barium, calcium, magnesium, aliphatic acids [204], esters of aromatic, phosphoric, and phosphorous acids [205], derivatives of phthalic anhydride [206], chelate compounds of titanium, in particular, triethanolamine titanate [207], etc., have been recommended as additives. Synergic action in the stabilization of polyvinyl chloride is also manifested in stabilizing systems in which certain components do not decelerate but, on the contrary, accelerate decomposition when used individually. It is known, for example, that zinc salts accelerate the thermal dehydrochlorination of polyvinyl chloride; however, their introduction in small amounts increases the stability of certain compositions [71].

2. Organometallic Compounds

Among the organometallic stabilizers, compounds of quadrivalent tin are most widespread. Monomer organotin compounds most often have the general formula

$$R_nSnX_{4-n}$$

where R is an organic radical bonded to the tin atom through the carbon atom; X is an organic or, rarely, an inorganic radical, bonded to the tin atom through an oxygen or sulfur atom or an acid group.

The high effectiveness of the stabilizing action of organotin compounds is due to their polyfunctionality. They contain a C−Sn bond, which can be broken during decomposition of the polymer, forming free radicals, selective with respect to the decomposition products of the polymer. Termination of the reaction chain of polyvinyl chloride decomposition according to such a mechanism has been demonstrated experimentally for dibutyltin derivatives [42]. In addition, thanks to the presence of the X-groups, which can be replaced by a chlorine atom, organotin compounds form mercaptans, dienophilic reagents, epoxy compounds, phenols, ethers, and other substances which themselves give a stabilizing effect, according to a reaction of exchange decomposition with hydrogen chloride, depending on the nature of the X-group. The mechanism of the stabilizing action of polymer derivatives of alkyltin, for which the structural formulas sometimes remain unelucidated, apparently does not differ in principle from the mechanism of the action of monomer organotin stabilizers.

As an example of organotin stabilizers in which all four valence bonds of tin represent C − Sn bonds, we might mention dibutyldiphenyltin [199, 208], as well as tetraalkylstannanes [199].

There is a patent recommending the activation of tetraalkylstannane radicals by the inclusion of some electronegative group − carbonyl, nitrile, etc., − in their composition [209].

Of the alkoxy derivatives, organotin radicals with a number of $-OR$ groups from one [210-212] to four [213] are recommended. It is indicated in the patent [213] that tetraalkoxy derivatives $Sn(OR)_4$ represent polymer products. The production and use as stabilizers of tin derivatives based on polyatomic aliphatic alcohols – ethylene glycol, diethylene glycol, pentaerythritol, etc., [214], as well as derivatives based on phenols [211], have been patented. It has been established that the organotin alcoholates described in the literature are polymer compounds

$$RO- \left[\begin{array}{c} R' \\ | \\ -Sn-O- \\ | \\ R' \end{array} \right]_n -R$$

with values of n from one to 10 [215].

Organotin compounds containing the mercapto group substantially surpass the corresponding alkoxy derivatives in their stabilizing action [77]. The high activity of sulfur-containing organotin stabilizers has also been shown in [216], where alkoxy and mercapto derivatives of dialkylstannanes with an alkyl group containing eight carbon atoms in the chain were investigated. Organotin stabilizers containing up to three mercapto groups have been proposed in the patents [217-219]. Tetramethyl-, ethyl-, butyl-, octyl-, dodecyl-, phenyl-, tolyl-, and tetrabenzylmercaptides of tin with the general formula $Sn(SR)_4$ have been proposed in the patent [218].

Organotin stabilizers with a complex mercapto group structure were recommended in the patent [219], where the X-group in the compound R_2SnX_2 possesses the composition:

$$-S-CH_2-CH_2-O-\overset{\overset{\displaystyle O}{||}}{C}-(CH_2)_4-\overset{\overset{\displaystyle O}{||}}{C}-O-CH_2-CH_2-S-$$

in the patent [220], where the X-group represents

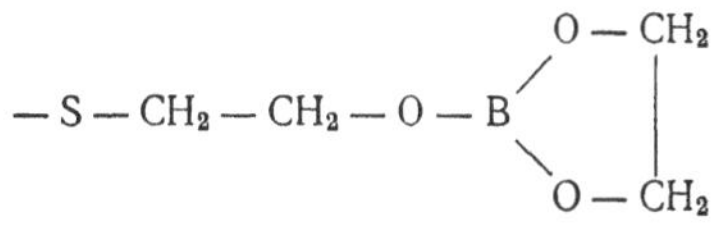

$$-S-CH_2-CH_2-O-B \overset{\displaystyle O-CH_2}{\underset{\displaystyle O-CH_2}{<}} \Big|$$

as well as in the patent [221], where the R'-group in stabilizers with the general formula $R_nSn(SR')_{4-n}$ represents

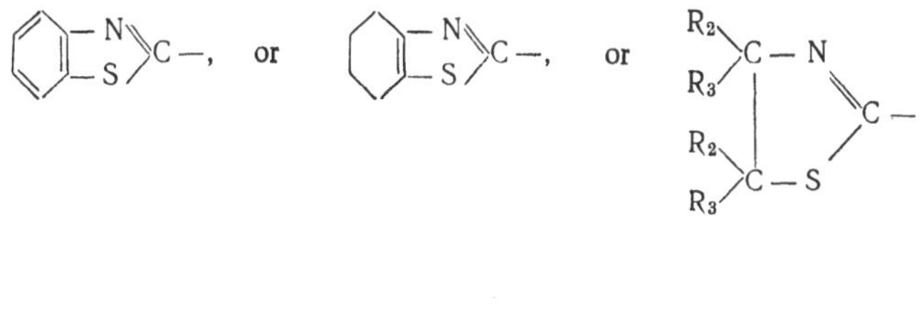

In [222], polymer organotin compounds with sulfur bridges were proposed as stabilizers of suspension polyvinyl chloride.

Acid derivatives of alkyl- and arylstannanes have received the most widespread practical use. In attempting to create polyfunctional effective stabilizers, many authors propose compounds with a complicated structure of the acid radical. Thus, together with dibutyltin salts of maleic, fumaric, or crotonic acids [223], salts of dialkyl- or diaryltin with acids possessing the composition:

$$
\begin{array}{c}
\mathrm{HO-\overset{\displaystyle O}{\overset{\|}{C}}-(CH_2)_m} \\
\diagdown \\
\mathrm{HO-\underset{\underset{O}{\|}}{C}-(CH_2)_n}\diagup S_x
\end{array}
$$

where m and n possess values from one to eight, while x = 1 or 2 [224]:

$$
\begin{array}{c}
\mathrm{HO-\overset{\displaystyle O}{\overset{\|}{C}}-CH_2} \\
\diagdown CH-S-\overset{\displaystyle O}{\overset{\|}{C}}-C_6H_5 \\
\mathrm{HO-\underset{\underset{O}{\|}}{C}-CH_2}\diagup
\end{array}
$$

$$
\begin{array}{c}
\mathrm{HO-\overset{\displaystyle O}{\overset{\|}{C}}-CH=CH-O-\overset{\displaystyle O}{\overset{\|}{C}}-CH_2} \\
\mathrm{HO-\underset{\underset{O}{\|}}{C}-CH=CH-O-\underset{\underset{O}{\|}}{C}-CH_2}
\end{array}
$$

$$
\mathrm{HO-\underset{\underset{O}{\|}}{C}-CH=CH-\underset{\underset{O}{\|}}{C}-O-C_{12}H_{25}}\;[225]
$$

as well as with epoxy acids [226, 227], with dibasic acids in which one group forms a dibutyltin salt, while the other forms an ester with propylene glycol [228], and with esters of xanthic acids [229] have been proposed.

Of the organotin compounds with inorganic acids, salts of tributyltin and diphenyltin with the dioctyl ester of orthophosphoric acid or the dihexyl ester of pyrophosphoric acid [230]:

$$
\mathrm{(C_4H_9)_3-Sn-O-\overset{\displaystyle O}{\overset{\|}{P}}\!\!\begin{array}{l}\diagup OC_8H_{17}\\ \diagdown OC_8H_{17}\end{array}}
$$

as well as salts of dibutyltin and boric acid [220]:

$$(C_4H_9)_2 - Sn \underset{O}{\overset{O}{<}} B - O - \underset{\underset{C_4H_9}{|}}{\overset{\overset{C_4H_9}{|}}{Sn}} - O - B \underset{O}{\overset{O}{>}} Sn - (C_4H_9)_2$$

have been proposed as stabilizers.

The condensation products of the oxide of dibutyl- or diphenyltin with aldehydes, ketones, oximes, amidooximes, and with complex esters of polybasic acids [119, 231-234] or directly with polybasic acids and polyatomic alcohols [235] have been recommended as stabilizers of chlorine-containing high-molecular compounds. High effectiveness of the action in the stabilization of polyvinyl chloride has been noted among polymer products representing derivatives of alkylstannic and alkylthiostannic acids [236]. It is extremely possible that the high activity of such condensation products is related to the effectiveness of the stabilizing action of the initial oxides of alkyl- and arylstannanes [237, 238], as well as the sulfides of alkyl- and arylstannanes [239].

The activity of organotin stabilizers can be substantially increased by using them in a mixture with additives that give a synergic effect. Esters of orthoformic or orthosilic acids [240], diesters of maleic or fumaric acids [241], salts of phosphoric acids [242, 243], stabilizers containing lead [208], epoxy compounds, barium and cadmium salts, chelate compounds, and other substances [62, 93] have been recommended as the second component in stabilizing systems based on organotin compounds. In view of the relatively high cost of organotin stabilizers, their use in a mixture with additives that intensify the effectiveness of the action is of extremely great significance, since it permits a substantial reduction of the consumption of the basic stabilizer.

In the literature the toxicity of certain organotin stabilizers, as well as the unpleasant odor characteristic of sulfur-containing compounds, are noted. There are data [216, 244] that dialkylstannane derivatives with alkyl radicals containing eight carbon atoms in the chain are nontoxic, but are somewhat inferior to dibutyltin derivatives in effectiveness. The mercaptide based on dioctyltin and the ester of thioglycolic acid, possessing the composition:

$$(C_8H_{17})_2Sn \; (S - CH_2 - \overset{\overset{O}{\|}}{C} - O - iso\text{-}C_8H_{17})_2$$

is distinguished by a weak odor and reduced toxicity.

The high activity of organotin stabilizers is responsible for their use in the production of colorless transparent objects from nonplasticized compositions, as well as in the reprocessing of polyvinyl chloride by the extrusion method. In both cases the necessary condition is high thermal stability of the compositions.

A more detailed survey of the patents and periodical literature on the use of organotin compounds to stabilized chlorine-containing high-molecular compounds is given in [80, 154].

In addition to organotin compounds, the survey work [199] mentions organolead compounds, recommended as stabilizers of chlorine-containing high-molecular compounds — hydroxides of trialkyl- and tri-aryllead, alkyl- and aryllead salts of aliphatic acids, and triethyllead hexylmaleate. Alkyl- and arylplumbanes have not found wide use in the plastics industry on account of their high toxicity.

The use of organotitanium compounds as stabilizers of polyvinyl chloride was recommended in the patent [245]. Certain other alkyl-titannates give good indices when used together with calcium, cadmium, barium, tin, and zinc salts of aliphatic acids.

3. Epoxide Compounds

Until comparatively recently, the stabilizing action of epoxide compounds was explained by researchers by their ability to react with hydrogen chloride, forming the chlorohydrin:

$$-CH-CH- \quad +HCl \rightarrow -CHOH-CHCl-$$
$$\diagdown \diagup$$
$$O$$

However, in such an interpretation of the mechanism of the action of epoxy stabilizers, the following phenomena could not be explained: the lack of correspondence of the high stabilizing activity and comparatively weak accepting activity of epoxy compounds in the decomposition of polyvinyl chloride, conservation of the concentration of epoxide rings during the thermal processing of the composition, and, finally, a three-to-five-fold intensification of the stabilizing action of the epoxy compounds in the presence of salts of certain metals, in particular, barium and cadmium. The opinion was expressed that the stabilizing action of epoxy compounds is related to reactions not only of opening, but also of closing of the three-membered ring: the hypothesis was advanced that the three-membered ring is opened in the reaction with hydrogen chloride, and the chloro-hydrin is formed, from which the epoxy-ring structure is reestablished under the influence of metal salts [246].

It was shown somewhat later that epoxy compounds inhibit the radical decomposition of hydroperoxides [12, 247]. In an investigation of the decomposition of tertiary butylhydroperoxide under the action of cobalt octoate, it was established that the combination of the epoxy resin "Epone 834" with a cadmium salt is a better inhibitor of peroxide decomposition than each of the components individually; this mixture is close to dibutyltin dilaurate in effectiveness.

The basic types of epoxy stabilizers are glycidyl esters – condensation products of epichlorohydrin with phenols and aliphatic alcohols, epoxided esters of higher aliphatic acids, and alicyclic epoxy esters of aliphatic and aromatic acids and alcohols [246, 247, 251-256].

Of the glycidyl esters, phenoxypropeneoxide, pentachlorophenoxy-propene oxide, β-naphthoxypropene oxide [63, 55], as well as liquid and solid aromatic epoxy resins with various molecular weights [248] have found use as stabilizers of polyvinyl chloride. Pentachlorophenoxy-propene oxide guarantees good thermal stability of crystalline polymers and copolymers of vinylidene chloride [249]. The results of an investigation of "epones" – condensation products of phenols with epichlorohydrin – as stabilizers of polyvinyl chloride and vinyl chloride copolymers with vinyl acetate have been published [247]; the activity of epoxy esters has been compared with the activity of cadmium and strontium soaps; it has been established that epoxy esters are more effective stabilizers for the homopolymer and for the copolymer than the cadmium and strontium soaps; mixtures of epoxy resins with soaps have given very good results for the homopolymer of vinyl chloride.

Aromatic glycidyl esters with various substituents in the nucleus – chlorine, nitro group, alkyl radicals, as well as hydroquinone bis-glycidyl and resorcinol bis-glycidyl esters – have been investigated as stabilizers of the copolymer of vinyl chloride with vinylidene chloride. It has been established that the stabilizing action of aromatic glycidyl esters depends on the nature and position of the substituents in the aromatic nucleus; esters with a chlorine atom or nitro group in the 4-position prove most effective [21].

The condensation products of epichlorohydrin with hydroxybenzo-phenones, for example, with 2,4-dihydroxybenzophenone, have been proposed as thermal stabilizers for the copolymer of vinyl chloride with vinylidene chloride [250].

The shortcomings of low-molecular glycidyl esters are their volatility and their ability to cause dermatitis; hence preference is given to the nonvolatile and less toxic epoxide resins in the practice of stabilizing high-molecular compounds.

Epoxided complex esters of higher aliphatic acids in compositions based on polyvinyl chloride exert a plasticizing action together with their stabilizing action [257-260]. Long-chain epoxy compounds with the epoxy group in the middle of the chain are more effective stabilizers and plasticizers than compounds with epoxy groups situated at the ends of short chains [255]. The shortcoming of epoxide stabilizer-plasticizers, representing epoxided fats and oils, is their not very good compatibility with polyvinyl chloride.

In an investigation of various epoxy compounds, including epoxided animal fats, epoxy esters of unsaturated fatty acids, epoxided cotton and soy oils, as stabilizer-plasticizers of polyvinyl chloride, it was found that the presence of free hydroxyl or carboxyl groups in the esters leads to a sharp reduction of the compatibility with the polymer. The insufficient compatibility and rapid bleeding of epoxy plasticizers during the aging of objects of polyvinyl chloride may also be due to residual unsaturation in the molecules [246]. In the case of exhaustive epoxidation, followed by reduction of the residual unsaturation, epoxide stabilizer-plasticizers show good compatibility with the polymer. Epoxided fats and oils are used in practice as secondary stabilizers in compositions containing metallic salts [61, 98, 247, 261, 262].

In compositions based on the copolymer of vinyl chloride with vinyl-acetate, three series of alicyclic epoxy compounds have been tested: 3,4-epoxycyclohexylmethyl esters of lauric, palmitic, 9,10-epoxystearic, 9,10,12,13-diepoxystearic, oxalic, maleic, terephthalic, sebacic, and other acids of the general formula

$$O\!\!<\!\!\bigcirc\!\!- CH_2 - O - \overset{\overset{\displaystyle O}{\|}}{C} - R$$

complex esters of 3,4-epoxycyclohexylcycloformic acid and diethylene glycol, 3-methyl-1,5-pentanediol, 2-ethyl-1,3-hexanediol, and 2-ethyl-hexyl alcohol, with the general formula

$$O\!\!<\!\!\bigcirc\!\!- \overset{\overset{\displaystyle}{C}}{\underset{\underset{\displaystyle O}{\|}}{}} - OR$$

and 4,5-epoxycyclohexane-1,2-dicarboxylates of methyl, ethyl, butyl, 2-ethylhexyl, n-decyl, and certain other alcohols of the general formula [256]

$$O\!\!<\!\!\bigcirc\!\!\begin{matrix} \overset{\overset{\displaystyle O}{\|}}{C} - OR \\ \underset{\underset{\displaystyle O}{\|}}{C} - OR \end{matrix}$$

The authors have established that alicyclic epoxy compounds are effective stabilizers of the polymer, in certain cases surpassing epoxided soy bean oil. Analogous stabilizer-plasticizers of polyvinyl chloride, representing epoxyhexahydrophthalates of butyl, hexyl, iso-octyl, and certain other aliphatic alcohols, were described in [254]; epoxided esters of 3,4-epoxycyclohexane-1,1-dimethanol and their alkyl-substituted derivatives [263], as well as 3,4-epoxy-1 (or 6)-cyclohexylmethyl esters of

9,10-epoxystearic acid or 5,6-epoxylauric have been proposed for stabilizing chlorine-containing high-molecular compounds. These compounds give synergic mixtures with barium or cadmium salts [264].

In addition to the three basic types of epoxide types enumerated, the literature contains data on the use of certain other substances containing epoxide rings in the molecules to stabilize polyvinyl chloride. Salts of lead, barium, calcium, cadmium, and aliphatic epoxy acids, with 11-22 carbon atoms in the chain, have been described. It has been shown that in contrast to salts of nonepoxided aliphatic acids, salts of epoxy acids give no synergic effect when they are used together [71].

Soaps, metallic salts of organic acids, esters of phosphoric and phosphinic acids, organotin compounds, chelate and certain other substances are customarily used to increase the activity of epoxide stabilizers [24, 63, 265].

A special region of application of epoxide stabilizers in mixtures with barium and cadmium salts is the preparation of elastic, transparent, colorless objects from polyvinyl chloride.

4. Nitrogen-Containing Organic Compounds

Of the nitrogen-containing organic compounds, aromatic and, to a considerably lesser degree, aliphatic amines, amides of organic and inorganic acids, mainly derivatives of urea and thiourea, as well as heterocyclic compounds of the aromatic and aliphatic series, have found use for stabilizing chlorine-containing polymers and copolymers.

The most complete idea of the mechanism of the stabilizing action of amines can be given by studies devoted to the protection of rubbers, including halogen-containing rubbers. It is known from these studies that amines play a dual role in the stabilizing of high-molecular compounds. They are extremely active decelerators of the oxidative decomposition of polymers and, moreover, suppress the catalytic action of polyvalent metals, accelerating the decomposition of the polymers. Aromatic and heterocyclic amines have been investigated as agents for bonding metals into complexes; it has been established that the introduction of complex-formers prevents the negative influence of metals on the decomposition of polymers [82].

Many amines, as experiments show, accelerate the dehydrochlorination of polyvinyl chloride. This is apparently related to the high reactivity of amines or their conversion products with respect to the undecomposed polymer. As a result of the peculiarity of the influence on chlorine-containing high-molecular compounds noted, nitrogen-containing substances have not found widespread use as stabilizers of polymers and copolymers of vinyl chloride. They are used, as a rule, in small concentrations in the form of additives, which intensify the action of the basic stabilizers.

Of the amines, esters of β-aminocrotonic and p-aminobenzoic acids, p-aminobenzenesulfamide [63, 266], hexamethylenetetramine [267], and certain others have been used for the stabilization of polyvinyl chloride. Phenyl-β-naphthylamine, ethylphenylethanolamine [269], dicyanalkyl-amines [270], caprolactam [271], and N-(p-aminophenyl)hexamethylene-imine [268] have been recommended in the patent literature for the stabil-ization of polyvinyl chloride and vinyl chloride copolymers.

Amides of organic acids are used more often than amines for the stabilization of polyvinyl chloride. The patent and periodical literature pertaining to the stabilization of the vinyl chloride polymer and copoly-mers by amides, in particular, by derivatives of urea and thiourea, is presented rather fully in the survey [154]. Of interest are the recom-mendations on the use of substituted amides of phosphoric [272] and stearic [273] acids as stabilizers of polyvinyl chloride. These amides contain three-membered ethyleneimine rings, which open readily when they react with HCl.

Of the heterocyclic nitrogen-containing compounds, α-phenylindole is used to stabilize latex polymers and copolymers of vinyl chloride [99]. Melamine has found use for the stabilizing of suspension polymers, where α-phenylindole proves inactive [63]. Derivatives of morpholine [274], symmetrical triazine [275], pyrazole [276], as well as derivatives of imidazole, imidazoline, oxazole, oxazoline, thiazole, and thiazoline [277] are recommended in the patent literature for stabilization.

Heterocycles containing these compounds possess the structures:

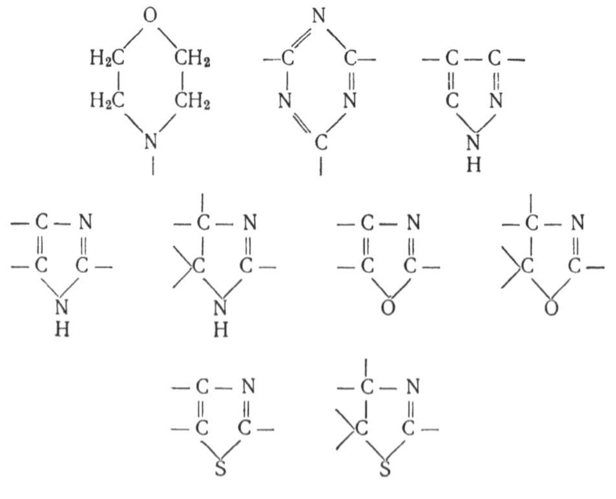

We should mention the use of nitro compounds as stabilizers of polyvinyl chloride. In spite of the fact that the activating influence of the nitro group upon introduction into the molecule of an inhibitor of radical chain processes is well known, there are no data in the literature on the use of substances of this type in the plastics industry. The information available on stabilizers, whose molecules contain nitro groups, pertains to organotin compounds [152] and glycidyl esters [21].

5. Sulfur-Containing Organic Compounds

Such compounds are used, as a rule, in a mixture with stabilizers that bond hydrogen chloride. Moreover, salts of metals that form colored sulfur-containing compounds are not recommended for use as acceptors.

The stabilizing action of sulfur-containing compounds is related to their ability to decompose peroxide and hydroperoxide groups without the formation of free radicals, as well as to their property of terminating the reaction chains in the decomposition of polymers. Sulfur-containing compounds are used in practice as thermal stabilizers, since many of them prove to be insufficiently stable to the action of ultraviolet light and can cause sensitized decomposition of the polymer when the composition is irradiated. There is information that the activity of mercapto derivatives, in particular, dodecyl mercaptan, in the process of thermal decomposition of polyvinyl chloride, does not decrease when the temperature is raised to 190-200°C [278].

As has been indicated in a number of studies, sulfur-containing compounds give a synergic effect with amines and phenols in the stabilization of high-molecular compounds [89, 279]. The following have been proposed and are used as stabilizers of polyvinyl chloride: mercaptides of antimony [280, 281], condensation products of aldehydes or ketones with mercaptans [63], thioesters [282], salts of thioacids [283], aromatic esters of aliphatic sulfonic acids [284], esters of xanthic acids [285]; the use of the polysulfide of the composition

$$
\begin{array}{ccc}
 & X & X \\
H_2C\diagup\diagdown CH_2 & \quad & H_2C\diagup\diagdown CH_2 \\
H_2C\diagdown\diagup CH_2 & & H_2C\diagdown\diagup CH_2 \\
 & N\!\!-\!\!\!-\!\!Sn\!\!-\!\!\!-\!\!N &
\end{array}
$$

(X = atom of sulfur, oxygen, or methylene group; n can take values from one to six [286]) is recommended.

6. Phosphorus-Containing Stabilizers

Of the phosphorus-containing stabilizers, aromatic esters of phosphoric acid, such as tricresyl phosphate, have received the most widespread use. In many of them, stabilizing action is combined with plasticizing action. The properties of a number of esters of phosphoric acid
used as stabilizers and plasticizers of polyvinyl chloride are discussed
in the collection [287].

In recent years, aliphatic and aromatic esters of phosphoric acids –
phosphites – have been recommended more and more often for the stabilization of polyvinyl chloride and vinyl chloride copolymers. The patent
[288] recommends aryl phosphites with long-chain alkyl substituents in
the aromatic nuclei: it is indicated that such phosphites are complex
formers, and hence can be used as agents suppressing the harmful influence of polyvalent metals. The patent [289] recommends phosphites
with different radicals in the ester groups:

$$P \begin{array}{l} \diagup OR_1 \\ - OR_2 \\ \diagdown OR_3 \end{array}$$

and indicates that the radicals R_1, R_2, and R_3 contain halogen on the
carbon atom next to that through which the ester bond with the phosphorus atom is effected. Mixed alkyl-aryl phosphites surpass esters
containing only aliphatic or only aromatic radicals in their activity.

Esters of phosphinic acids with a C – P bond in the molecule have
been proposed for the stabilization of halogen-containing high-molecular
compounds [189]. In tests of the stabilizing action of esters of iso-
propoxyphosphinic acid, it was shown that their activity increases with
increasing temperature. The experimental data give a basis for assuming that the investigated esters, of the general formula

$$\begin{array}{c} RO \\ \diagdown \\ RO \end{array} \overset{O}{\underset{\|}{P}} - \overset{CH_3}{\underset{O}{C}} - CH_2$$

are bifunctional compounds in the decomposition of polyvinyl chloride.
They contain the epoxide group, the stabilizing action of which is manifested at temperatures of the order of 170-180°C. The high activity of
the esters at 190-200°C is explained by the possibility of their decomposition at the C–P bond, forming free radicals capable of reacting with
the decomposition products of polyvinyl chloride [79]. The probability
of such a mechanism of the stabilizing action of alkylphosphinic acid es-

ters is confirmed by the data on the high strength of the C–P bond in esters containing epoxide groups [290]. The possibility is not excluded that the high stabilizing activity of the phosphites is related to their conversion, according to the reaction of A. E. Arbuzov, to esters of alkylphosphinic acids, under the action of alkyl halides:

$$P\,(OR)_3 + R_1X \rightarrow R_1 - \underset{\underset{\displaystyle OR}{\displaystyle \diagdown}}{\overset{\overset{\displaystyle O}{\displaystyle \|}}{P}} \hspace{-0.3em}\diagup^{\displaystyle OR} + RX$$

As has already been mentioned in the discussion of metallic salts of organic and inorganic acids, the acid radicals of such stabilizers can represent esters of phosphoric or alkylphosphinic acids. In addition to salts of alkylphosphinic acids, the use of sodium, aluminum, barium, and calcium salts of aliphatic esters of phosphoric acid is recommended [291].

Phosphorus-containing stabilizers are used in practice in a mixture with other stabilizers – hydrogen chloride acceptors, phenols, organotin compounds, or epoxide compounds.

7. Aromatic Compounds

Of the aromatic compounds, phenols and phenolates have found use for the stabilization of chlorine-containing high-molecular compounds. The use of phenolates provides for accepting of hydrogen chloride, forming free phenols, which in turn can slow down the thermal decomposition of the polymer on account of termination of the reaction chain of decomposition, forming stable radicals In . The use of barium di-(nonylphenolate), barium di-nonyl-o-cresolate, and strontium di-(octylphenolate) [292], the phenolates of barium, strontium, calcium, cadmium, zinc, lead, and substituted aryl-, alkyl-, and arylalkylphenols, in which the substituents can contain from four to 24 carbon atoms [293], as well as zirconium tetraphenolate [294] has been proposed. Phenolates are recommended for use in mixtures with other stabilizers, for example, with triphenyl phosphate [292] or with hydrogen chloride acceptors of the type of metallic salts [293]. Of the phenols, diphenylolpropane, which is usually used in a mixture with hydrogen chloride acceptors, has found widespread use for the stabilization of polyvinyl chloride [295, 296].

In recent years, esters of aromatic acids and phenols, as well as aromatic hydroxyketones, have been finding wide use as stabilizers.

These compounds, as has already been mentioned, are good absorbers in the ultraviolet region of the spectrum. The studies [32, 297–304] are devoted to a description of their light-stabilizing properties.

Of the aromatic hydroxyketones, 2,4-di-hydroxybenzophenone and
2-hydroxy-4-methoxybenzophenone have found practical use. The use of
derivatives of 2-hydroxybenzophenone, of the general formula

in which the substituents X and Y represent alkyl, alkylene, alkoxy,
alkyleneoxy, and aryl radicals, halogen or hydrogen atoms, is recommended in the patents [305-309]. Aromatic hydroxyketones, in the molecules of which the ketone groups are connected by methylene groups, are
recommended in the patents [310, 311].

Of the esters of aromatic acids, the disalicylate of hydroquinone
[312], the monosalicylates of hydroquinone, resorcinol, and catechol
[313], the di-(2-hydroxy-5-chlorobenzoate) of resorcinol [314], dibenzoyl-
resorcinol [315], and others [316] have been recommended as stabilizers.
The patent [317] recommends the use of diesters of aliphatic dicarboxylic
acids and hydroxybenzophenones of the general formula

where X is a hydrogen atom or alkyl, Y is a hydrogen atom or hydroxyl
group, n = 2-8.

8.   Aliphatic Compounds

Of the aliphatic compounds, alcohols have found use as stabilizers;
aliphatic esters and alcoholates are also recommended for the stabilization of polyvinyl chloride. Of the alcohols, ethylene glycol, glycerin,
and hexanetriol in a mixture with urea [318], alcohols with unconjugated
double bonds, for example, 3,7-dimethyloctadiene-1,6-ol-3 [319], and
alcohols of the acetylenic series, for example, butynediol [63] are used
to stabilize the chlorine-containing high-molecular compounds. Of the
esters, butylacetyl ricinoleat and propylene glycol diricinoleate [63], as

well as complex monoesters of high-molecular aliphatic acids and poly-atomic alcohols [320] are recommended.

Of the alcoholates, zirconium tetraalcoholates $Zr(OR)_4$ [294], in which the radical R represents an aliphatic or alicyclic radical containing up to 18 carbon atoms, are recommended. Aliphatic alcohols, esters, and alcoholates are used, as a rule, in a mixture with hydrogen chloride acceptors.

9. Chelate Compounds

Chelate derivatives used as stabilizers of high-molecular com-pounds have been discussed in sufficient detail in the survey article [154]. It is indicated that the stabilizing action of such substances is based on their ability to give complexes with metal chlorides, formed in the de-composition of chlorine-containing polymers [321], and thereby to sup-press the catalytic action of salts of metals of variable valence [94]. Some of the chelate compounds can act as dienophilic reagents, decom-posing polyene structures formed as a result of dehydrochlorination of the polymer [24]. Compounds of calcium with 1,3-dicarbonyl derivatives, capable of keto-enol tautomerism, for example, with the ethyl ester of acetoacetic acid [199], have been proposed for the stabilization of poly-vinyl chloride:

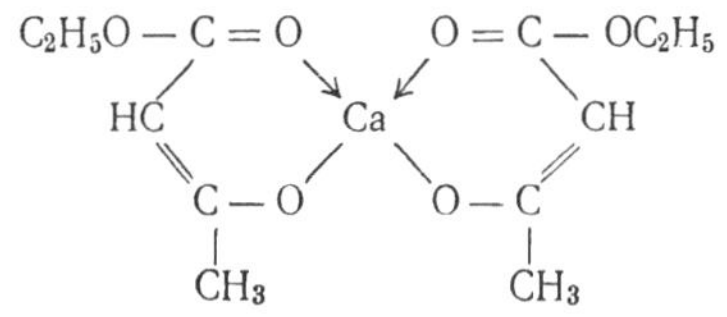

as well as the product obtained from hydroxydibutyltin and 2,4-pentane-dione [322]:

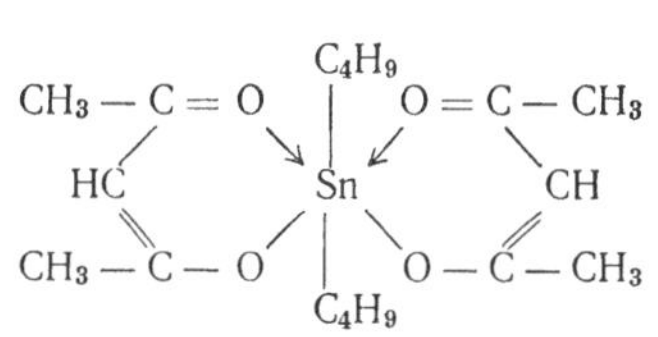

derivatives of copper or nickel [323]:

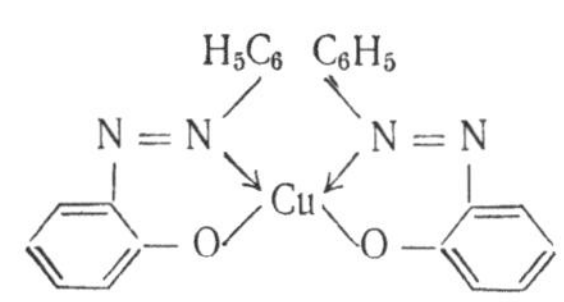

as well as compounds based on alkylenepolyamino-polyacetic acid [324]:

$$O=\underset{|}{C}-O-M-O-\underset{|}{C}=O$$

(structure)

$$M'-O-\overset{O}{\overset{\|}{C}}-CH_2-\underset{|}{N}-(CH_2)_n-\underset{|}{N}-CH_2-\overset{O}{\overset{\|}{C}}-O-M'$$

where M represents lead, cadmium, tin, or zinc; M' represents an alkali or alkaline earth metal.

10. Polymer Compounds with Conjugated Bonds

Polymer compounds with an acyclic conjugated chain – polyphenylacetylene, the copolymer of equimolar amounts of phenylacetylene and p-diethynylbenzene, and the dehydrochlorination products of polyvinyl chloride have been proposed as stabilizers of high-molecular compounds [325]. The stabilizing action of the indicated compounds is related to their ability to form relatively inactive biradicals upon excitation to the triplet state. The distinguishing feature of the action of polymers with an aromatic conjugated chain is an increase in the effectiveness of the stabilizing action with increasing temperature. The use of polymers with a system of conjugated bonds is also possible in the case of their individual use and in a mixture with stabilizers of other types. They give a synergic effect with certain stabilizers.

BIBLIOGRAPHY

1. C. S. Marvel, J. H. Sample, and M. F. Roy, J. Am. Chem. Soc. 61:3241, 1939.
2. C. S. Marvel, Org. Chem. 1(8):754, 1943.
3. D. D. Druesedow and C. F. Gibbs, Natl. Bur. Std. 525:69, 1953; Mod. Plastics 30:123, 1953.
4. H. Mark and A. V. Tobolsky, Physical Chemistry of High Polymeric Systems, Interscience Publishers, New York, 1950, p. 403.
5. J. D. Cotman, N. Y. Acad. Sci. 57:117, 1953.
6. V. M. Yur'ev, A. N. Pravednikov, and S. S. Medvedev, Doklady Akad. Nauk SSSR 124:2, 1959.
7. A. H. Willbourn, J. Polymer Sci. 34:569, 1959.
8. E. J. Arlman, J. Polymer Sci. 12:543, 1954.
9. G. Talamini and G. Pezzini, Makromol. Chem. 39:26, 1960.
10. B. Baum and L. H. Wartman, J. Polymer Sci. 28(118):537, 1958.
11. E. J. Arlman, J. Polymer Sci. 12:547, 1954.
12. D. E. Winkler, J. Polymer Sci. 35(128):1959.
13. A. Cittadini and R. Paolillo, Materie Plastiche 4:314, 1960.
14. B. Baum, SPE Journal 17(1):71, 1961.

15. U. P. A. Groll and G. W. Hearne, Ind. Eng. Chem. 31:1530, 1939.

16. L. M. Porter and F. F. Rust, J. Am. Chem. Soc. 78:5571, 1956.

17. A. M. Goodall and K. E. Howlett, J. Chem. Soc. p. 2596, 1954.

18. D. H. R. Barton and K. E. Howlett, J. Chem. Soc. p. 155, 1949.

19. V. W. Fuchs and D. Louis, Makromol. Chem. 22:1, 1957.

20. E. W. R. Steacie, Atomic and Free Radical Reactions, Vol. 1, Reinholt, New York, 1954.

21. G. Ya. Gordon, Summaries of Reports at the Conference on the Aging and Stabilization of Polymers, Moscow, Academy of Sciences USSR Press, 1961.

22. G. Smets, Angew. Chem. 67:2, 57 (1955).

23. A. A. Berlin, Successes of Polymer Chemistry and Technology, Collection 1, Moscow, State Press for Chemical Literature, 1955, p. 67.

24. H. V. Smith, Brit. Plastics 25:304, 1952.

25. A. L. Scarbrough, W. L. Kellner, and P. W. Rizzo, Natl. Bur. Std. (U. S.), Circ. 525, 1953.

26. R. R. Stromberg, S. Straus, and B. G. Achhammer, J. Polymer Sci. 35(129):355, 1959.

27. C. F. Bersch, M. R. Harvey, and B. G. Achhammer, J. Res. Natl. Bur. Std. 60:481, 1958.

28. C. E. Schildknecht, Vinyl and Related Polymers, J. Wiley and Sons, Inc., 1952.

29. G. A. Razuvaev and K. S. Minsker, Zhur. Obshchei Khim. 28:983, 1958.

30. R. R. Stromberg and S. Straus, J. Res. Natl. Bur. Std. 60:147, 1958; Ekspress-Inform. CBM, 27(119-124):1958.

31. S. Otani, J. Chem. Soc. Japan, Ind. Chem. Sect. 61(10):1324,1958.

32. F. Kirchhof, Gummi Asbest No. 10:614, 1958.

33. V. N. Kondart'ev, Photochemistry ("Successes in Physics"), Moscow-Leningrad, State Press for Technical and Theoretical Literature, 1933.

34. A. N. Terenin, Photochemistry of Dyes and Related Organic Compounds, Moscow-Leningrad, Academy of Sciences USSR Press, 1947.

35. K. Bongeffer and P. Gartek, Fundamentals of Photochemistry, Moscow, United Scientific and Technical Press, 1935.

36. L. A. Matheson and R. F. Boyer, Ind. Eng. Chem. 44:867, 1952.

37. G. P. Mack, Kunststoffe 43:94, 1953.

38. L. H. Wartman, Ind. Eng. Chem. 47:1013, 1955.

39. V. A. Kargin and M. N. Shteding, Khim. Prom. No. 3:137, 1955.

40. V. W. Fox, J. G. Hendrics, and H. J. Ratti, Ind. Eng. Chem. 41(8):1774, 1949.

41. A. M. Bocharova, Dissertation, Moscow, 1953.

42. A. S. Kenyon, Natl. Bur. Std. (U. S.), Circ. 525:81, 1953.

43. T. Imoto and I. Ago, Kogyo Kagaku Zasshi 61:97, 1958; C. A. 53:16582, 1959.
44. T. Matsuda, J. Minematsu, and M. Joshioka, C. A. 51:17228, 1957.
45. T. Matsuda, J. Minematsu, and E. Nakajama, C. A. 52:5028, 1958.
46. J. Minematsu and I. Asada, Referat. Zhur. Khim. 1960, 28854, 28855.
47. A. Chapiro, J. chim. phys. 53:895, 1956.
48. S. H. Pinner, Nature 183:1108, 1959.
49. Z. Kuri, H. Ueda, and S. Shida, J. Chem. Phys. 32(2):371, 1960.
50. G. J. Atchison, J. Polymer Sci. 49:385, 1961.
51. B. R. Loy, J. Polymer Sci. 50(153):245, 1961.
52. A. Konishi, Nippon Kagaku Zasshi 78:1517, 1957.
53. Kh. S. Bagdassar'yan, Theory of Radical Polymerization, Moscow, Academy of Sciences USSR Press, 1959, p. 230.
54. A. A. Berlin, Z. V. Popova, and D. M. Yanovskii, Summaries of Reports at the Conference on the Aging and Stabilization of Polymers, Moscow, Academy of Sciences USSR Press, 1961.
55. G. Ya. Gordon, Vinylidene Chloride and Its Copolymer, Moscow, State Press for Chemical Literature, 1957.
56. K. Thinius, E. Schröder, and V. Waurick, Plaste Kautschuk 6(1):11, 1959.
57. B. I. Fedoseev, Z. V. Popova, and D. M. Yanovskii, Plast. Mass. No. 1:35, 1963.
58. B. I. Fedoseev, Z. V. Popova, and D. M. Yanovskii, Vysokomolekulyarnye Soedineniya 5(5):659, 1963.
59. G. P. Belonovskaya, S. E. Bresler, B. A. Dolgoplosk, A. T. Os'-minskaya, and A. G. Popov, Doklady Akad. Nauk SSSR 128(6):1179, 1959.
60. M. V. Vol'kenshtein, Uspekhi Fiz. Nauk 67(1):131, 1959.
61. G. H. Taft, Mod. Plastics 34(9):170, 1957.
62. E. Parker, Kunststoffe 8:443, 1957.
63. H. V. Smith, Brit. Plastics 27(8):307, 1954.
64. F. Chevassues and R. de Broutelles, Mat. Plast. p. 854, 1958.
65. H. Gibello, Le Chlorure de Vinyle et ses Polymères, Paris, 1959.
66. W. F. Fisher and B. M. Vanderbilt, Mod. Plastics 33(9):165, 1956.
67. A. S. Danyushevskii and E. A. Godzevich, Plastmassy No. 1:46, 1960.
68. A. S. Danyushevskii, Plast. Mass. No. 3:35, 1961.
69. A. Frye and R. W. Horst, J. Polymer Sci. 40(137):419, 1959.
70. C. H. Fuchsman, SPE Tech. Papers 5(30):5, 1959.
71. D. M. Yanovskii, A. A. Berlin, E. N. Zil'berman, and N. A. Rybakova, Zhur. Priklad. Khim. 32:1575, 1959.
72. Z. V. Popova and D. M. Yanovskii, Zhur. Priklad. Khim. 33(1):186, 1960.
73. Z. V. Popova and D. M. Yanovskii, Zhur. Priklad. Khim. 34(6):1324, 1961.

74. Z. V. Popova and D. M. Yanovskii, Vysokomolekulyarnye Soedineniya 2(2):210, 1960.

75. Z. V. Popova, D. M. Yanovskii, E. N. Zil'berman, N. A. Rybakova, and V. I. Ganina, Zhur. Priklad. Khim. 34(4):874, 1961.

76. Z. V. Popova and D. M. Yanovskii, Intern. Symp. Macromol. Chem. III:372, 1960.

77. A. A. Berlin, Z. V. Popova, and D. M. Yanovskii, Zhur. Priklad. Khim. 33:871, 1960.

78. Z. V. Popova, D. M. Yanovskii, N. V. Kozlova, and A. I. Krymova, Zhur. Priklad. Khim. 35:164, 1962.

79. Z. V. Popova, D. M. Yanovskii, P. A. Kirpichnikov, A. S. Kapustina, and V. M. Davydova, Zhur. Priklad. Khim. 36:187, 1963.

80. E. Ferraris, Materie Plastiche 22(11):400, 1956.

81. A. S. Kuz'minskii, N. N. Lezhnev, and Yu. S. Zuev, Oxidation of Raw and Cured Rubbers, Moscow, 1957.

82. A. S. Kuz'minskii, V. D. Zaitseva, and N. N. Lezhnev, Doklady Akad. Nauk SSSR 125:1057, 1959.

83. F. Zeitz, Catalysis, Problems of the Theory and Methods of Investigation [Russian Translation], Moscow, Foreign Literature Press, 1955.

84. Z. V. Popova and D. M. Yanovskii, Vysokomolekulyarnye Soedineniya 3(12):1782, 1961.

85. Z. V. Popova, D. M. Yanovskii, G. O. Tatevos'yan, and O. A. Shtekker, Plast. Mass. No. 5:3, 1962.

86. Z. V. Popova, A. A. Berlin, and D. M. Yanovskii, Zhur. Priklad. Khim. 36:1091, 1963.

87. A. A. Berlin, Z. V. Popova, and D. M. Yanovskii, Doklady Akad. Nauk SSSR 131(3):563, 1960.

88. A. A. Berlin, Z. V. Popova, and D. M. Yanovskii, Vysokomolekulyarnye Soedineniya 4(8):1172, 1962.

89. A. A. Berlin, Z. V. Popova, and D. M. Yanovskii, SPE Transactions 3(1):27, 1963.

90. E. Huckel, Z. Elektrochem. 43:752, 1937.

91. M. E. Dyatkina and Ya. K. Syrkin, Izvest. Akad. Nauk SSSR, Otdel. Khim. Nauk, No. 6:534, 1945.

92. V. A. Kargin and T. I. Sogolova, Zhur. Fiz. Khim. 31:1328, 1957.

93. H. W. Smith, Brit. Plastics 29(10):373, 1956.

94. G. Bonfiglio, Materie Plastiche 20(4):293, 1954.

95. J. G. Hendricks, Ind. Eng. Chem. 43(10):2335, 1951.

96. G. P. Mack, Ind. Plastiques Mod. (Paris) 5(3):34, 1953.

97. R. E. Lally, Paint and Varnish Production 42(10):18, 64, 1952.

98. A. A. Berlin, E. N. Zil'berman, N. A. Rybakova, A. M. Sharetskii, and D. M. Yanovskii, Zhur. Priklad. Khim. 32(4):863, 1959.

99. K. Thinius, Chem. Tech. No. 2:81, 1952.

100. M. Chamberlain, R. A. Delap, and C. L. Stacy, Ind. Eng. Chem. 48(7):1209, 1956.

101. G. Bonfiglio, Materie Plastiche 23(3):175, 1957.

102. J. C. Bauwens and G. Homes, Compt. rend. 250:1853, 1960.

103. G. F. Bush, Am. Dyestuff Reptr.49(3):33, 1960; Brit. Plastics Federation,10, 61/6963.

104. J. Minematsu and T. Jamada, Makromol. Chem. 47(2-3):249, 1961.

105. B. Kinlay, Fibres Plastiques 33(3):73, 1961.

106. H. J. Klingner, Plaste Kautschuk 8(10):532, 1961.

107. G. P. Mack, Ind. Plastique Mod. (Paris) 5(4):43, 1953.

108. G. P. Mack, Mod. Plastice 31(3):150, 218, 1953.

109. A. Scarbrough, Mod. Plastics 29(9):111, 1952.

110. M. Lisi and S. Varga, Chem. Zvesti 14(1):19, 1960.

111. A. S. Danyushevskii, N. Ya. Parlashkevich, Z. N. Frolova, and I.S. Shentsis, Plastmassy No. 2:9, 1961.

112. A. Hartmann, Kolloid-Z. 139(3):146, 1954.

113. E. B. Snyder and R. Fantini, ASTM Bull. 59:218, 1956.

114. B. Schmitt and F. Heck, Kunstoffe 46(12):555, 1956.

115. H. A. Reehling, U. S. Patent 2773746, 1957.

116. T. Jamada, Referat. Zhur. Khim. 1957, 11945, 19449.

117. L. G. Angert and A. S. Kuz'minskii, The Role and Use of Antioxidants in Raw and Cured Rubbers, Moscow, State Press for Chemical Literature, 1957.

118. Shiramatsu, Toytaro, Hashimoto, Zhuzaburo, and Naraba Kozo, Repts. Elect. Commun. Lab. 2(11):12, 1954.

119. A. Guyot and J. P. Benevise, Ind. Plastiques Mod. (Paris) 13(5):37, 1961; Brit. Plastics Federation 8, 61/5347.

120. C. Wippler, Nucleonics 18(8):69, 1960; Brit. Plastics Federation 10, 61/6958.

121. V. A. Kargin and M. N. Shteding, Khim. Prom. No. 2:74, 1955.

122. G. G. Himmler and F. R. Nissel, Plastics Technol. 3(4):280, 1957.

123. A. A. Berlin, G. S. Petrov, and V. F. Prosvirkina, Zhur. Fiz. Khim. 32(11):25, 1958.

124. Furudzava, Kondé, and Goto, Chem. High. Polymers 10(98):259, 1953.

125. Yamaguti, Otsu, and Imoto, J. Chem. Soc. Japan, Ind. Chem. Sect. 58(6):472, 1955.

126. G. M. Guzman and T. M. Tatou, Anales Real. Soc. Espan Fis. Quim. (Madrid) Ser. B 55(2):129, 1959.

127. Z. Mencik, Collection Czech. Chem. Commun. 24:3793, 1959.

128. H. Grohn and Hoang-Binh, Plaste Kautschuk 8(2):631, 1961.

129. S. Krimm, C. V. Liang, and G. B. Sutherland, J. Polymer Sci. 22(100):95, 1956; 22(101) 221, 1956.
130. S. Krimm, A. R. Berens, V. L. Folt, and J. J. Shipman, Chem. & Ind. No. 46:1512, 1958.
131. H. Sobue, Kogyo Kagaku Zasshi 61:106, 1958.
132. S. Khirosi, T. Ionéo, and T. Iosio, Enka Biniru Kikai Géppo No. 50:9, 1958.
133. S. Varga, Chem. Zvesti 11(2):72, 1957.
134. S. Narita, S. Ichinone, and S. Enomoto, J. Polymer Sci. 37(131): 273, 281, 1959.
135. Shimanouchi, Tsuchiya, and Mizushima, J. Chem. Phys. 30(5): 1365, 1959.
136. M. Asahina and S. Enomoto, J. Chem. Soc. Japan No. 81:1370, 1960.
137. R. A. Burley and W. J. Bennett, Appl. Spectroscopy 14:32, 1960; C. A. 13:13721, 1960.
138. K. Hukuda, Mem. Fac. Sci., Kyushu Univ. 133(1):41, 1960.
139. R. Gautron, J. chim. phys. 58(2):159, 1961.
140. D. Malcolm and G. Francis, J. Am. Chem. Soc. 83(11):2584, 1961.
141. J. C. Bevington, Fortschr. Hochpolymer.-Forsch. 2(1):1, 1960.
142. A. Szcstzpanek and R. Weber, Maschinenmarkt 5(8):27, 1960.
143. J. F. Salhofer, Gummi, Asbest. Kunstoffe 14(8):752, 1961.
144. D. Taylor, Plastics Inst., (London) Trans. J. 28(76):160, 1960; Brit. Plastics Federation 2, 61/996.
145. G. C. Nolla, Rev. Plasticos 10(55):14, 1959.
146. R. Hammond, Plastics Inst. Trans. J. 26(63):49, 1958.
147. A. Coen and P. Parrini, Materie Plastiche 27(8):760, 1961.
148. J. Schreiber, Plaste Kautschuk 8(9):454, 1961.
149. Späth, Gummi, Asbest, Kunststoffe 10(10):566, 1957.
150. J. Baumens and G. Homes, Compt. rend. 250(10):1853, 1960.
151. F. Chevassues and R. de Brouteilles, Stabilisation du Poly- chlorure de Vinyle, Editions Amphora. Rep. Francais, 1955.
152. S. Midzutani, J. Soc. Rubber Ind. Japan 29(8):692, 1956.
153. B. Henderson, Can. Plastics Nov., 66, 1957.
154. E. N. Zil'berman, Uspekhi Khim. i Tekhnol. Polimer., Coll. 3, Moscow, 1960.
155. V. A. Voskresenskii, S. V. Fridland, V. V. Atamanova, E. M. Orlova, and V. A. Byl'ev, Summaries of Reports at the Confer- ence on the Aging and Stabilization of Polymers, Moscow, Academy of Sciences USSR Press, 1961.
156. A. S. Danyushevskii, Dissertation, Moscow, 1961.
157. H. H. Jellinek, Degradation of Vinyl Polymers, 1955.

158. M. Bornengo, Materie Plastiche 24(7):606, 1958.

159. I. Tsunéyukisuké, Eng. Mater. 6(6):69, 1958.

160. E. E. Griesser and W. T. Higgins, Pensez Plast. No. 65:29, 1957.

161. H. J. M. Langshow, Plastics (London), 25:40, 1960; C. A. 9:9347, 1960.

162. P. Dubois, Kunststoffe 49(11):632, 1959.

163. H. Stäger, Kunststoffe 49(11):589, 1959.

164. N. L. Perry, SPE Journal 15(7):550, 1959.

165. C. F. Weider, Kunststoffe 43:102, 1953.

166. German Federated Republic Patent 950152, 1956.

167. U. S. Patent 2938883, 1960.

168. Niewiadomski and Czapiga, Przeglad Electrotechn. 34(7):348, 1958.

169. U. S. Patent 2882251, 1959.

170. U. S. Patent 2874145, 1959.

171. German Federated Republic Patent 1023220, 1958.

172. M. E. Pozin, Hydrogen Peroxide and Peroxide Compounds, Moscow-Leningrad, State Press for Chemical Literature, 1951, pp. 259-270.

173. U. Jacobson, Brit. Plastics 34:6, 328, 1961.

174. E. Rosental, Kunststoff-Rundschau 4:133, 1957.

175. H. V. Smith, Brit. Plastics 27:176, 1954.

176. U. S. Patent 2752319, 1956.

177. R. Reichherzer, Kuststoff—Plastics 6:165, 1959.

178. E. T. Denisov and N. M. Emanuél', Uspekhi Khim. 27:365, 1958.

179. W. G. Wennels, Trans. J. Plast. Inst. 23(51):44, 1955.

180. W. G. Wennels, Plastics 20(212):93, 1955.

181. U. S. Patent 2960490, 1960.

182. French Patent 1111551, 1956.

183. German Federated Republic Patent 950326, 1956.

184. H. Passedouet, Rev. Franc. Corps Gras 3(8-9):587, 1956.

185. Great Britain Patent 815875, 1959.

186. U. S. Patent 2782176, 1957.

187. U. S. Patent 2784171, 1957.

188. French Patent 1198168, 1960.

189. U. S. Patent 2959568, 1960.

190. Italian Patent 487734, 1959.

191. German Federated Republic Patent 1009728, 1957.

192. Japan Patent 293, 1956.

193. U. S. Patent 2902465, 1959.

194. U. S. Patent 2960491, 1960.

195. German Federated Republic Patent 946480, 1956.

196. U. S. Patent 2853466, 1958.

197. U. S. Patent 2958675, 1960.

198. Canadian Patent 508544, 1955.
199. H. V. Smith, Brit. Plastics 27:213, 1954.
200. N. L. Perry, Pract. Plastics 9(7):7, 1958.
201. N. L. Perry, Plast. and Paint 2(5):35, 1958.
202. N. L. Perry, Rubber Age 85(3):449, 1959.
203. A. Chansen, Rubber Age 76(5):715, 1955.
204. U. S. Patent 2744631, 1956.
205. Great Britain Patent 841890, 1960.
206. U. S. Patent 2863745, 1959.
207. U. S. Patent 2933465, 1960.
208. M. Imoto and T. Otsu, J. Inst. Polytech., Osaka City Univ. 4:124, 1953.
209. U. S. Patent 2745819, 1956.
210. Canadian Patent 510495, 1955.
211. Canadian Patent 513532, 1955.
212. Canadian Patent 546847, 1958.
213. A. Maillard, A. R. J. Deluzarche, and J. C. Maire, Bull. Soc. Chim. France No. 6:853, 1958.
214. U. S. Patent 2583084, 1951.

215. N. A. Rybakova, N. K. Taikova, and E. N. Zil'berman, Tr. po Khim. i Khim. Tekhnol. 2:183, 1959.
216. J. G. A. Luijten and S. Pezarro, Brit. Plastics 30(5):183, 1957.

217. Swedish Patent 148940, 1955.
218. German Federated Republic Patent 1008908, 1957.
219. U. S. Patent 2883363, 1959.

220. German Federated Republic Patent 1007327, 1957.
221. U. S. Patent 2731441, 1956.
222. L. Kamphenkel, Plaste Kautschuk 3(9):209, 1956.
223. G. J. M. Van der Kerk and J. G. A. Luijten, Ind. Chim. Belge 21:567, 1956.
224. German Democratic Republic Patent 16800, 1959.
225. Japan Patent 5983, 1954.
226. Great Britain Patent 791119, 1958.
227. USSR Patent No. 117676.
228. U. S. Patent 2938013, 1960.
229. U. S. Patent 2759906, 1956.
230. U. S. Patents 2630436 and 2630442, 1953.
231. German Democratic Republic Patent 14024, 1958.
232. U. S. Patents 2591657 and 2593267, 1952.
233. Great Britain Patent 718393, 1954.
234. USSR Patent No. 108007.
235. U. S. Patent 2715111, 1955.
236. USSR Patent No. 212023, 1961.

237. U. S. Patent 2763632, 1956.

238. German Federated Republic Patent 968827, 1957.

239. U. S. Patent 2746946, 1956.

240. Great Britain Patent 740203, 1956.

241. Japan Patent 9096, 1956.

242. U. S. Patent 2954362, 1960.

243. U. S. Patent 2868765, 1959.

244. H. H. van Reuss, Angew. Chem. 70(5):135, 1958.

245. U. S. Patent 2777826, 1956.

246. N. L. Perry, Ind. Eng. Chem. No. 6:862, 1958.

247. D. E. Winkler, Ind. Eng. Chem. No. 6:863, 1958.

248. E. S. Narrecott, Brit. Plastics 24:341, 1951.

249. German Federated Republic Patent 962833, 1957.

250. U. S. Patent 2922777, 1960.

251. E. Savelli, Materie Plastiche 23(1):27, 1957.

252. H. Passedouer, Pensez Plast. 8(225):5, 1961; Brit. Plastics
     Federation 4, 61/3847.

253. G. Burdess and C. Finlay, Fibres and Plastics 22(1):21, 1961;
     22(3):73, 1961; Brit. Plastics Federation 4, 61/2510.

254. Greenspan and Gall, Ind. Eng. Chem. No. 6:861, 865, 1958.

255. Brice and Budde, Ind. Eng. Chem. No. 6:868, 1958.

256. D. H. Mullins, Ind. Eng. Chem. No. 6:873, 1958.

257. U. S. Patent 2739161, 1956.

258. L. P. Witnauer, H. B. Knight, W. E. Polm, R. E. Koos, W. C.
     Ault, and D. Swern, Ind. Eng. Chem. No. 11:2304, 1955.

259. J. Dvorak and E. Nejedly, Chem. Przimyal 8(4):209, 1958.

260. E. W. Lines, Plastics Technol. 7(3):51, 1961.

261. A. A. Berlin, E. N. Zil'berman, N. A. Rybakova, A. M. Sharet-
     skii, and D. M. Yanovskii, Khim. Nauka i Prom. 2:809, 1957.

262. N. A. Rybakova, E. N. Zil'berman, and A. A. Berlin, Tr. po
     Khim. i Khim. Tekhnol. 1:679, 1958.

263. U. S. Patent 2924582, 1960.

264. Great Britain Patent 827986, 1960.

265. W. J. Marmion, Research 7:351, 1954.

266. K. Thinius, Kunststoffe 40:191, 1950.

267. M. Bomar, Chem. Prumysl 34:326, 1959.

268. U. S. Patent 2612500, 1952.

269. U. S. Patent 2989496, 1961.

270. German Federated Republic Patent 927536, 1955.

271. German Federated Republic Patent 1035895, 1960.

272. German Federated Republic Patent 879314, 1953.

273. USSR Patent No. 125037.

274. U. S. Patent 2721883, 1955.

275. U. S. Patent 2714057, 1955.

276. U. S. Patent 2946765, 1960.

277. German Democratic Republic Patent 5482, 1953.
278. Z. V. Popova and D. M. Yanovskii, Summaries of Reports at the Conference on the Aging and Stabilization of Polymers, Moscow, Academy of Sciences USSR Press, 1961.
279. P. I. Levin, A. F. Lukovnikov, M. B. Neiman, and M. S. Khloplyankina, Vysomolekulyarnye Soedineniya 3(8):1243, 1961.
280. German Federated Republic 1100437, 1960.
281. Great Britain Patent 739766, 1955.
282. Great Britain Patent 827393, 1960.
283. USSR Patent No. 131505.
284. U. S. Patent 2684955, 1954.
285. German Democratic Republic Patent 7282, 1954.
286. U. S. Patent 2962474, 1960.
287. Off. Matieres Plastiques 8(81):103, 1961; Brit. Plastics Federation 7, 61/4496.

288. French Patent 1176735, 1959.
289. Great Britain Patent 803082, 1958.

290. V. S. Abramov and A. S. Kapustina, Doklady Akad. Nauk SSSR 111(6):1243, 1956.
291. U. S. Patent 2935490, 1960.
292. U. S. Patent 2935491, 1956.
293. French Patent 1127671, 1956.
294. U. S. Patent 2932624, 1960.
295. U. S. Patent 2625521, 1953.
296. H. C. Murfitt, Brit. Plastics 33(12):578, 1960.
297. J. Weth and A. C. Signore, Am. Paint J. 42:6, 117, 1957.
298. J. T. Leonard, Chem. Prod. 21(5):173, 1958.
299. W. Arndt, Plastics World 16(8):8, 1958.
300. H. Stecher, Adhaesion 2(6):243, 1958.
301. P. Graham, J. Chem. Eng. Data 8(4):372, 1959.
302. H. L. Perry, Plast. Eng. J. 15(7):550, 1959.
303. C. Vale and W. Taylor, Chem. Ind. No. 9:268, 1961; Brit. Plastics Federation 4, 61/2583.
304. H. Gysling and H. J. Keller, Kunststoffe 1:13, 1961.
305. U. S. Patent 2876210, 1959.
306. U. S. Patent 2891996, 1959.
307. U. S. Patent 2904529, 1959.
308. U. S. Patent 2937157, 1960.
309. U. S. Patent 2947723, 1960.
310. U. S. Patent 2807605, 1958.
311. U. S. Patent 3001970, 1961.
312. Japan Patent 5992, 1959.
313. Japan Patent 10490, 1956.
314. Japan Patent 2190, 1959; C. A. 20, 19460, 1959.

315. U. S. Patent 2900361, 1959.
316. U. S. Patents 2858293, 2889295, 2898323, 2818404, 1959.
317. U. S. Patent 2943076, 1960.
318. German Democratic Republic Patent 7284, 1954.
319. U. S. Patent 2676940, 1954.
320. U. S. Patents 2595636, 1951; 2666752, 1954.
321. D. Swern, Paint Varnish Production 46(6):26, 1956.
322. U. S. Patent 2604483; Great Britain Patent 735030, 1955.
323. U. S. Patent 2665265; French Patent 1052357, 1955.
324. U. S. Patent 2724706, 1955.
325. USSR Patent No. 131085.

Chapter VII

AGING AND STABILIZATION OF POLYAMIDES

Materials made from polyamides age rapidly during use. Many factors play a role in this case: sunlight, heat, the oxygen of the air, moisture. Hence, from the very beginning of the industrial production of polyamides (1941-1942), the first patents on their stabilization appeared [1-4]. However, in spite of the large number of studies presently available in the field of the aging of polyamides, as well as patents on their stabilization, it is difficult to compile a distinct representation of the mechanism of the processes that occur in polyamides under the influence of various factors – heat, oxygen, light.

There are almost no studies at all elucidating the mechanism of the action of various classes of stabilizers, in spite of the abundance of the existing patents.

I. THERMAL AGING OF POLYAMIDES

This type of aging polyamides has been studied in the greatest detail. The changes that occur in polyamides upon heating have been the object of investigation of many authors during the last decade, both here and abroad.

Appreciable changes during thermal influence on polyamides in the absence of oxygen occur at temperatures above 200°C. However, at 150°C, nylon already losses 4% of its strength after heating for 18 hr in an atmosphere of nitrogen; in the case of heating under the same conditions in air, the strength is reduced by 25% [5].

1. Change in the Molecular Weight during Thermal Aging of Polyamides

The works of a number of authors have investigated the change in the molecular weight that occurs during thermal destruction of polyamides of various chemical structures.

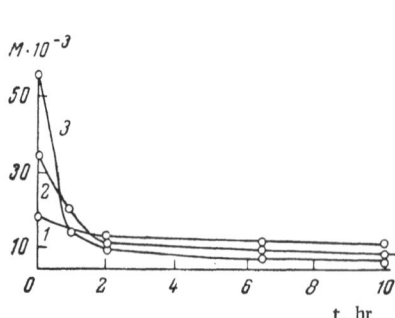
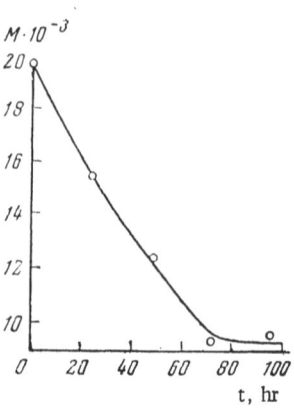

Fig. 103. Dependence of the molecular
weight of polyhexamethyleneadipamide on
the duration of heating at 300°C (in a
stream of nitrogen) for three samples with
initial molecular weights of 18,000 (1),
35,000 (2), and 55,000 (3).

Fig. 104. Dependence of the
molecular weight of poly-
caproamide on the duration
of its heating in sealed ampoules in nitrogen at 250°C.

When fused polyhexamethylenedipamide is heated in a stream of
nitrogen at 270°C the molecular weight does not decrease; at 300°C the
molecular weight with respect to viscosity decreases to a definite value,
independent of the molecular weight of the initial polymer [6]. Figure
103 shows the decrease in the molecular weight as a function of the time
of heating for three samples of polyhexamethyleneadipamide with different initial molecular weights (18,000, 35,000, and 55,000). As a result of four-hour heating, the molecular weights of all the samples decrease to approximately 9000, and then remain unchanged.

Since the drop in the molecular weight occurs independent of the
type of terminal groups (polyamides stabilized by acetic acid and unstabilized were studied) and proceeds more sharply in high-molecular
polyamides, the authors assume that the events of destruction occur not
only at the ends, but also in any sites of the chain, probably at any amide
bond. Destruction predominates at the beginning of heating. The establishment of a constant equilibrium molecular weight at 300°C is explained by an increase in the influence of structuring processes.

At 330°C, the process of structuring predominates over destruction;
hence, in addition to a decrease in the molecular weight of the soluble
portion, heating leads to the formation of a cresol-insoluble substance,
the amount of which increases as heating progresses and constitutes 97%
after 6 hr.

Thermal destruction of polycaproamide occurs at lower temperatures
than that of polyhexamethyleneadipamide. The curve in Fig. 104 shows

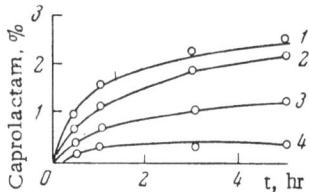

Fig. 105. Dependence of the amount of ε-caprolactam formed in the process of thermal depolymerization of polycaproamide in the presence of 0.037% water, on the duration of heating at 230°C for four samples with initial specific viscosity of the polymer 0.400 (1), 0.450 (2), 0.612 (3), and 0.786 (4).

the variation of the molecular weight as a function of the time of heating of polycaproamide in sealed ampoules in an atmosphere of nitrogen at 250°C. Here, just as for polyhexamethyleneadipamide at 300°C, first a decrease, and then the establishment of constant molecular weight is observed [7].

The thermal destruction of polynanthoamide [a polyamide obtained by polycondensation of ω-aminoenanthic acid $NH_2(CH_2)_6COOH$], as well as mixed polyamides containing the radical of ε-aminocaproic acid $- NH(CH_2)_5CO -$ (anid 40/60 – the product obtained by polycondensation of hexamethyleneadipamide and ε-caprolactam, taken in a 40 : 60 ratio, and anid G-669, produced by polycondensation of hexamethyleneadipamide, ε-caprolactam, and hexamethyleneazelaamide) at 300°C under vacuum is accompanied by a reduction of the molecular weight with respect to viscosity. This is explained by the authors of [8, 9] by hydrolytic cleavage of the amide bond on account of the water remaining in the polyamide even after its thorough drying.

Polycaproamide is characterized by a depolymerization reaction, related to the existence of labile equilibrium in the system

$$H\,[NH\,(CH_2)_5\,CO]_n\,OH \rightleftarrows n\,(CH_2)_5 - NH + H_2O$$
$$\underset{CO}{\diagdown\diagup}$$

Below are cited the equilibrium concentrations of the monomer and polymer corresponding to various temperatures [10]:

Temperature, °C	160	220	250	280
Concentration, %:				
Monomer	3	10	12.5	15
Polymer	97	90	87.5	85

The thermal depolymerization of polycaprolactam is explained by processes of hydrolysis [12], acidolysis, and aminolysis [11, 12].

The kinetic curves of the formation of caprolactam at 230°C for four samples of polycaproamide of various initial molecular weights are shown in Fig. 105 [11]. The increase in the molecular weight, i.e., the decrease in the concentration of terminal groups, is responsible for a decrease in the rate of depolymerization of polycaproamide.

The main direction of thermal depolymerization is interaction of the free terminal amino and carboxyl groups with the last amide bond of the same molecule, splitting out ε-caprolactam (intramolecular acidolysis and aminolysis) [11, 12]:

$$\sim NH - (CH_2)_5CO - HN (CH_2)_5COOH \rightarrow \sim NH (CH_2)_5COOH + NH - (CH_2)_5$$
$$CO$$

$$\sim CO (CH_2)_5NH - CO (CH_2)_5NH_2 \rightarrow \sim CO (CH_2)_5NH_2 + NH - (CH_2)_5$$
$$CO$$

In addition to intramolecular processes of aminolysis and acidolysis, intermolecular processes are also possible [12]:

Intermolecular Acidolysis

$$\sim NHCO (CH_2)_5 - NH \mid CO (CH_2)_5NH_2$$
$$+$$
$$HO \mid CO (CH_2)_5 - NH - CO (CH_2)_5 \sim \rightarrow$$
$$\rightarrow \sim NHCO (CH_2)_5NHCO (CH_2)_5 - NHCO (CH_2)_5 \sim + HOOC (CH_2)_5 NH_2$$
$$HOOC (CH_2)_5NH_2 \rightarrow H_2O + (CH_2)_5 - NH$$
$$CO$$

Intermolecular Aminolysis

$$\sim NH (CH_2)_5CO \mid NH (CH_2)_5 COOH$$
$$+ \rightarrow \sim NH (CH_2)_5CONH (CH_2)_5CONH (CH_2)_5 + NH_2 (CH_2)_5COOH$$
$$H_2 \mid N (CH_2)_5CONH (CH_2)_5 \sim$$

$$NH_2 (CH_2)_5COOH \rightarrow H_2O + (CH_2)_5 - NH$$
$$CO$$

The curves of Fig. 105 actually show that the decrease in the molecular weight of the initial polymer, i.e., increase in the concentration of terminal groups, leads to an increase in the rate of formation of caprolactam.

Substitution of the terminal amino groups by acetylation leads to a decrease in the rate of depolymerization. Increasing the water content in the polymer considerably accelerates the depolymerization reaction.

The depolymerization of polycaprolactam can occur not only from the ends of the chain, but also at any portion of the chain, as a result of reactions of intramolecular exchange [8, 9]:

$$\sim \overset{O}{\overset{\|}{C}} \quad \overset{H}{\overset{|}{N}} \sim \rightarrow \sim \overset{O}{\overset{\|}{C}} - \overset{H}{\overset{|}{N}} \sim + (CH_2)_5 - NH$$
$$\overset{|}{NH} \quad \overset{|}{C} = O \qquad\qquad CO$$
$$(CH_2)_5$$

When polycaproamide is heated in an open system in a stream of dry nitrogen at 300°C and higher, continuous depolymerization occurs, liberating ε-caprolactam, the decrease in the molecular weight with respect to viscosity being proportional to the amount of monomer distilled off.

Enant, which depolymerizes at 350°C, liberating ω-enantholactam, also behaves analogously.

The possibility of the splitting out of ε-caprolactam not only from the ends of the chain, but also in any portion of the macromolecule is confirmed by experiments on the destruction of mixed polyamides, containing ε-aminocaproic acid radicals in the chain. When mixed polyamides (anid 40/60 and anid G-669) are heated in a stream of inert gas at 300°C, ε-caprolactam distills off, as in the case of homogeneous polycaproamide [8, 9].

2. Chemical Processes That Occur in the Thermal Aging of Polyamides

The products obtained in the pyrolysis of polyamides have been investigated by many authors. The experimental data are frequently contradictory, and hence the principal schemes of the reactions that occur, proposed by various researchers, also differ from one another.

There is a considerable discrepancy in the values of the molecular weights determined according to viscosity and according to terminal carboxyl groups for polyamides heated at 230-250°C or produced at too high temperatures [7, 13, 14]. Thus, for example, in the production of polyhexamethyleneadipamide at 250°C, its molecular weight, determined according to viscosity, is 19,950, while that calculated according to the titration of terminal carboxyl groups is 22,200 [13].

The data cited below [7] show that after brief supplementary heating of polycaproamide when it is produced at 230-250°C in an atmosphere of nitrogen, the molecular weights determined for three samples according to the viscosity proved considerably lower than the molecular weights according to the concentration of terminal carboxyl groups:

Molecular weight, calculated according to:			
Viscosity	7,600	8,300	12,200
Concentration of terminal groups	11,200	15,500	27,600

The authors explain such a discrepancy in the molecular weights by the splitting out of carboxyl groups during heating.

In the case of pyrolysis under vacuum at 400°C of the mixed polyamide produced by polycondensation of caprolactam and hexamethyleneadipamide, as well as the mixed polyamide obtained from hexamethyleneadipamide, ε-caprolactam, and hexamethylenesebacamide [15, 16], the
following products were detected in the gas phase by mass spectroscopic
analysis: carbon monoxide, carbon dioxide, cyclopentanone, water,
various hydrocarbons (methane, ethane, propane, butane, ethylene, but-
enes). Ammonia and other nitrogen compounds were not detected in the
gas phase. All the nitrogen remains in the solid residue [15-18]. The

authors proposed that the weakest bond in the chain $-\overset{|}{\underset{\overset{||}{O}}{C}}-\overset{|}{\underset{\overset{|}{H}}{N}}-$ is broken;

this leads to the reaction

$$R-\overset{|}{\underset{|}{NH}}-CO\,(CH_2)_4CO-\overset{|}{\underset{|}{NH}}-R \rightarrow -CO\,(CH_2)_4CO--\rightarrow \begin{matrix} CH_2-CH_2 \\ \underset{CH_2 \quad CH_2}{|\qquad |} \\ \underset{CO}{\diagdown\diagup} \end{matrix} +CO$$

At the end of the molecule, such a cleavage of an amide bond gives

$$R-NHCO\,(CH_2)_4COOH \rightarrow RNH_2 + CO_2 + \begin{matrix} CH_2-CH_2 \\ \underset{CH_2 \quad CH_2}{|\qquad |} \\ \underset{CO}{\diagdown\diagup} \end{matrix}$$

These reactions explain the formation of CO and CO_2 in the pyrolysis
of polyamides; however, the amount of carbon dioxide obtained considerably exceeds the number of carboxyl groups in the initial polyamide.

Most of the water liberated during pyrolysis is adsorbed by the initial polymer, in the opinion of the authors. A small amount of water can
be formed according to the reaction [15, 16]

$$\sim (CH_2)_x NHCO\,(CH_2)_y \sim \qquad \sim (CH_2)_x - N = \underset{\underset{OH}{|}}{C} - (CH_2)_y \sim$$

$$\rightarrow \qquad\qquad\qquad\qquad \xrightarrow{-H_2O}$$

$$\sim (CH_2)_y CONH\,(CH_2)_x \sim \qquad \sim (CH_2)_y - \overset{\overset{OH}{|}}{C} = N\,(CH_2)_x \sim$$

$$\sim (CH_2)_x - N = \underset{\underset{O}{|}}{C} - (CH_2)_y \sim$$

$$\rightarrow \qquad\qquad\qquad\qquad$$

$$\sim (CH_2)_y - \overset{\overset{O}{|}}{C} = N\,(CH_2)_x \sim$$

The formation of large amounts of CO_2 can be explained by hydrolysis of the amide bond, thanks to the presence even of traces of water,
followed by decarboxylation [17, 18]:

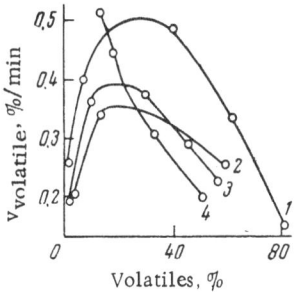

Fig. 106. Dependence of the rate of formation of volatile destruction products on the amount of volatiles formed during heat at 345°C of the copolymer of polyhexamethyleneadipamide and polycaproamide. 1) Pure sample; 2) in the presence of 1% 1,5-diaminoanthraquinone; 3) in the presence of 1% H_3PO_4; 4) in the presence of 3% H_2SO_4.

$$R - \overset{\overset{\text{O}}{\|}}{C} - \underset{\underset{\text{H}}{|}}{N} - R' + H_2O \rightarrow \underset{\downarrow}{RCOOH} + R'NH_2$$
$$RH + CO_2$$

The activation energies calculated for various samples lie in the range from 15 to 42 kcal/mole, which the authors relate to the sensitivity of the process of hydrolysis to traces of catalyst.

Figure 106 shows that in the presence of sulfuric and phosphoric acids, as well as 1,5-diaminoanthraquinone, the rate of formation of volatile products decreases. The action of 1,5-diaminoanthraquinone as an inhibitor of thermal destruction indicates the possibility of a radical mechanism of the process; however, in the opinion of the authors, it is masked by the hydrolytic process. Actually, thorough purification and drying of one of the samples of polycaproamide gives an increase in the preexponential factor from 10^7 to 10^{10} and an increase in the activation energy from 34 to 43 kcal/mole.

The authors of [17, 18] propose that still more thorough purification can give a value of the preexponential factor of 10^{13}-10^{16} and an activation energy of 50-60 kcal/mole. This value of the activation energy corresponds to the energy of thermal destruction of polyamides, proceeding according to a purely radical mechanism in the absence of moisture and other impurities.

Thermal decomposition of amides of the type of RCONHR' and biamides, derivatives of hexamethylenediamine, $RCONH(CH_2)_6NHCOR$ $(R - C_5H_{11}, C_6H_5)$, was conducted at 350°C in nitrogen to elucidate the causes of the formation of large amounts of carbon dioxide [19]. It was found that all these amides decompose very slowly; after 5 hr, more than 95% of the starting material remains unchanged. On the other hand, under the same conditions N,N'-di-n-butylamides $RNHCO-(CH_2)_xCONHR$ (where $R = C_4H_9$, x = 3, 4, 5, 8) decompose in considerable amounts, while in the case of N,N'-dibutyladipamide (when x = 4), which simulates polyhexamethyleneadipamide, the starting material decomposes almost entirely.

Analysis gives the following content of volatiles (in moles per mole of amide) for the pyrolysis of N,N'-dibutyladipamide: n-butylamine,

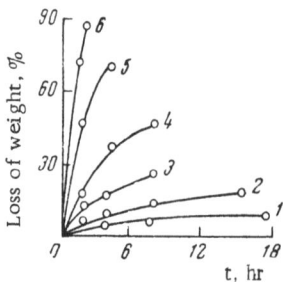

Fig. 107. Dependence of the loss in weight on the duration of heating of polyhexamethyleneadipamice in nitrogen: 1) 280°; 2) 305°; 3) 330°; 4) 350°; 5) 375°; 6) 400°C.

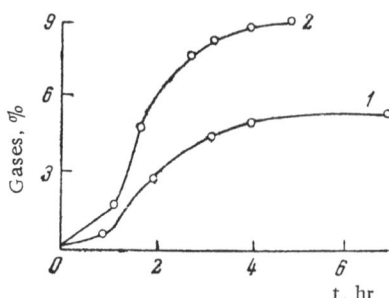

Fig. 108. Dependence of the amount of gases formed on the duration of heating of polyhexamethyleneadipamide at 375°C. 1) Ammonia; 2) carbon dioxide.

0.967; CO_2, 0.466; CO, 0.008; C_4H_{10}, 0.03; C_4H_8, 0.01; H_2, 0.003; NH_3, 0.000005. Other dibutylamides (x = 3, 5, 8) give n-butylamine, small amounts of hydrocarbons, and carbon monoxide as the volatiles, carbon dioxide is not formed. Decomposition of di-n-hexyladipamide at 350°C gives n-hexylamine and CO_2. Thus, the reaction forming carbon dioxide during heating under anhydrous conditions in the absence of three carboxyl groups is a specific reaction of N-substituted adipamides.

In contrast to Achhammer [15, 16] and Straus and Wall [17, 18], many authors have detected not only carbon dioxide among the gaseous destruction products, but also large amounts of ammonia [8-10, 20]. Ammonia can be formed [10] in the reaction of the terminal amino groups, this reaction proceeding even in the production of polyamides:

$$\sim CO - NH(CH_2)_6NH_2 + H_2N(CH_2)_6NHCO \sim \rightarrow \ \sim CO - NH(CH_2)_6 - NH(CH_2)_6NHCO \sim + NH_3$$

In addition to secondary amino groups, tertiary amino groups can also be obtained [9]:

$$\begin{matrix} \sim CO - NH\,(CH_2)_6 \\ \qquad\qquad\qquad\Large\diagdown NH + NH_2\,(CH_2)_6NH \sim \rightarrow NH_3 + N \diagdown \\ \sim CO - NH\,(CH_2)_6 \end{matrix} \begin{matrix} (CH_2)_6NHCO \sim \\ - (CH_2)_6NHCO \sim \\ (CH_2)_6NHCO \sim \end{matrix}$$

The amount of ammonia obtained in the destruction of polyamides also exceeds the number of terminal amino groups in the initial polymer.

The large amounts of ammonia and carbon dioxide are explained [9, 21] by hydrolysis on account of the water remaining even after thorough drying of the polyamide, as a result of which new terminal amino and carboxyl groups are formed; they are split out in the form of ammonia and carbon dioxide, decarboxylation proceeding more readily than de-amination.

In [20] the kinetics of the thermal decomposition of polyhexamethyl-eneadipamide was studied. The curves in Fig. 107 show the loss in weight as a function of the time of heating at various temperatures (within the interval 280-400°C). At 280-305°C, the loss in weight is 10-20% after 16 hr; at 400°C, 85% of the initial sample is already liberated in the form of volatile products after 2 hr. Ammonia has been detected among the gaseous products, together with carbon dioxide. The kinetic curves of the formation of these gases, cited in Fig. 108, exhibit an induction period.

In contrast to Achhammer [15, 16], in [20] the cleavage not of the $-N-C-$ bond, but of the CH_2-CO- bond is proposed.

A detailed analysis of the products of thermal destruction of poly-hexamethyleneadipamide and polycaproamide is given in [22]. The poly-amides were heated at 300-305°C in a stream of dry nitrogen. The gase-ous pyrolysis products contain large amounts of NH_3, CO_2, H_2O, small amounts of n-hexylamine, n-pentylamine, and cyclopentanone. Analysis of the hydrolysis products of the residue obtained after pyrolysis of the polyamides permitted the authors to propose a scheme for the process.

The process of destruction begins with cleavage of the $-CH_2-\overset{|}{\underset{|}{N}}H-$ bond in the β-position to the carbonyl group (according to Achhammer, the $-CO-\overset{|}{\underset{|}{N}}H-$ bond in the α-position to the carbonyl group is broken). The radicals formed in this case are stabilized by the formation of a double bond:

$$\sim C-C-(CH_2)_5-C-N-CH_2-(CH_2)_4-C-NH\sim \rightarrow$$
$$\rightarrow \sim C-N-(CH_2)_5-C-NH_2+CH_2=CH-(CH_2)_3-C-NH\sim$$
$$\downarrow$$
$$\sim C-N-(CH_2)_5-C\equiv N+H_2O$$

Nitrile groups have actually been detected by means of the infrared spectra, in amounts of 5% of the initial amide groups.

Hydrolysis of the amide bonds on account of the water formed as a result of the last reaction can explain the large amounts of CO_2 and NH_3 obtained in the pyrolysis of polyamides.

Polyhexamethyleneadipamide at 305°C readily forms cross-linked structures. The product remaining after 6 hr heating is entirely insoluble in formic acid.

For polycaproamide, the formation of cross-linked structures occurs more slowly; at 280°C the polymer is converted to an infusible, insoluble product after 12 days. The authors explain the cross-linking of polyamides by the presence of secondary amino groups, formed in the interaction of the terminal amino groups, as was indicated above, as well as ketone groups, formed according to the following reaction (interaction of carboxyl groups):

$$R - C - NH \ (CH_2)_6 COOH + HOOC \ (CH_2)_5 NHCOR' \rightarrow$$
$$\overset{\|}{O}$$

$$\rightarrow RCONH \ (CH_2)_5 - C \ (CH_2)_5 NHC - R' + CO_2 + H_2O$$
$$\overset{\|}{O} \qquad \overset{\|}{O}$$

Further interaction of the secondary amino groups (for example, with the terminal carboxyls) leads to the formation of three-dimensional structures:

$$>NH + HOOC \ (CH_2)_4 CONH \sim \rightarrow >N - C - + H_2O$$
$$\overset{\|}{O}$$

At high temperatures (380–400°C), the destruction of polyhexamethyleneadipamide is autoaccelerated on account of the water formed in the self-condensation of cyclopentanone and the condensation of cyclopentanone with α-ethylpyrrolidone [9, 23].

The following formulas are proposed for the condensation products obtained:

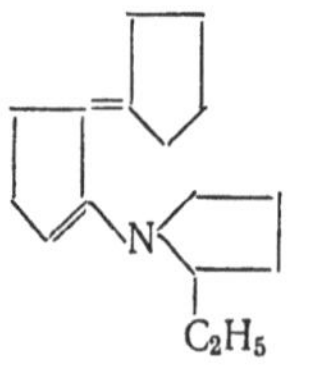

condensation product of two molecules of cyclopentanone with α-ethylpyrrolidone

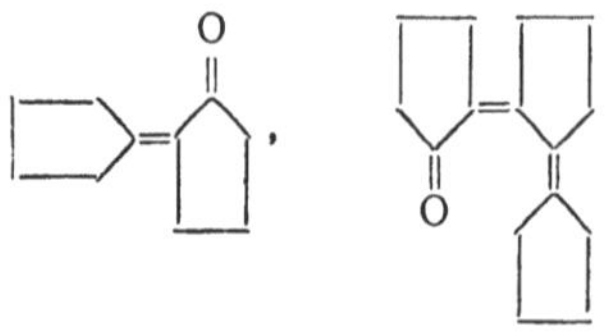

possible products of self-condensation of cyclopentanone

Complex cyclic structures have also been detected in the pyrolysis of N, N'-dibutyladipamide, modeling polyhexamethyleneadipamide [24].

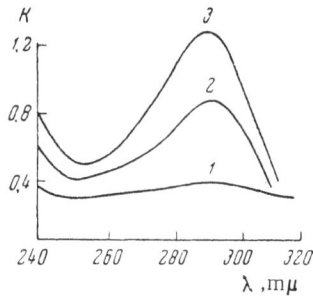

Fig. 109. Absorption spectra of 1% solutions of polyhexamethylene-dipamide in 60% sulfuric acid. 1) Initial sample; 2) after 4 hr of heating at 275°C; 3) after 8 hr of heating at 275°C; K) absorption coefficient.

At 290°C, derivatives of cyclopentanone, pyridine, and pyrrole of the following structure are formed:

— C_4H_9 2-n-butyl-cyclopentanone

C_4H_9 — — C_4H_9 2,5-di-n-butylcyclopentanone

— C_3H_7 2-n-propyl-3,4,5,6-bis-trimethylene-pyridine

— CH_3 2-methyl-N-butylpyrrole

C_4H_9

The formation of pyrrole rings as a result of the thermal destruction of polyamides is also indicated by the positive Erlich reaction (red color with N,N'-dimethylaminobenzaldehyde hydrochloride); the initial polyamide does not give this reaction [8, 9, 19, 24].

A number of authors [24, 25] have studied the absorption spectrum in the ultraviolet region for polyhexamethyleneadipamide, subjected to pyrolysis.

It has been shown [24] that the heating of polyhexamethyleneadipamide, as well as the model substance dibutyladipamide, at 290°C in an atmosphere of nitrogen leads to the appearance of a band at 280 mμ.

Polyhexamethyleneadipamide [25], preliminary heated at 275°C in an inert medium, gives an absorption band at 290 mμ in solution in 60% sulfuric acid. The spectra of the initial samples and samples heated for various periods of time, cited in Fig. 109, show that the intensity of this band increases with time of preliminary heating of polyhexamethyleneadipamide.

Attempts to assign the band at 290 mμ to cyclopentanone have given a negative result (the spectra of cyclopentanone exhibit a band at 2730 A on account of the carbonyl group). The band at 280-290 mμ may be explained by the formation of substituted indoles in the pyrolysis of polyhexamethyleneadipamide [24].

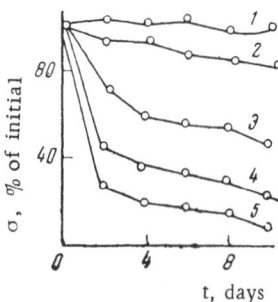

Fig. 110. Dependence of the breaking strength of polycapro-amide fiber on the duration of heating in air at various temperatures: 1) 80°; 2) 100°; 3) 120°; 4) 140°; 5) 160°C.

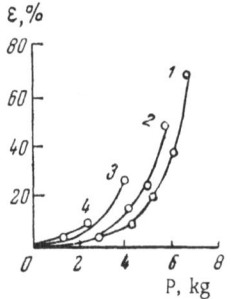

Fig. 111. Dependence of the stretching on the load for poly-amide film heated in air at 140°C. 1) Initial film; 2) after four weeks of heating; 3) after eight weeks of heating; 4) after 16 weeks of heating.

II. THERMAL OXIDATION
OF POLYAMIDES

The heating of polyamides in the presence of oxygen or air leads to considerable changes in their chemical composition and physicochemical properties. This is accompanied by a loss of valuable physicomechanical properties in the polyamide materials: decrease in the breaking strength and breaking elongation for fibers and films, intensification of brittleness.

The curves cited in Fig. 110 show [26] that polycaproamide fiber loses 0, 14.5, 50.8, 76, and 89.3% of its original strength as a result of eight-hour heating in air at 80, 100, 120, 140, and 160°C, respectively.

The curves of Fig. 111 [27] show a substantial reduction of the breaking strength and breaking elongation for polyamide film after heating at 140°C for a prolonged time (up to 16 weeks). Figure 112 represents curves of the dependence of the stress on the deformation for unoxidized capron film and capron film oxidized at 140°C for 8 hr. Such capron film losses 40% of the breaking strength in comparison with the initial value and practically does not stretch at all.

The thermograms [28] cited in Fig. 113 show a substantial difference in the thermal effects that occur during heating of poly-hexamethyladipamide fiber in nitrogen and in air. On both curves a weak endothermic effect, related to the removal of moisture, is detected around 100°C. An exothermic reaction begins in air at about 185°C, being interrupted at about 255°C by a weak endo-thermic effect, related to melting of the polymer. In nitrogen the exo-thermic reaction is entirely absent, which permits us to ascribe it to the oxidation of the polyamide.

When various polyamides [9, 29] (polycaproamide, polyhexamethyl-eneadipamide, polyenanthoamide, and the mixed polyamide obtained by polycondensation of hexamethyleneadipamide and ε -caprolactam, taken in a 40 : 60 ratio) are heated in the presence of oxygen, gels that swell

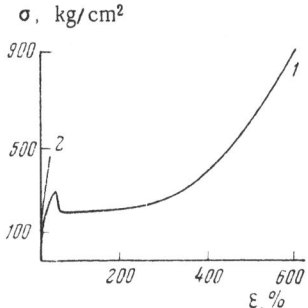

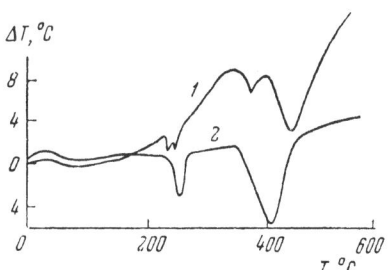

Fig. 112. Dependence of the stress (σ) on the deformation (ε). 1) Initial capron film; 2) film oxidized in air at 140°C for eight hours.

Fig. 113. Thermogram of polyhexamethyleneadipamide. 1) In air; 2) in nitrogen.

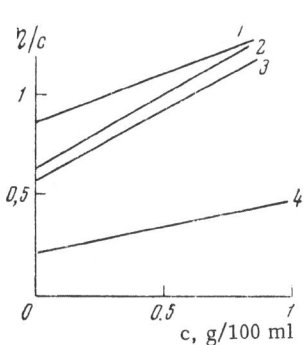

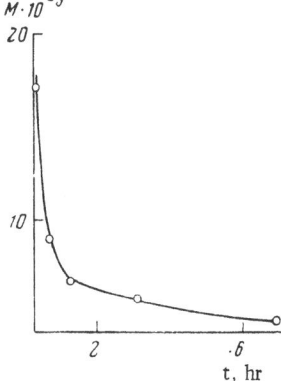

Fig. 114. Dependence of the viscosity in m-cresol on the concentration of the solution for the initial capron film (1) and film oxidized for 8 hr at 140°C (2), 170°C (3), and 200°C (4).

Fig. 115. Dependence of the molecular weight of polycaproamide, determined according to the viscosity in 85% formic acid, on the duration of heating in air at 200°C.

up in cresol are formed even at temperatures of 160-170°C; a nonswelling gel fraction is formed at 200-205°C. The viscosity of the soluble portion drops, while the Haggins constant increases (Fig. 114).

The curve of Fig. 115 [30] shows the dependence of the molecular weight of polycaproamide fiber on the time of heating in air at 200°C. After only 2 hr, the molecular weight of the polyamide, determined according to the viscosity in 85% formic acid, drops almost to half its value; thereafter, the molecular weight changes negligibly. As it follows from Fig. 116 [30], structuring processes occur together with the destruction processes; the amount of the fraction insoluble in 50% sul-

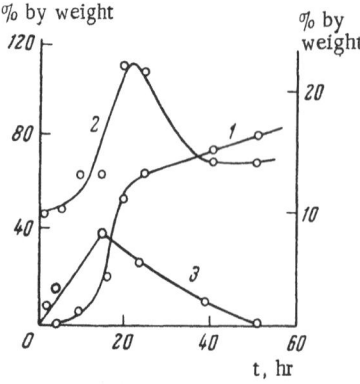

% by weight

Fig. 116. Amount of insoluble and
soluble fractions as a function of the
duration of heating of polycaproamide in air at 200°C. 1) (left-hand
scale) amount of product insoluble in
50% H_2SO_4; 2) (right-hand scale)
amount of product soluble in water;
3) (right-hand side) amount of product soluble in chloroform.

furic acid increases. After 10 hr the
insoluble fraction constitutes 4% of the
weight of the initial polycaproamide,
while after 20 hr it reaches 60%. The
amount of the substance soluble in water
and chloroform first increases, and then
drops.

The maximum amounts of water-
and chloroform-soluble products are
formed after 20 hr of heating, i.e., at
the time when the structuring processes
are substantially accelerated.

The chemical structure of the polyamide does not influence the general
direction of thermooxidation. Great influence is exerted by the degree of ori-
entation. Thus, for example, in weakly
oriented capron film, structuring processes occur at an appreciable rate at
only 160°C, while a nonswelling polymer
with a dense network is formed at 170°C
[29].

For highly oriented capron fiber, structuring is not observed even
in the case of oxidation for 8 hr at 170°C. At temperatures of 230-240°C,
polyhexamethyleneadipamide fiber gives an infusible, insoluble, and nonswelling black product after only an hour [30-33].

Side processes of purely thermal destruction and structuring [29]
take place at high temperatures together with the oxidative processes.
Thus, for example, caprolactam distills off when polycaproamide and
mixed polyamides containing ε-aminocaproic acid in the chain are oxidized (350°C). In this case the degree of polymerization of the polycaprolactam is sharply reduced, and no structuring occurs.

The mixed polyamide (anid G-669) is converted to a three-dimensional product at this same temperature. Such a behavior of polycaproamide is apparently explained by the fact that at high temperatures destruction of the caprolactam polymer predominates over thermal and
thermooxidative structuring (at comparatively low temperatures – below
200°C – capron is structured at an even greater rate than anid G-669).

Extremely interesting is a study of the oxidation of polycaproamide
by means of the ultraviolet absorption spectra of solutions of polycaproamide in 50% H_2SO_4 [30, 34]. Figure 117 shows the variation of the spec-

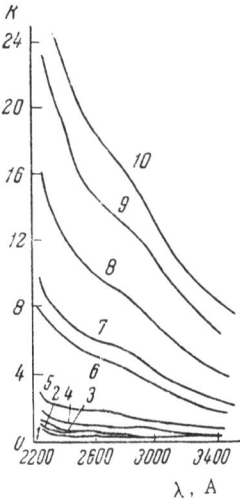

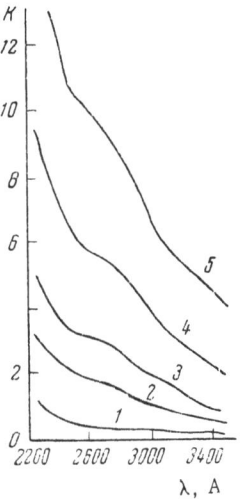

Fig. 117. Absorption spectra of solutions of polycaproamide in 50% H_2SO_4 after heating at 200°C for various periods of time; K) Absorption coefficient. 1) Unheated sample; 2) after heating for 8 hr under vacuum; 3) after heating for 8 hr at a nitrogen pressure of 100 mm Hg; 4) 8 hr at an oxygen pressure of 10 mm Hg; 5) 8 hr at an oxygen pressure of 30 mm Hg; 6) 8 hr at an oxygen pressure of 50 mm Hg; 7) 8 hr at an oxygen pressure of 100 mm Hg + 20 mm Hg of water vapors; 8) 7 hr at an oxygen pressure of 200 mm Hg; 9) 7 hr at an oxygen pressure of 200 mm Hg; 10) 8 hr at an oxygen pressure of 200 mm Hg.

Fig. 118. Absorption spectra of solutions of polycaproamide oxidized for 8 hr at an oxygen pressure of 100 mm Hg, in 50% H_2SO_4 at various temperatures: 1) 120°; 2) 140°; 3) 160°; 4) 180°; 5) 200°C; K) absorption coefficient.

tra in the region of 2200-3400 A during heating of polycaproamide fiber with a molecular weight of 16,500 at 200°C under vacuum and oxygen. When the oxygen pressure is increased, the absorption increases substantially within the entire investigated region, especially sharply in the region of 2200-2400 A.

Figure 118 presents the absorption spectra of solutions of polycaproamide fiber heated for 8 hr in the presence of oxygen at various temperatures. Here, as in the case of differing oxygen pressures, a quantitative correlation exists between the intensity of the absorption in the region of 2200-3400 A and the oxidation temperature.

Figure 119 shows that a direct relationship exists between the variation of the spectra and the amount of absorbed oxygen.

Apparently the oxidized polycaprolactam contains a group or groups of atoms with characteristic absorption in the region of 2200-3400 A.

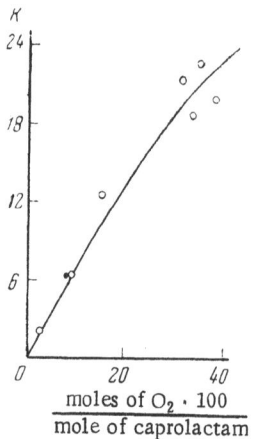

Fig. 119. Dependence of the absorption coefficient (K) at 2400 A on the amount of absorbed oxygen.

The absorption in this region can serve as a sensitive measure of the degree of oxidation of polycaproamide.

The curves of Fig. 120 show the variation of the absorption coefficient at 2400 A with time for polycaproamide fiber, oxidized in air at various temperatures. For small oxygen pressures, the dependence of the rate of the oxidation of polycaprolactam on the oxygen pressure to the power 0.75 is expressed by a straight line (Fig. 121). This permits us to derive an empirical equation for the dependence of the rate of absorption of oxygen on the pressure [30]:

$$-\frac{d[O_2]}{dt} = k p_{O_2}^{n}$$

where n = 0.75. The straight lines in Fig. 122, expressing the dependence of the logarithm of the maximum rates on 1/T for various oxygen pressures, show that when the oxygen pressure increases, not only the oxidation rate, but also the temperature coefficient of the rate of the process increases [30].

The kinetic curves of the oxidation of polyamides, cited in Fig. 123 exhibit induction periods only at low temperatures or small partial pressures of oxygen [35]. The same pattern was also detected in an investigation of the oxidation of polycaproamide by means of the ultraviolet spectra [30].

It was shown in [93] that water, carbon dioxide, carbon monoxide, acetaldehyde, formaldehyde, and methanol are formed in the oxidation of polycaproamide at 190°C. Nitrogen, ammonia, and volatile amines were not detected. In the oxidation of one unit of polycaproamide (113 g) in 1 hr, 0.12 mole of oxygen is absorbed, and the following are liberated (in moles): water, 3.5×10^{-2}; CO_2, 2.2×10^{-2}; CO, 0.98×10^{-2}, acetaldehyde, 3×10^{-4}; formaldehyde, 3×10^{-4}; methanol, 2.7×10^{-4}.

The authors propose a mechanism for the thermal oxidation of polyamides, which gives a good explanation for the formation of all these products. Initiation of the chain process of oxidation occurs by stripping of the most labile hydrogen atom by a molecule of oxygen.

According to studies of the photooxidation and radiation irradiation of low-molecular amines and polyamides [38, 95, 96], the most labile hydrogen atom is situated at the carbon atom next to the NH– group.

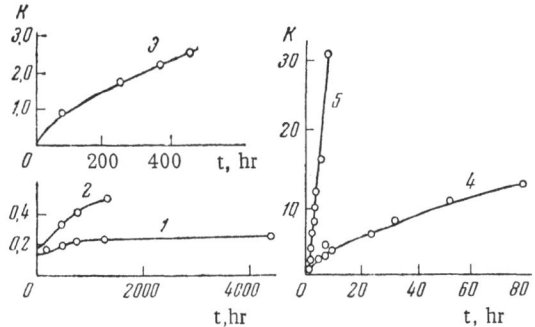

Fig. 120. Variation of the absorption coefficient (K) at 2400 A of solutions of polycaproamide in 50% H_2SO_4 as a function of the time of its oxidation in air at various temperatures: 1) 50°; 2) 60°; 3) 100°; 4) 150°; 5) 200°C.

$$\sim CH_2CONHCH_2CH_2\sim \; + \; O_2 \; \rightarrow \; \sim CH_2CONH\overset{\cdot}{C}HCH_2\sim \; + \; HO_2^{\cdot}$$

Then a peroxide radical and hydroperoxides are formed at the site of stripping of the hydrogen atom. Water is obtained when the hydroperoxides decompose (see p. 16). In the thermal oxidation of polyamides, the liberation of water may lead to hydrolysis of the polymer and an increase in the number of terminal carboxyl groups. Decarboxylation gives carbon dioxide. In addition to decomposition of hydroperoxides, decomposition of peroxide radicals may also occur:

$$\sim CH_2CONH\underset{\underset{O-O\cdot}{|}}{C}H - CH_2 - CH_2 \sim \; \rightarrow \; \sim CH_2CONH\overset{\overset{O\dotplus O}{\overset{|\;\;|}{}}}{\underset{\cdot}{C}}HCH_2-$$

$$\rightarrow \; \sim CONHC\underset{H}{\overset{\diagup O}{\diagdown}} \; + \; \overset{O\cdot}{\underset{|}{CH_2}} - CH_2 \sim$$

$$\qquad I \qquad\qquad\qquad II$$

Carbon monoxide can be obtained in the decomposition of compound I:

$$\sim CH_2CONHC\underset{H}{\overset{\diagup O}{\diagdown}} \; \rightarrow \; \sim CH_2CONH_2 + CO$$

Decomposition of the radical II produces formaldehyde:

$$\overset{O\cdot}{\underset{|}{CH_2}} - CH_2\sim \; \rightarrow CH_2O + \overset{\cdot}{C}H_2 - CH_2\sim$$

$$\qquad\qquad III$$

Isomerization and further transformations of the radical III leads to the formation of acetaldehyde:

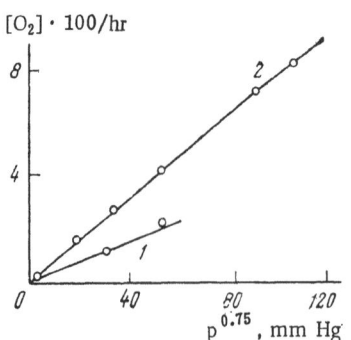

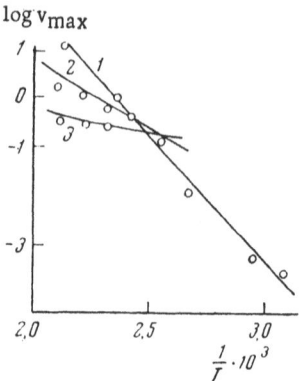

Fig. 121. Rate of absorption of oxygen at various pressures (p) and various temperatures: 1) 160°C; 2) 200°C; [O_2]) moles of O_2 per mole of caprolactam.

Fig. 122. Dependence of the maximum rate of oxidation on the temperature in the coordinates log v_{max} vs. 1/T at various temperatures: 1) 30; 2) 100; 3) 156 mm Hg.

$$\cdot CH_2 - CH_2 - CH_2 \sim\ \rightarrow CH_3 - \overset{\cdot}{C}H - CH_2\ \sim\ \overset{O_2}{\rightarrow}\ CH_3 - \overset{\overset{\displaystyle O - O\cdot}{|}}{C}H - CH_2 \sim$$

$$\rightarrow CH_3 - \underset{\cdot}{\overset{\overset{\displaystyle O + O}{|\ \ \ |}}{C}}H\quad CH_2\ \sim\ \ \rightarrow CH_3 - C\overset{O}{\underset{H}{\diagdown}}\ \ + CH_2 - CH_2 \sim$$

Radical III can add oxygen and isomerize according to a different scheme:

$$\overset{\cdot}{C}H_2 - CH_2 - CH_2 \sim\ \rightarrow\ \underset{O \overset{|}{\underset{|}{+}} O}{\overset{\overset{\displaystyle O - O\cdot}{|}}{C}H_2 - CH_2 - CH_2 \sim}\ \rightarrow\ \underset{O\diagdown}{\overset{\overset{\displaystyle O - O}{|\qquad |}}{\cdot CH_2\qquad CH_2 - CH_2 \sim}}$$

$$\rightarrow\ CH_3\quad \cdot CH - CH_2 \sim\ \rightarrow CH_3O\cdot\ +\quad \underset{H}{\overset{O\diagdown}{\diagup}}C - CH_2 \sim$$

$$\downarrow RH$$
$$CH_3OH$$

Thus, all the volatile products detected by the authors in the thermal oxidation of polycaproamide are produced as a result of the decomposition of peroxides and peroxide radicals.

The induction periods and S-shaped form of the kinetic curves of oxidation at low temperatures and low oxygen pressure, and the detection in the oxidation products of water, acetaldehyde, formaldehyde, and methanol, formed in the decomposition of peroxides and peroxide radicals – all this indicates an autoaccelerated character of the process of oxidation of polyamides with degenerate branches to hydroperoxides.

Polyamides heated in air at 80-90°C for prolonged periods exhibit an intense reaction for peroxides. At higher temperatures the peroxides are rapidly decomposed, and they cannot be detected [29].

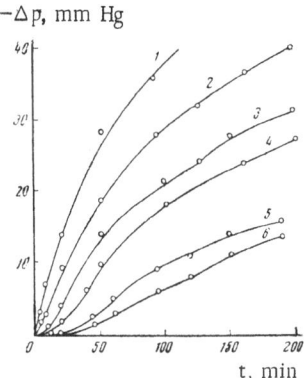

−Δp, mm Hg

t, min

Fig. 123. Variation of the oxygen pressure as a function of the duration of thermooxidative destruction of polyhexamethylenesebacamide at various temperatures and oxygen pressures: 1) 200°C, 200 mm Hg; 2) 180°C, 200 mm Hg; 3) 160°C, 200 mm Hg; 4) 200°C, 100 mm Hg; 5) 140°C, 200 mm Hg; 6) 200°C, 50 mm Hg.

CO and N_2 have been detected among the gaseous products obtained in the oxidation of polycaproamide at 200°C [30].

Many authors have investigated the causes of the yellowing of polyamides [36, 37]. Polyhexamethyleneadipamide, heated in air, yellows to a substantially greater degree than when it is heated in nitrogen under vacuum [5, 36]. Photocolorimetric determination of the color change of polyhexamethyleneadipamide during heating [36] showed that it yellows after 10 hr at 110°C in oxygen, but not under vacuum or in nitrogen. In oxygen containing water, the coloration decreases. Acetylation of the terminal amino groups increases the stability of the fiber to yellowing. Polyhexamethyleneadipamide treated with acetic anhydride remains white when heated at 140°C in oxygen. Only yellowed polyhexamethyleneadipamide gives a red color, denoting the presence of the pyrrole ring, with Ehrlich's reagent.

The yellow destruction products are apparently the result of interaction of amino groups with the oxidation products of polyamides, resulting in the formation of various pyrrole derivatives [36].

Decomposition of the hydroperoxide can lead to the formation of methylol products [29]:

$$RCHNHCOR \rightarrow RCHNHCOR \rightarrow RCHNHCOR$$
$$\underset{O-OH}{|} \qquad \underset{O\cdot}{|} \qquad \underset{OH}{|}$$

Cross-links can be obtained in the recombination of macroradicals:

$$\begin{array}{cc} R-CH-NH-COR & R-CH-NH-COR \\ | & | \\ R-CH-NH-COR & O \\ & | \\ & R-CH-NH-COR \end{array}$$

At temperatures above 240°C, three-dimensional structures can form on account of the reaction with the methylol portions of the oxide chains:

$$\underset{R-CH-NHCOR}{\overset{OH}{|}} + R-NHCO-R \rightarrow \underset{R-N-COR}{\overset{R-CH-NHCOR}{|}} + H_2O$$

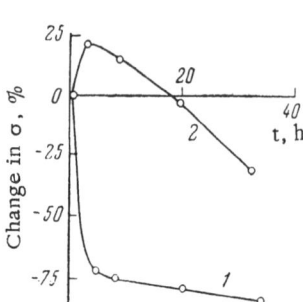

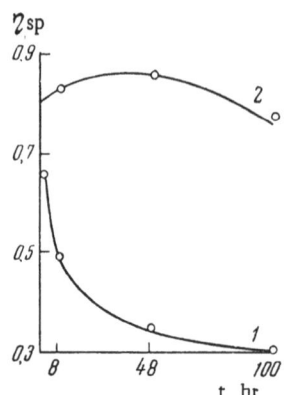

Fig. 124. Variation of the breaking strength as a function of the
duration of heating in air at 180°C.
1) Capron fiber; 2) capron fiber
stabilized by 0.05% CuCl₂.

Fig. 125. Variation of the viscosity of 0.5% solutions in tricresol as a function of the duration of heating in air at 150°C.
1) Capron fiber; 2) capron fiber
with an addition of DNPDA (the
additive was introduced by powdering).

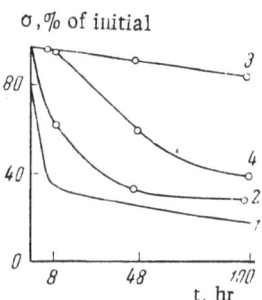

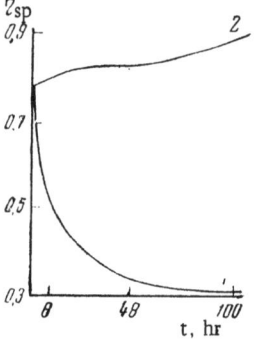

Fig. 126. Variation of the
strength of capron fibers powdered with DNPDA, as a function
of the duration of heating. 1)
Initial fiber heated at 150°C;
2) the same, at 180°C; 3) with an
addition of DNPDA, heated at
150°C; 4) the same at 180°C.

Fig. 127. Variation of the viscosity of 0.5% solution in tricresol as a function of the duration of heating in air at 150°C.
1) Capron fibers; 2) capron fiber
with an addition of DNPDA (the
additive is introduced during the
process of polymerization of caprolactam).

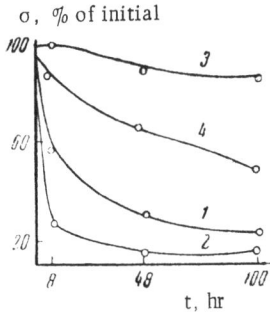

σ, % of initial

Fig. 128. Variation of the strength of capron fibers as a function of the duration of heating (DNPDA introduced during the process of polymerization of caprolactam). 1) Initial fiber, heated at 150°; 2) the same, at 180°; 3) with an addition of DNPDA, heated at 150°; 4) the same at 180°C).

The methylol products can also decompose, forming molecules with terminal aldehyde and amide groups, which can also give branching:

$$R-CH-NH-COR \rightarrow RC\begin{subarray}{l}O\\\diagup\\\diagdown\\H\end{subarray} + NH_2-C\begin{subarray}{l}O\\\diagup\\\diagdown\\R\end{subarray}$$
$$\quad\quad\mid$$
$$\quad OH$$

and further:

$$RC\begin{subarray}{l}O\\\diagup\\\diagdown\\H\end{subarray} + \begin{subarray}{l}RNHCOR\\ \\RNHCOR\end{subarray} \rightarrow \begin{array}{l}R-N-COR\\\mid\\R-CH\\\mid\\R-N-COR\end{array}$$

III. STABILIZATION OF POLYAMIDES AGAINST THERMAL OXIDATION

The patent literature contains very many indications of the possibility of stabilizing polyamides against thermal oxidation. However, the literature contains almost no research works discussing the mechanism of the action of antioxidants for polyamides.

A number of survey works consider the action of inorganic [39-41], as well as organic [35, 42-44] additives to polyamides.

In all these works a comparison was made of the mechanical properties of stabilized and unstabilized polyamide fibers, subjected to thermal oxidation.

Figure 124 [41] shows the variation of the breaking strength (in % of the original) of capron fiber, stabilized by a small addition of $CuCl_2$, and unstabilized capron fiber during heating in air at 180°C. During the initial period of heating (up to 18 hr), even an increase in the breaking strength is observed for the stabilized fiber, while for the unstabilized fiber the strength drop by almost 75% after only 2 hr.

A large number of antioxidants of organic structure, of the type of amines and phenols, were investigated in [42, 43]. The additives were introduced by powdering the capron in the form of crumbs before spinning of the fiber, as well as into the fused caprolactam before polymerization. The antioxidants belonging to the amine class, such as N,N'-di-β-naphthyl-p-phenylenediamine, N,N-phenylcyclohexyl-p-phenylenediamine, N,N'-diphenyl-p-phenylenediamine, and phenyl-β-naphthylamine, preserve the strength of the fiber by 70-95% when it is heated for 2 hr

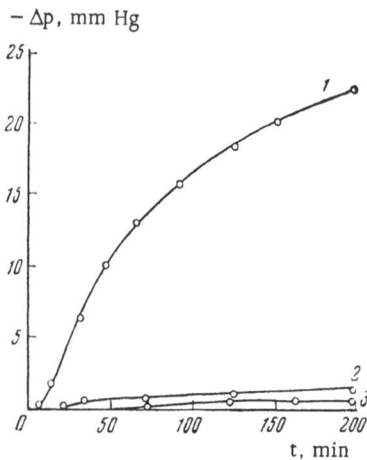

Fig. 129. Variation of the oxygen pressure as a
function of the duration of thermooxidative destruction of polyhexamethylenesebacamide at
160° and an oxygen pressure of 200 mm Hg. 1)
Polyhexamethylenesebacamide without inhibitors;
2) polyhexamethylenesebacamide in the presence
of 0.5% DNPDA; 3) in the presence of 0.5% neozone D.

at 200°C. Under these conditions, the unstabilized fiber retains only 22%
of its strength. Inhibitors of the phenol type (2,2-methylene-bis-4-
methyl-6-tert-butylphenol, 2,6-di-tert-butyl-4-methylphenol, etc.) give
essentially no protective effect when the fiber is heated in an air medium.

Aromatic diamines: N,N'-di-β-naphthyl-p-phenylenediamine (DNPDA)
and N,N'-phenylcyclohexyl-p-phenylenediamine, proved most effective.
Figure 125 and 126 show the variation of the viscosity of 0.5% solutions
in tricresol and the variation of the strength of capron fibers, stabilized
by DNPDA and unstabilized during prolonged heating (the stabilizer is introduced by powdering). Figures 127 and 128 show the same dependences
after the introduction of DNPDA before polymerization. Independent of
the method of introduction, DNPDA is a rather effective stabilizer, which
substantially increases the thermal stability of capron fiber.

In the stabilization of polyhexamethylenesebacamide, aromatic
amines also proved most effective: DNPDA and phenyl-β-naphthyl-
amine (neozone D).

The kinetic curves of the absorption of oxygen, cited in Fig. 129,
show a substantial deceleration of the process of oxidation for hexamethylenesebacamide, stabilized with DNPDA and neozone D [35].

Mixed polyamides [44], produced by polycondensation of hexamethyleneadipamide, hexamethylenesebacamide, and caprolactam (nylon 66/610/6),

as well as a polymer with aromatic rings in the chain (nylon 66/6-F/6; F – residue of isophthalic acid) are well stabilized by various amines (di-β-naphthyl-p-phenylenediamine, phenyl-β-naphthylamine, diphenyl-p-phenylenediamine); phenols are also relatively ineffective in this case.

Mixed polyamides containing radicals of the cyclic diamine piperazine (nylong 66/610/pip-10 and nylon 66/610-pip-6) are stabilized with considerably more difficulty that nylon 66/610/6 and nylon 66/6-F/6.

Certain compounds that improve the light stability of polyamides are also stabilizers against thermal oxidative destruction. For example, hydroxyphenylbenzoxazole, which exhibits an appreciable effect as a photostabilizer [45], is also an inhibitor of thermal oxidation [42].

In [94] the kinetics of the oxidation of polyamides stabilized by phenyl-β-naphthylamine was compared with the stabilizer consumption. The oxidation process becomes appreciable when more than half (at 200°C) of the initial amount of phenyl-β-naphthylamine remains unconsumed in the polymer. A considerable content of the inhibitor at the beginning of appreciable oxidation of the polyamide leads to a substantial deceleration of the absorption of oxygen. Hence the effectiveness of stabilizers of the type of aromatic amines should be evaluated not according to the value of the induction period, but according to the decrease in the rate of absorption of oxygen. It is interesting to note that stabilizers of the type of free stable radicals are almost entirely consumed during the induction period (Fig. 129).

1.   Some Stabilizers of Thermal Oxidation for Polyamides

Inorganic Stabilizers

Among the inorganic inhibitors of oxidation, the greatest number of patents belong to copper and its compounds, to bromides and iodides of the alkali metals and to various inorganic and organic derivatives of phosphoric acids.

Two-component and three-component mixtures of these compounds exhibit a synergic effect.

1.   One-Component Stabilizers:

a) metallic copper, mono- and divalent copper salts of inorganic and organic acids (copper naphthenate is especially effective) [46];

b) bromides and iodides of the alkali metals [47];

c) orthophosphoric, pyrophosphoric, phosphorous acids and their aryl esters in amounts of 0.3-1% by weight [48]; hypophosphorous acid,

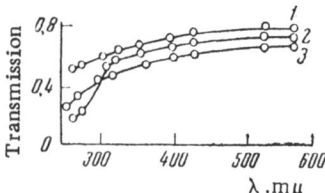

Fig. 130. Spectra of various polyamides. 1) Polyhexamethylenesebacamide; 2) polyhexamethyleneadipamide; 3) polycaproamide.

sodium hypophosphite, diphenyl phosphinate, butylphenyl phosphinate, etc.,
[49];

d) metals capable of being oxidized:
Mg, Al, Sr, Ca [50].

2. Two-Component Mixtures:

a) copper salts (copper acetate,
CuCl$_2$, CuSO$_4$) + a salt of an organic base
(benzyltrimethylammonium iodide, mor-
pholinium iodide, tetra-n-butylammoni-
um iodide, trimethylammonium bromide)
[51];

b) copper salt + halogen derivative of NH$_4^+$, alkali and alkaline earth
metals [52];

c) phosphorus compound with inorganic halide (H$_3$PO$_3$, NaH$_2$PO$_4$,
methyl phosphite, butyl phosphite, butyl phosphate with KI, KBr) [53].

3. Three-Component Mixtures:

a) copper compound, halogen compound, and phosphorus compound
[54];

b) mixtures including an organic antioxidant (for example, phosphorus compound, halogen compound, and 2-mercaptobenzimidazole [55]).

The mechanism of the action of inorganic additives to polyamides
has so far not been studied at all. The possibility of the formation of a
complex and a chelate compound of copper with the amide groups, leading to an increase in their stability, is proposed [40]:

$$\sim CH_2CH_2C - N - CH_2CH_2 \sim$$
$$\underset{O \ldots Cu \;\; O}{\overset{\parallel \quad\;\; |}{}}$$
$$\sim CH_2CH_2 - N - C - CH_2CH_2 \sim$$

Phosphorus compounds may bond various impurities contained in the
polyamides, which might catalyze the process of polyamide destruction
at high temperatures [39].

Organic Stabilizers

The usual antioxidants, also used for other polymers, of the type of
aromatic amines and phenols, as well as certain special inhibitors, are
recommended as organic stabilizers for polyamides.

TABLE 18. Variation of the Physicochemical Properties of Capron Film during Its Irradiation by Ultraviolet Light under Vacuum

Duration of irradiation, hr	Solubility, %	$[\eta]$	Haggins constant
0	100	0.930	0.34
2	100	0.957	0.61
5	98	1.160 •	0.45
10	87	1.178 •	0.51
120	69	—	—

• Data for the soluble part of the polymer.

1. Aromatic Amines: β-naphthylamine, phenyl-α-naphthylamine, diphenylguanidine, phenthiazine [3]; phenyl-β-naphthylamine [35], N,N'-diphenyl-p-phenylenediamine [56]; N,N'-di-β-naphthyl-p-phenylenediamine [42, 44]; N,N'-phenylcyclohexyl-p-phenylenediamine [35, 42].

2. Aromatic Hydroxy Compounds:

a) containing one OH group: β-naphthol, dibenzylphenol [57]; 2,6-di-tert-butyl-p-cresol [58];

b) containing several OH groups and cyclohexyl groups: cyclohexylpyrocatechol, dicyclohexylpyrocatechol, cyclohexylhydroquinones, dicyclohexylpyrogallol [59];

c) halogenated di-(hydroxyphenyl)methanes [60], di-(2-hydroxy-4-chlorophenyl)methane

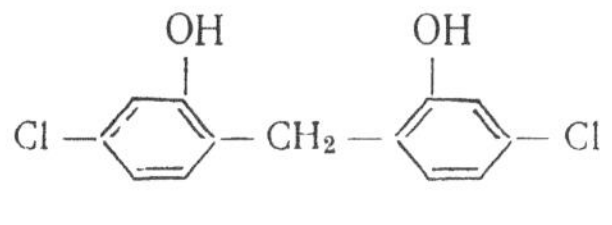

di-(4-hydroxy-3,5-dibromophenyl)methane, etc.

3. Amines Possessing Condensed Rings, which may contain substituents (alkyl, alkoxy, amino, or OH groups) [61].

4. Aliphatic Substances Containing Several Hydroxyl Groups: hydroxy acids, glycols (mixture of stearic acid and glycol, ω-hydroxycaproic acid) [62].

IV. PHOTOAGING OF POLYAMIDES

Objects made from polyamides (fibers, films, etc.) change properties sharply under the action of ultraviolet rays: they become brittle, infusible, turn yellow.

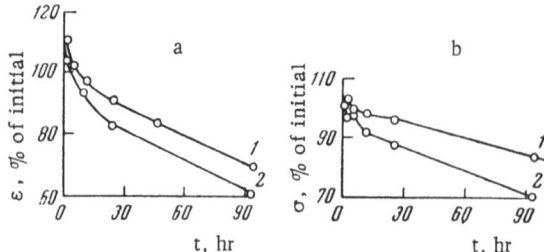

Fig. 131.   Variation of the breaking elongation (a) and breaking strength (b) of polyamide fibers as a function of the time
of irradiation in an atmosphere of nitrogen. 1) Polyhexamethyleneadipamide; 2) polycaproamide.

The action of heat and oxygen is intensified during simultaneous ir-
radiation [17, 18, 63-66]. The $-\overset{\shortmid}{\underset{\underset{\mathrm{O}}{\|}}{C}}-\overset{\shortmid}{\underset{\underset{\mathrm{H}}{\shortmid}}{N}}-$ bond is the weakest in the

chain and requires 53 kcal/mole for its cleavage. The energy of the

$-\overset{\shortmid}{\underset{\shortmid}{C}}-\overset{\shortmid}{\underset{\shortmid}{C}}-$ bond is equal to 80 kcal/mole. The amide bonds $-\overset{\shortmid}{\underset{\underset{\mathrm{O}}{\|}}{C}}-\overset{\shortmid}{\underset{\underset{\mathrm{H}}{\shortmid}}{N}}-$
should be broken first.

Chemical changes occur during irradiation if the incident light rays
are absorbed by the substance. The spectra of polycaproamide, polyhexamethyleneadipamide, and polyhexamethylenesebacamide, cited in
Fig. 130, show that waves shorter than 350 mμ are most sensitive for
polyamides, since there is little absorption above 350 mμ [67, 68].

When polyamides are irradiated, the viscosity of the solutions decreases [16, 69]. The amount of the substance that dissolves in water
(determined after 4 hr extraction in a Soxhlet apparatus [69]) increases
approximately in proportion to the time of irradiation of polycaproamide, which can be explained by cleavage of the molecular chains according to the laws of change.

Under the action of short-wave (253.7 mμ) and long-wave (300-
400 mμ) radiation on various polyamides (polycaproamide, polyhexamethyleneadipamide, and polyhexamethylenesebacamide), in all cases of
irradiation (in nitrogen, air, and oxygen), products incompletely soluble
in m-cresol are obtained [67, 68].

Table 18 presents data obtained in the irradiation of capron film
under vacuum by the full spectrum of the PRK-2 lamp at 30°C [70].

The increase in the characteristic viscosity of the solution and increase in the Haggins constant indicate the formation of branched struc-

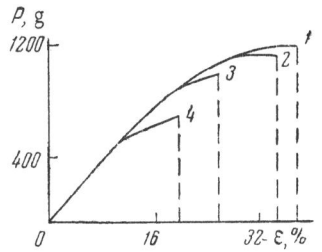

Fig. 132. Dependence of the load on the elongation for nonirradiated and irradiated polyhexamethylene-adipamide fiber. 1) Nonirradiated fiber; 2) irradiated in nitrogen for 96 hr at 40°C; 3) the same, in air; 4) the same, in oxygen.

tures and an increase in the average value of the molecular weight. After 5 hr a gel fraction is formed, the amount of which increases upon further irradiation.

Irradiation of the mixed polyamide G-669 (copolymer of caprolactam, hexamethyleneadipamide, and hexamethylene-azelaamide) in the presence of oxygen leads to an appreciable reduction of the viscosity of the solution [70, 71]. However, determination of the molecular weight by the method of light scattering for the polyamide irradiated for 50 hr at 70°C showed that the average value of the molecular weight increases from 13,000 to 21,300, while the characteristic viscosity decreases from 0.384 to 0.270. Such a lack of agreement of the change in the molecular weight determined by the method of light scattering and the characteristic viscosity is explained by the formation of more symmetrical branched molecules. Thus, during photolysis (irradiation under vacuum) and during photooxidation, the molecular weight increases, which leads to the formation of insoluble three-dimensional structures in the case of irradiation under vacuum.

The change in the mechanical properties of polyamides depends greatly on the conditions of irradiation. Thus, in the case of photolysis in a medium of dry nitrogen under the action of short-wave radiation (2537 A), the curves cited in Fig. 131 were obtained, showing a significant drop in the breaking elongation and breaking strength during irradiation of the polyamides at 40°C [68]. For polycaproamide fibers, the breaking elongation drops by 42% as a result of 92 hr irradiation, while for polyhexamethyleneadipamide, the value drops by 30%. In this case the breaking strength for polycaproamide drops by 28%, while that for polyhexamethyleneadipamide drops by 10%.

Figure 132 shows that irradiation of polyamide fibers by the spectrum in the region of 300-400 mμ in an atmosphere of dry nitrogen gives very negligible changes in the mechanical properties. In air and oxygen under such irradiation, the mechanical properties deteriorate sharply [68]. Such an absence of significant changes in the mechanical properties of polyamides during irradiation by ultraviolet light without oxygen is also noted in other studies [15, 63, 70].

In the photolysis and photooxidation of polyamides, the number of terminal carboxyl and amino groups changes, the shape of the kinetic curves differing for the content of $-COOH$ and $-NH_2$ groups.

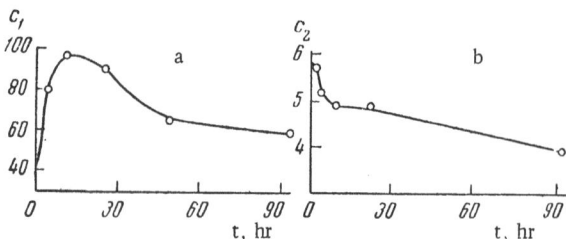

Fig. 133. Variation of the concentration of terminal NH_2 groups (a) and COOH groups (b) as a function of the time of irradiation of polycaproamide in an atmosphere of nitrogen. c_1) Concentration of NH_2 groups, equiv $\cdot$ 10^{-6}/g; c_2) concentration of COOH groups, equiv $\cdot$ 10^{-5}/g.

The dependence of the variation of the concentration of terminal amino groups on the time of irradiation in nitrogen, cited in Fig. 131, is expressed by a curve passing through a maximum. The content of carboxyl groups, as it follows from Fig. 133, drops the entire time in this case [68]. The lack of correspondence in the variation of the numbers of terminal groups: increase in the content of amino groups during photolysis and decrease in carboxyl groups, is apparently explained by

the following. The cleavage of an amide bond $-\overset{\displaystyle |}{\underset{\displaystyle O \ \ H}{C-N}}-$ leads to the for-

mation of radicals $-\overset{\displaystyle C^{\cdot}}{\underset{\displaystyle O}{\|}}$ and $-\overset{\displaystyle N^{\cdot}}{\underset{\displaystyle H}{|}}$, the radical $-N^{\cdot}H$ giving an amino

group as a result of secondary processes, while the radical $-\overset{\displaystyle C^{\cdot}}{\underset{\displaystyle O}{\|}}$ does not

lead to the formation of a carboxyl group. The decrease in the number of carboxyl groups can be explained by decarboxylation [71].

In the photooxidation of polyamides, a different path of the curves of the content of terminal groups as a function of the time of irradiation is observed [72]. As is shown in Fig. 134, the content of carboxyl groups increases along an S-shaped curve, while the content of amino groups passes through a maximum for polycaproamide, and decreases during the first four hours of irradiation, subsequently remaining unchanged, for polyhexamethyleneadipamide.

The increase in the number of carboxyl groups is explained by their formation as a result of oxidative processes in the presence of oxygen. The number of amino groups can decrease as a result of their interaction with the aldehyde groups formed during photooxidation [70, 71].

When polyamides are irradiated by ultraviolet light under vacuum [68, 71], as well as in the presence of oxygen [71], the samples turn

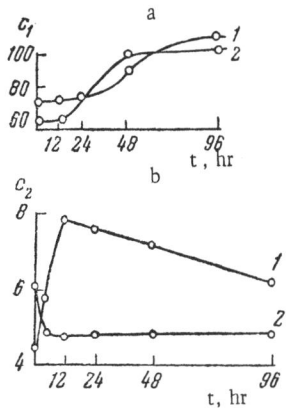

Fig. 134. Variation of the concentration of COOH groups (a) and
NH_2 groups (b) as a function of the
time of irradiation in oxygen. 1)
Polycaproamide; 2) polyhexamethyleneadipamide; c_1) concentration
of COOH groups, equiv $\cdot$ 10^{-6}/g;
c_2) concentration of NH_2 groups,
equiv $\cdot$ 10^{-5}/g.

yellow, the color intensity increasing
with increasing duration of irradiation.
The appearance of a yellowish-brown
color is apparently explained by the formation of hetero-chain compounds of the
pyrrole series, as in the case of the
thermal destruction of polyamides [71].
Actually, samples that have turned yellow during irradiation give a positive
Erlich reaction for pyrrole rings. The
irradiation of polyamides in air and in
oxygen not by the entire spectrum of the
PRK-2 lamp, but only by the near ultraviolet, does not lead to yellowing [68,
71]; in this case the Erlich reaction is
negative.

It is interesting that samples irradiated under these conditions turn
yellow after prolonged storage in the
dark or after brief heating. Thus, for
example [68], a sample irradiated in
the dark for 200 hr turns yellow after
several weeks of storage in the dark at 20°C or after 2 hr heating at
100°C. The color of fibers not preliminarily irradiated does not change
in this case. The author of [68] cites no explanation for this phenomenon.

An absorption band in the region of 2870 A is detected in the ultraviolet spectra of polycaproamide films that have turned yellow as a result of irradiation and give a positive Erlich reaction [74, 75]. The authors relate this to the appearance of heterocyclic compounds of the
pyrrole series in the irradiated films.

Films irradiated by near ultraviolet in the presence of oxygen and
containing no compounds of the pyrrole series give spectra possessing
no absorption band at 2870 A.

In the work of R. Ford [72], the ultraviolet spectra in the region
of 250-350 mμ of polyamide films irradiated in air by the unfiltered light
of a mercury lamp were studied. Films of a mixed polyamide, produced by joint condensation of ε-caprolactam, hexamethyleneadipamide,
and hexamethylenesebacamide (nylon 6/66/610) were investigated. Figure
135 presents the spectra of irradiated and nonirradiated polyamide films.
The initial nonirradiated film (spectrum 1) absorbs in the region of wavelengths shorter than 2400 A (this absorption is ascribed to the amide

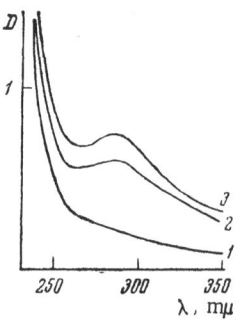

Fig. 135. Spectra of films of the copolymer of polycapro-amide, polyhexamethylene-adipamide, and polyhexamethyl-enesebacamide. 1) Before ir-radiation; 2) directly after ir-radiation in air for 80 hr; 3) ir-radiated film 2 after storage in the dark for 24 hr; D) optical density.

bonds) [25, 75, 76]. In the case of irradia-tion (spectrum 2) a band appears around 2900 A; the absorption in the entire region of 250-350 mμ is also somewhat increased. When the irradiated films are stored in the dark, the absorption at 2900 A continues to increase. Spectrum 3 of Fig. 135 pertains to irradiated film after its storage in the dark for 24 hr, before the cessation of the change in the absorption intensity.

Figure 136 shows the variation of the optical density of the film at 2900 A during irradiation and storage in the dark. The in-tensity of the band at 2900 A increases during storage in the dark, at an initial rate great-er than the rate of change of the intensity of this band directly before the light was turned off. After the following turning on of the light, the band intensity drops to values cor-responding to the intensity before the stor-age of the film in the dark.

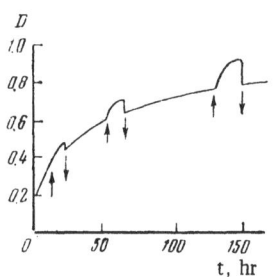

Fig. 136. Variation of the op-tical density (D) at 2900 A for polyamide film during irradia-tion and storage in the dark. Arrows pointing up, light turned on; arrows pointing down, light turned off.

According to the data of Table 19, the increase in the optical density of the band at 2900 A during storage of the film in the dark (Δ_1) proves to be directly proportional to the increase in the optical density of this band at the moment when the light is turned off, in comparison with the optical density of the nonirradiated film (Δ_2).

These experimental data show that the ratio $\Delta_2 / \Delta_1 = R$ is approximately constant for various periods of irradiation of the film. The author explains all these facts by the follow-ing scheme of the radical chain process of polyamide photooxidation:

$$A + h\nu_1 \xrightarrow{k_{\varepsilon_1}} B^{\cdot} + C^{\cdot} \quad \text{at the rate} \quad w_1 = k_{\varepsilon_1} J_1 \qquad (1)$$

$$B^{\cdot} + O_2 \xrightarrow{k_2} BO_2^{\cdot} \quad \text{at the rate} \quad w_2 = k_2 [B^{\cdot}] [O_2] \qquad (2)$$

$$BO_2^{\cdot} + h\nu_3 \xrightarrow{k_{\varepsilon_3}} B^{\cdot} + O_2 \quad \text{at the rate} \quad w_3 = k_{\varepsilon_3} J_3 [BO_2^{\cdot}] \qquad (3)$$

TABLE 19. Variation of the Optical Density of the Band at 2900 A
as a Function of the Duration of Irradiation of Polyamide Film

Time of irradiation, hr	Δ_1	Δ_2	$\Delta_2/\Delta_1 = R$
30	0.065	0.265	4.08
60	0.095	0.365	3.84
70	0.098	0.382	3.90
100	0.120	0.500	4.17
180	0.170	0.670	3.94

A is the polyamide molecule; B and C are radicals formed in the photochemical decomposition of the polyamide molecule; $BO_2^{\cdot}$ is the group responsible for the absorption at 2900 A; and J_1 and J_3 are light intensities
at the frequencies ν_1 and ν_3, respectively.

The concentration $[BO_2^{\cdot}]$ is proportional to the increase in the band
intensity in the irradiation of the film (Δ_2), while $[B^{\cdot}]$ is proportional
to the further increase in this band during storage of the film in the dark.

An analysis of the kinetic equations of the reactions cited above gives
the following expression:

$$\frac{[BO_2^{\cdot}]}{[B^{\cdot}]} = \frac{k_2\,[O_2]}{k_{\varepsilon_s}\,J_3} = R$$

The concentration $[O_2]$ can be considered constant; R is a constant. The
experimentally obtained constant values of the quantities R (Table 19)
confirm this scheme.

Of the three reactions cited above, the slowest is the reaction (1) of
primary photolysis. The rate of reaction (2) is greater than the rate of
reaction (1), since after the light is turned off, the intensity of the band
at 2900 A increases more rapidly than in the light. The rate of reaction
(3) is higher than the rate of reaction (1), since when the light is again
turned on, the concentration $[BO_2^{\cdot}]$ drops. Thus, the author ascribes
the band at 2900 A to a peroxide radical, unstable to irradiation with frequency ν_3.

The spectra in Fig. 137 for polyamide irradiated under high vacuum
show a general increase in the intensity in the region of 2500–3000 A,
which also continues in the dark; however, the peak at 2900 A is absent
in this case.

Obviously these data do not agree with the data cited in the work of
Hsü Chi-P'ing [71], where the appearance of a peak around 2900 A was
also observed in the irradiation of films under vacuum. Achhammer [15]

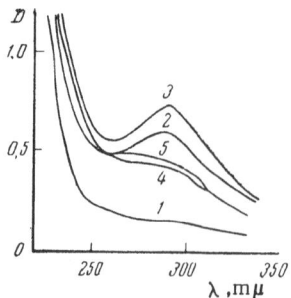

Fig. 137. Spectra of polyamide films. 1) Before irradiation; 2) directly after irradiation in air for 90 hr; 3) after 24 hr storage of film 2 in the dark; 4) after irradiation of film 1 under vacuum for 90 hr; 5) after 24-hr storage of film 4 in the dark; D) optical density.

detected a decrease in the absorption in the region of 250-340 mμ in the irradiation of polyamide films by ultraviolet light under vacuum.

The difference in the absorption spectra in the ultraviolet region obtained by different authors can be explained by insufficient purity of the starting materials. As has been indicated, the band at 2900 A also appears in the case of thermal destruction of polyamides [24, 25]. It can also be present in the initial polyamide as a result of the products of side reactions that occur during the production of the polyamides. The intensity of this band increases with increasing time of polycondensation [25]. The presence or absence in the initial polyamide of a product giving absorption at 2900 A may be responsible for the different behavior of the polymer during subsequent irradiation.

The products obtained in the photolysis and photooxidation of polyamides have been investigated in many studies [15, 16, 68, 70, 71, 74]. Thus, in the irradiation of films of a mixed polyamide [15] with ultraviolet light at 95°C, gaseous products: CO, CO_2, H_2O, and hydrocarbons were detected mass spectroscopically [15]. Just as in thermal destruction, the author proposes cleavage of the $-\overset{\overset{\displaystyle O}{\|}}{C}-\overset{\overset{\displaystyle H}{|}}{\underset{|}{N}}$ bond next to the carbonyl group:

$$-\overset{\overset{\displaystyle H}{|}}{N}-\overset{\overset{\displaystyle O}{\|}}{C}-(CH_2)_x-\overset{\overset{\displaystyle O}{\|}}{C}-\overset{\overset{\displaystyle H}{|}}{N}--\to-\overset{\overset{\displaystyle O}{\|}}{C}-(CH_2)_x-\overset{\overset{\displaystyle O}{\|}}{C}-\to CO + \text{hydrocarbons}$$

The formation of the remaining products (CO_2, H_2O) is explained in the same way as in pyrolysis [15]. In the photolysis of polycaproamide at 30°C by the complete spectrum of the lamp PRK-2, hydrogen, carbon monoxide, and small amounts of hydrocarbons (ethane, ethylene, propane, propylene, n-butylene) were detected by the method of gas chromatography [71, 73, 74]. In 120 hr of irradiation, less than two cleavages of the basic chain at the $-\overset{\overset{\displaystyle O}{\|}}{C}-\overset{\overset{\displaystyle H}{|}}{N}-$ bond, followed by splitting out of

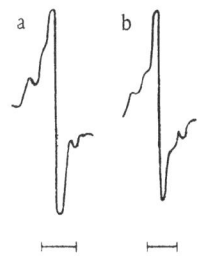

a   b

40 oerst   40 oerst

Fig. 138. EPR spectra of polycaproamide irradiated by ultraviolet radiations. a) EPR spectrum at the temperature of liquid nitrogen; b) EPR spectrum after thawing of the irradiated film.

a CO molecule, occur, along with about 18 cleavages of C – H bonds, liberating hydrogen. The formation of small amounts of hydrocarbons indicates that cleavage of the C – H and C – N bonds is also accompanied by cleavage of C – C bonds. Even if only part of the radicals formed in the cleavage of C – H bonds recombine, processes of structuring should predominate over destruction of the basic chain.

Actually, as has been indicated, the viscosity of the solution of irradiated polyamide in tricresol first increases during the photolysis of polycaproamide, and then a gel fraction is formed.

In the irradiation of polycaproamide film under vacuum not by the entire spectrum of the PRK-2 mercury-quartz lamp, but only by the near ultraviolet (filtration of the far ultraviolet, up to 3000 A, with Pyrex glass), the hydrogen content in the gaseous photolysis products drops sharply; the characteristic viscosity is simultaneously reduced, and no gel fraction is formed.

The following scheme of the processes that occur during irradiation under vacuum is proposed [71] on the basis of a study of the composition of the gases formed during photolysis of polyamides:

1) Destruction of the basic polyamide chains as a result of homolytic cleavage of the weak C – N bonds, followed by the liberation of carbon monoxide:

$$\sim (CH_2)_n CONH (CH_2)_m \sim \xrightarrow{h\nu} \sim CH_2CH_2 \dot{C}O + \dot{N}HCH_2CH_2 \sim$$

$$\sim CH_2CH_2\dot{C}O \rightarrow \sim CH_2\dot{C}H_2 + CO$$

The further transformations of the $\sim CH_2CH_2NH$ and $\sim CH_2CH_2$ radicals can lead to the appearance of new terminal amino groups, double bonds in the polyamide molecules, etc;

2) In addition to cleavage of the polymer chains, there is a splitting out of a hydrogen atom, which, as for lower amides [38], is proposed in the α-position to the NH group:

$$\sim CH_2CH_2CONHCH_2CH_2 \sim \xrightarrow{h\nu} \sim CH_2CH_2CONH\dot{C}HCH_2 \sim + H\cdot$$
$$H\cdot + H\cdot \rightarrow H_2$$
$$H\cdot + RH \rightarrow H_2 + R\cdot, \text{ etc.}$$

Cross-links are formed in the recombination of macroradicals:

$$
\begin{array}{l}
\sim CH_2CH_2CONHCHCH_2 \sim \\
+ \qquad\qquad\qquad\qquad\qquad\quad \rightarrow \\
\sim CH_2CH_2CONHCHCH_2 \sim
\end{array}
\qquad
\begin{array}{l}
\sim CH_2CH_2CONHCHCH_2 \sim \\
\qquad\qquad\qquad | \\
\sim CH_2CH_2CONH\overset{|}{C}HCH_2 \sim
\end{array}
$$

In the photolysis of polycaproamide, hydrogen and carbon monoxide are liberated at a practically constant rate, which indicates the un-branched character of the radical-chain processes [74].

In the case of irradiation by near ultraviolet, cleavage of the weak C – N bonds predominate; this leads to a decrease in the characteristic viscosity and a reduction of the hydrogen content in the gaseous photolysis products.

The EPR spectra of polycaproamide irradiated at the temperature of liquid nitrogen, cited in Fig. 138, represent a triplet with intensity ratio (1 : 6 : 1), which shows the presence of the triplet (1 : 2 : 1) super-imposed upon a singlet.

When the temperature is raised, the triplet is converted to a quinti-plet with an intensity ratio 1 : 4 : 6 : 4 : 1, superimposed upon the singlet. The quintiplet disappears rapidly, and the singlet remains, lasting for no less than two weeks.

The authors of [70, 71] ascribe the triplet to the radical $CH_2-CH_2-\overset{\bullet}{C}O$, the conversion of the triplet to a quintiplet – to the transformation of this radical to the radical $-CH_2-\overset{\bullet}{C}H_2$ by the splitting out of CO. The singlet is ascribed to the radical formed when an H atom is split out in the α-position to the NH group.

The quantum yield of the photolysis of polycaproamide, calculated according to the summary mass evolution (H_2 and CO) in the case of monochromatic radiation (2537 A) is equal to $6\text{-}7 \times 10^{-4}$ [74].

In the photooxidation of polycaproamide, the same gaseous products were found [75] as in the case of irradiation without oxygen. The amount of hydrogen formed is the same as in photolysis; ten times as much car-bon monoxide is liberated as in the absence of oxygen. The formation of large amounts of CO is explained by the appearance of new carbonyl groups in the oxidation of the polyamide.

Peroxide groups are detected in polyamides irradiated in the pres-ence of oxygen [63, 71]. Schwemmer [68] attempted to analyze the solid residue after the irradiation of polyamides by ultraviolet light. How-ever, the paper chromatography of the hydrolysis products of the ir-radiated polyamides that he performed gave the same results as for the initial nonirradiated polyamide.

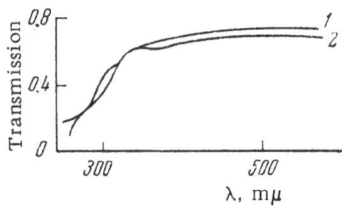

Fig. 139. Spectra of unstabilized poly-hexamethyleneadipamide (1) and the polymer stabilized by chromium (2).

Under the simultaneous action of oxygen and water vapors, ultraviolet irradiation leads to a deterioration of the mechanical properties, smaller in comparison with the oxidative process in the absence of moisture, which can be explained by the interaction of water with the $R-\overset{.}{N}H$ and $R-\overset{.}{C}O$ radicals formed and chain termination [71, 73].

The main factor promoting a change in the chemical composition and physico-mechanical properties of polyamides during photoaging consists of the photosensitized oxidative processes.

V. PHOTOSTABILIZATION OF POLYAMIDES

A number of organic and inorganic stabilizers are recommended for increasing light stability.

As the inorganic additives to polyamides recommended in the literature, we should mention primarily salts of trivalent chromium [77, 78]. The use of compounds of di- and hexavalent chromium is also possible. However, during the process of treatment of the polyamide with the stabilizer, divalent chromium is oxidized, while hexavalent chromium is reduced to the trivalent form. Thus, for example, polyamide fibers are treated first with potassium bichromate and then with sodium thio-sulfate [77]. Certain chromium dyes also exhibit a protective action against sunlight [79].

Light-stable polyamides are obtained by adding an aqueous suspension of carbon black containing water-soluble chromium salts, for example, CrF_3, before polycondensation [78]. In addition to increasing the photostability of polyamides, chromium also increases their stability to thermal oxidation [68, 41]. The spectra of polyamide samples treated with chromium, cited in Fig. 139, practically do not differ from the spectra of the untreated samples. The action of chromium apparently reduces to a strengthening of the amide bond, which is important both in photoaging and in thermal oxidation of polyamides [68].

The action of other metals – Al, Mn, Cu, Co, Ni (chromium, manganese, copper, and aluminum salts of phenolic acids and anthranilic acid [86], various inorganic and organic manganese salts [1, 2, 4]), polyphosphate complexes of the heavy metals [81-83], for example, Mn hexametaphosphate, produced by adding a solution of $(NaPO_3)_6$ to manganese acetate – apparently also reduces to strengthening of the amide bond. A mixture of manganese and copper compounds exhibits a synergic effect [41].

Inorganic additives: cerium hydroxide or oxide [41], compounds of pentavalent niobium and tantalum [84], are also used as photostabilizers.

A substantial increase in the light stability of polyamides is achieved by introducing additives of organic phosphorus compounds into the polymer [43]. Various complete esters of phosphorus, pyrocatecholphosphorus, α-naphthylphosphorous, and phenylphosphorous acids effectively reduce the drop in the mechanical properties of capron fiber during irradiation in comparison with the unstabilized form. The drop in the breaking strength as a result of 20-hr irradiation at 40 °C in air medium is 2.1% in the case of the addition, for example, of the dodecyl ester of pyrocatecholphosphorous acid; the drop in the breaking elongation is 0.5%; under these conditions the unstabilized fiber loses 41% of its strength and 45% of its breaking elongation.

Protection from photoaging is usually accomplished by the introduction of organic substances capable of absorbing ultraviolet light to a considerably greater degree than the polymer. In this case the stabilizer should not form active radicals capable of initiating the chain process of polymer oxidation. Otherwise, the ultraviolet absorber is a sensitizer, which accelerates the process of photoaging.

The absence of a protective action of an additive can be explained by its ability to give rays upon luminescence with a greater wavelength than those absorbed, however, possessing energy sufficient for the cleavage of weak bonds in the polymer chain (for example, the amide bond).

We should mention the following as ultraviolet absorbers used as photostabilizers for many polymers [85]:

1) compounds of the series of hydroxybenzophenones [86], for example, 2-hydroxy-4-methoxybenzophenone:

$$\text{benzene ring} - \overset{\overset{\displaystyle O}{\|}}{C} - \text{benzene ring} - OCH_3$$
$$\underset{\displaystyle OH}{}$$

2) benzotriazoles [87], for example:

$$\text{benzotriazole} = \overset{N}{\underset{N}{\diagdown}} N - \text{benzene ring with } OH \text{ and } CH_3$$

2-(2'-hydroxy-5'-methyl)phenylbenzotriazole

3) salol and other derivatives of salicylic acid [87], for example, 4-tert-butylphenyl salicylate:

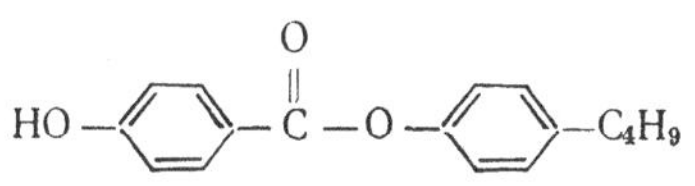

$$HO-\bigcirc-\overset{\overset{\displaystyle O}{\|}}{C}-O-\bigcirc-C_4H_9$$

Carbazole and its derivatives [for example, 3-(4-hydroxyphenyl)-2-aminocarbazole] [88], N,N'-polymethylene-bis-(o-hydroxybenzamides) [89, 90], are especially recommended for polyamides.

It is extremely important to elucidate the question of the relationship between the protective action of an additive and its ability for luminescence, as well as the correspondence of its absorption spectrum to the spectrum of the polyamide [91]. Additives of substances differing in their ability to luminesce, as well as possessing different spectra in the ultraviolet region have been used for this purpose [45, 91].

Luminophores whose absorption spectra correspond to the absorption spectrum of the polyamide:

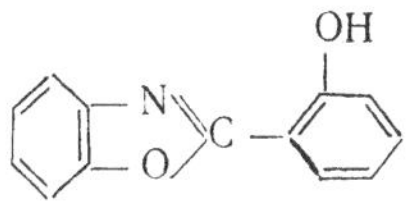

2-(o-hydroxyphenyl)benzoxazole

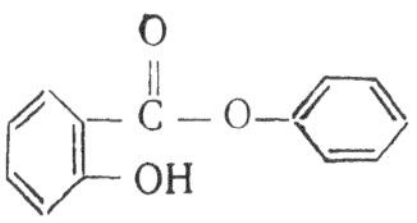

salol

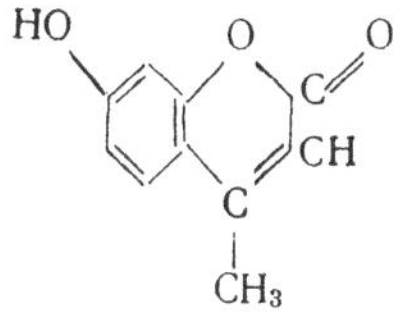

7-hydroxy-4-methylcoumarin

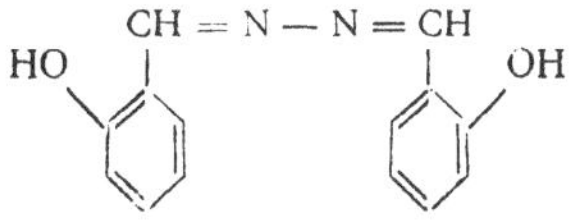

o-hydroxybenzaldazine

Luminophores possessing an absorption spectrum not corresponding to the spectrum of the polyamide:

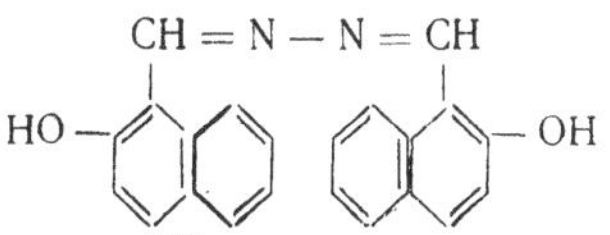

o-hydroxynaphthaldazine

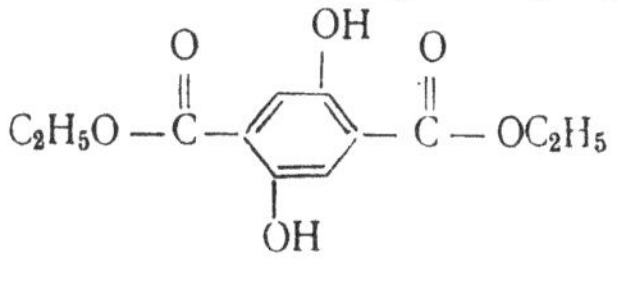

diethyl ester of 2,5-dihydroxyterephthalic acid

Organic substances that are not luminophores, but possess a spectrum corresponding to the spectrum of the polyamides:

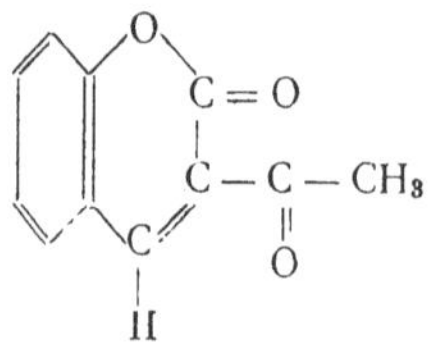

3-acetylcoumarin

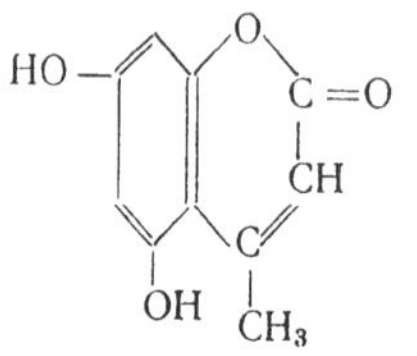

5,7-dihydroxy-4-methylcoumarin

Organic substances that are not luminophores and do not possess spectra corresponding to the spectrum of the polyamides:

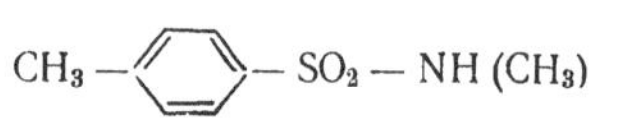

$CH_3 - \langle \rangle - SO_2 - NH\,(CH_3)$ N-methyl-p-toluenesulfamide

An investigation of the mechanical properties of the irradiated polyamide films showed that the process of photochemical destruction is decelerated by the introduction of such luminophores as 2-(o-hydroxyphenol)benzoxazole, salol, 7-hydroxymethylcoumarin, the spectra of which correspond to the spectrum of the polyamide. On the other hand, the luminophore o-hydroxybenzaldazine possessing an absorption spectrum corresponding to the spectrum of the polyamides, does not decelerate the photochemical destruction of polyamides, but is a strong sensitizer. The diethyl ester of 2,5-dihydroxyterephthalic acid, which does not possess a spectrum corresponding to the spectrum of the polyamide, still exerts a protective action in the case of irradiation by filtered ultraviolet light (2900-3200 A).

Substances that are not luminophores, but possess a spectrum overlapping with the spectrum of the polyamides (3-acetylcoumarin, 5,7-dihydroxy-4-methylcoumarin) manifest a stabilizing action in the case of irradiation by filtered light.

N-Methyl-p-toluenesulfamide, which is not a luminophore and does not possess a spectrum corresponding to the spectrum of the polyamide, possesses sensitizing properties.

Thus, the ability for luminescence is not directly related to the stabilizing properties of the additive. Correspondence of the absorption spectra of the additive and the polyamide exerts a greater influence, but also is not the deciding factor, such as, for example, in the case of o-hydroxybenzaldazine and the diethyl ester of 2,5-dihydroxyterephthalic acid.

Protection from photoaging can be accomplished by the introduction of antioxidants that suppress the processes of photosensitized oxidation. For example, effective stabilizers of polyamide film are the 2,6-di-tert-butyl-4-methylphenyl ester of pyrocatecholphosphorous acid, 2,6-di-tert-butylhydroquinone, and a mixture of potassium iodide with copper naphthenate. All these additives exhibit considerable protective action in the thermal oxidation of polyamides as well [92].

BIBLIOGRAPHY

1. German Patent 737943, 1941; C. A. 39:5599, 1945.
2-3. G. Hopff, A. Müller, and F. Wenger, Polyamides [Russian Translation], Moscow, State Press for Chemical Literature, 1958, pp. 154, 165, 203.
4. French Patent 955259; C. A. 45:9276, 1951.
5. S. P. Gundavda, J. Textile Ind. 47(5):289, 1956.
6. V. V. Korshak, G. L. Slonimskii, and E. S. Krongauz, Izvest Akad. Nauk SSSR, Otdel.Khim. Nauk, p. 221, 1958.
7. M. Staudinger and H. Schnell, Makromol. Chem. 1:49, 1947.
8. S. R. Rafikov and R. A. Sorokina, Vysokomolekulyarnye Soedineniya 1(4):549, 1959.
9. R. A. Sorokina, Dissertation, Institute of Heteroorganic Compounds, Academy of Sciences, USSR, 1961.
10. R. Hill, Chem. & Ind. 33:1083, 1954.
11. N. D. Katorzhnov and A. S. Strepikheev, Zhur. Priklad. Khim. 32:625, 1363, 1959.
12. S. Smith, J. Polymer Sci. 30:459, 1958.
13. V. V. Korshak and V. A. Zamyatina, Izvest. Akad. Nauk SSSR, Otdel. Khim. Nauk, p. 480, 1945.
14. V. V. Korshak and V. A. Zamyatina, Izvest. Akad. Nauk SSSR, Otdel, Khim. Nauk, p. 609, 1945.
15. B. G. Achhammer, F. W. Reinhart, and G. M. Kline, J. Appl. Chem. 1:301, 1951.
16. B. G. Achhammer, F. W. Reinhart, and G. M. Kline, J. Res. Natl. Bur. Std. 46:391, 1951.
17. S. Strauss and L. Wall, J. Res. Natl. Bur. Std. 60:39, 1958.
18. S. Straus and L. Wall, J. Research 63A:269, 1959.
19. J. Goodman, J. Polymer Sci. 13:175, 1954.
20. T. Hesselstrom, H. Coles, C. Balmer, H. Hanigen, and M. Keller, J. Text. Res. 22:742, 1952.

21. G. Tylor, J. Am. Chem. Soc. 69:635, 1947.

22. B. Kemerbik, Zh. Kroz, and V. Groll', Khim. i Tekhnol. Polimer. No. 4:53, 1961.

23. S. R. Rafikov, G. N. Chelnokova, and R. A. Sorokina, Vysoko-molekulyarnye Soedineniya 4(4):1637, 1962.

24. J. Goodman, J. Polymer Sci. 17:587, 1955.

25. A. Liquori, A. Mele, and V. Karelli, J. Polymer Sci. 10:510, 1953.

26. A. Agster, Melliand Textilber. 37:1338, 1956.

27. C. Reimer, Kunstoffe 45:367, 1955.

28. R. F. Schwenker, Textile Res. J. 30(8):624, 1960.

29. S. R. Rafikov and R. A. Korokina, Vysokomolekulyarnye Soedineniya No. 1:21, 1961.

30. W. Sbrolli and T. Capaccioli, Chim. Ind. 42(12):1325, 1960.

31. V. V. Korshak and S. R. Rafikov, Zhur. Obshchei Khim. 14, 974, 1944.

32. V. V. Korshak and S. R. Rafikov, Doklady Akad.Nauk SSSR 48:36, 1945.

33. V. V. Korshak and S. R. Rafikov, Doklady Akad.Nauk SSSR 56:597, 1947.

34. W. Sbrolli, T. Capaccioli, and E. Bertolli, Chim. Ind. 42(4):357, 1960.

35. I. I. Levantovskaya, M. P. Yazvikova, M. K. Dobrokhotova, B. M. Kovarskaya, and K. N. Vlasova, Plast. Mass. No. 3:19, 1963.

36. P. Rochas and J. C. Martin, Bull. Inst. Textile France No. 83:41, 1959.

37. F. H. Steiger, J. Text. Res. 27(6):459, 1957.

38. W. H. Sharkey and W. E. Mochel, J. Am. Chem. Soc. 81:3000, 1959.

39. Yu. A. Voitelev and N. D. Katorzhnov, Khim. Volokna No. 3:3, 1960.

40. G. I. Kudryavtsev, N. D. Katorzhnov, Yu. A. Voitelev, and V. V. Golubeva, Khim. Volokna No. 5:16, 1960.

41. Yu. A. Voitelev and N. D. Katorzhnov, Khim. Volokna No. 4:3, 1960.

42. L. G. Tokareva, N. V. Mikhailov, Z. I. Potemkina, and M. V. Kovaleva, Vysokomolekulyarnye Soedineniya No. 11:1728, 1960.

43. N. V. Mikhailov, L. G. Tokareva, Z. I. Potemkina, K. K. Burav-chenko, B. V. Petukhov, G. M. Terekhova, and P. A. Kirpichinkov, Summaries of Reports at the Conference on the Aging and Stabiliza-tion of Polymers, Moscow, Academy of Sciences USSR Press, 1961, p. 48.

44. M. Epstein and C. W. Hamilton, Mod. Plastics 37(7):142, 144, 146, 148, 150, 151, 154, 155, 190, 192, 1960.

45. V. V. Korshak, K. M. Mozgova, and V. P. Lavrishchev, Vysoko-molekulyarnye Soedineniya No. 8:1159, 1165, 1959.

46. German Patent 4752, 1943; Great Britain Patent 652947, 1951; C. A. 46:2912, 1952.
47. Announcement 75102, 1943; G. Hopff, A. Müller, and F. Wenger, Polyamides [Russian translation], Moscow, State Press for Chemical Literature, 1958.
48. Great Britain Patent 793196, 1958; Referat. Zhur. Khim. 1960, 20189, U. S. Patent 2493597, 1950; C. A. 44:27799, 1950.
49. German Federated Republic Patent 1078323 and 1069380; Zbl. 49: 16610, 1960.
50. U. S. Patent 2733162; Mod. Plastics No. 7:142, 1960.
51. Great Britain Patent 839067, 1960; C. A. 25:989, 1960.
52. Great Britain Patent 722724, 1955; Referat. Zhur. Khim. 1956, 21046, U. S. Patent 2705227, 1955; Referat. Zhur. Khim. 1956, 23948.
53. Great Britain Patent 715364, 1954; Referat. Zhur. Khim. 1955, 56899.
54. U. S. Patent 2705227, 1955; C. A. 49:10666, 1955.
55. U. S. Patent 2630421, 1953; Great Britain Patent 708029, 1954; Referat. Zhur. Khim. 1956, 14431.
56. French Patent 906892, 1944.
57. G. Hopff, A. Müller, and F. Wenger, Polyamides [Russian Translation], Moscow, State Press for Chemical Literature, 1958, p. 250.
58. U. S. Patent 2849446, 1958; Referat. Zhur. Khim. 1961, 3P 357.
59. German Federated Republic Patent 1002524, 1957; German Federated Republic Patent 1002534, 1958; Referat. Zhur. Khim. 1959, 80389.
60. German Federated Republic Patent 1001819, 1957; Referat. Zhur. Khim. 1958, 55853; German Federated Republic Patent 1001819, 1957; C. A. 52:23104, 1958.
61. German Democratic Republic Patent 4980, 1954; Referat. Zhur. Khim. 1955, 5998.
62. Italian Patent 500035, 1954.
63. A. Agster and O. Holzinger, Textil-Praxis 11:825, 1956.
64. G. S. Egerton, Tex 11(1):28, 30, 1952.
65. J. Boulton and D. L. C. Jackson, J. Soc. Dyers Colourists 59:21, 1943.
66. A. Sippel, Melliand Textilber. 38:898, 1957.
67. M. Schwemmer, Textil-Rundschau 11:70, 136, 1956.
68. M. Schwemmer, Eidgenoess. Materialpruefunger. Versuchsanstalt Ind. Bauw. Gewerbe, Zuerich No. 180, 1955.
69. T. Hashimoto, Bull. Chem. Soc. Japan 30:950, 1957.
70. S. R. Rafikov and Hsü Chi-P'ing, Vysokomolekulyarnye Soedineniya 3(1):56, 1961.
71. Hsü Chi-P'ing, Dissertation, Institute of Heteroorganic Compounds, Academy of Sciences, USSR, 1961.

72. R. A. Ford, J. Colloid Sci. 12(3):271, 1957.

73. Hsü Chi-P'ing and S. R. Rafikov, Vysokomolekulyarnye Soedineniya 4:851, 1962.

74. Hsü Chi-P'ing and S. R. Rafikov, Vysokomolekulyarnye Soedineniya 4(10):1474, 1962.

75. A. R. Goldfarb, J. Biol. Chem. 193:397, 1951.

76. S. Ham and J. R. Platt, J. Chem. Phys. 20:335, 1952.

77. Great Britain Patent 649481, 1951; C. A. 45:6851, 1951.

78. Swiss Patent 347979, 1960; Referat. Zhur. Khim. 1961, 13P 397.

79. O. Newsome, J. Soc. Dyers Colourists 66:277, 1950.

80. Great Britain Patent 688629, 1951; Referat. Zhur. Khim. 1956, 11415.

81. German Federated Republic Patent 1063378, 1960; Referat. Zhur. Khim. 1961, 17P 372.

82. German Federated Republic Patent 1069378, 1958; Zbl. 28: 9398, 1960.

83. Great Britain Patent 862577, 1961; Plast. Abstr. No. 11:1725,1961.

84. Great Britain Patent 802085, 1958; C. A. 53:3726, 1959.

85. R. A. Coleman and J. A. Welckseb, Mod. Plastics 36:117, 1959.

86. U. S. Patent 28335523, 1958; Referat. Zhur. Khim. 1961, 10L 167; Great Britain Patent 726792, 1955; Referat. Zhur. Khim. 1957, 67525; U. S. Patent 2552551, 1951; C. A. 45:73800, 1951.

87. R. G. Schmitt and R. C. Hirt, J. Polymer Sci. 45:35, 1960.

88. German Democratic Republic Patent 764663, 1957; Referat. Zhur. Khim. 1958, 75749.

89. Great Britain Patent 54326, 1942; C. A. 40:52615, 1945.

90. German Democratic Republic Patent 5084, 1954; Referat. Zhur. Khim. 1955, 53715.

91. V. P. Lavrishchev, Dissertation, Institute of Heteroorganic Compounds, Academy of Sciences, USSR, 1961.

92. M. B. Neiman, B. M. Kovarskaya, I. I. Levantovskaya, G. V. Dralyak, M. P. Yazvikova, V. A. Sidorov, V. N. Kochetkov, G. M. Trossman, G. O. Tatevos'yan, and I. B. Kuznetsova, Plast. Mass. No. 10:6, 1962.

93. I. I. Levantovskaya, B. M. Kovarskaya, M. B. Neiman, and G. V. Dralyak, Vysokomolekulyarnye Soedineniya 6(9):1569, 1964.

94. I. I. Levantovskaya, B. M. Kovarskaya, M. B. Neiman, E. G. Rozantsev, and M. P. Ivikova, Plast. Mass. No. 3:14, 1964.

95. E. T. Burell, J. Am. Chem. Soc. 83:74, 1961.

96. J. Shinohara and D. Ballantine, J. Chem. Phys. 36:3042, 1962.

Chapter VIII

AGING OF CERTAIN CONDENSATION POLYMERS

The mechanism of the processes of decomposition of a number of condensation polymers under the influence of heat and oxygen has received little study, and data on their stabilization are almost entirely absent. The stabilizers known to us are effective at temperatures no higher than 200°C; hence they cannot be used for polymers characterized by a high softening point and subjected to prolonged and strong heating during use.

However, the introduction of stabilizers is essential, since processes of thermal and thermooxidative destruction develop extremely intensively during the use and reprocessing of such polymers, leading to a sharp deterioration of their physicomechanical and dielectric properties. Thus, an extremely urgent problem at the present time is a detailed investigation of the processes of decomposition of condensation polymers for developing a theory of their stabilization. Works in the field of the study of the thermal and thermooxidative destruction of certain condensation polymers (epoxide, phenol-formaldehyde resins, polycarbonate, and polyarylates) are outlined below.

I. EPOXIDE RESINS

Epoxide resins are condensation products of substances containing the epoxy group (epichlorohydrin) with polyatomic phenols (4,4'-dihydroxy-diphenyl-2,2'-propane, resorcinol, etc.) and represent thermoplastic products with various softening points, depending on the ratio of the initial components and the conditions under which the condensation was conducted:

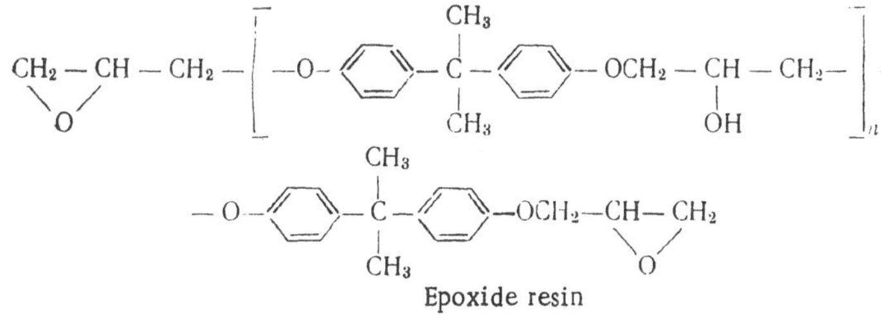

Epoxide resin

These resins can be hardened under the influence of certain substances (diamines, anhydrides of dicarboxylic acids, etc.), i.e., be converted to an infusible and insoluble state, characterized by the formation of a three-dimensional structure.

Below is depicted the proposed structure of the molecule of an epoxide resin, hardened with maleic anhydride:

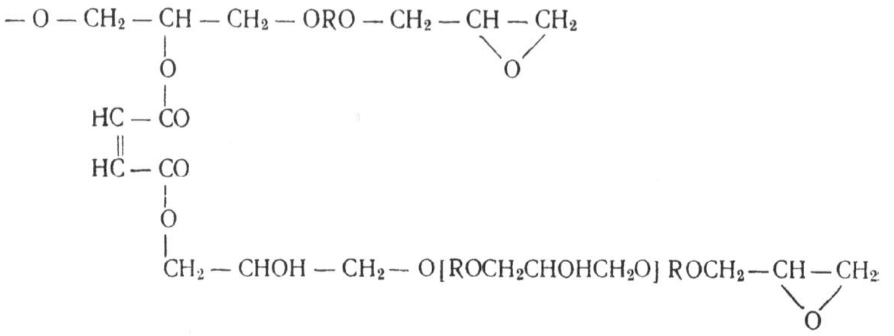

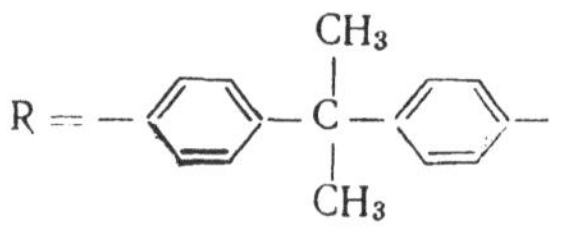

$$- O - CH_2 - CH - CH_2 - ORO - CH_2 - CH - CH_2$$

$$HC - CO$$
$$HC - CO$$

$$CH_2 - CHOH - CH_2 - O[ROCH_2CHOHCH_2O] ROCH_2 - CH - CH_2$$

$$R = - \bigcirc - \overset{\overset{\displaystyle CH_3}{|}}{\underset{\underset{\displaystyle CH_3}{|}}{C}} - \bigcirc -$$

The thermal destruction of resins based on 4,4'-dihydroxydiphenyl-2,2'-propane (resin ED-6, unhardened and hardened with polyethylene-polyamine and maleic anhydride) was studied in [1-4]. The thermal decomposition of these resins begins at temperatures above 200°C and is characterized by substantial gas evolution (the kinetics of the gas evolution was studied on a special setup, described in [1]).

Figure 140 presents the kinetic curves of gas evolution for the unhardened resin and the resin hardened with maleic anhydride, taken within a broad range of temperatures. Chromatographic analysis of the gaseous destruction products of these resins showed the presence of considerable amounts of methane, carbon monoxide, formaldehyde, acetaldehyde, and acrolein.

Table 20 presents the composition of the basic gaseous products of the thermal destruction of epoxide resins at 405°C.

Gas evolution practically ceases 5-15 min after the beginning of the reaction. However, the process of thermal destruction continues, liquid products of comparatively low molecular weight distilling off from the polymer. In the case of unhardened epoxide resins, the liquid destruction products, according to the data of infrared spectroscopy, represent a mixture of low-molecular fractions of the resin, capable of being converted to the infusible and insoluble state under the influence of hardeners.

TABLE 20. Composition of the Basic Gaseous Products of the Thermal Destruction of Epoxide Resins at 405°C

(moles of gas/mole of resin)

Resin	H_2	CO	CH_4	CO_2	C_2H_6	C_2H_4	C_3H_8	C_3H_6
Epoxide ED-6	0.01	0.3	0.1	0.08	0.01	0.02	0.005	0.1
ED-6, hardened with polyethylenepolyamide	0.01	0.1	0.1	0.05	0.03	0.03	0.02	0.1
ED-6, hardened with maleic anhydride	0.01	0.32	0.09	1.71	0.01	0.02	0.01	0.08

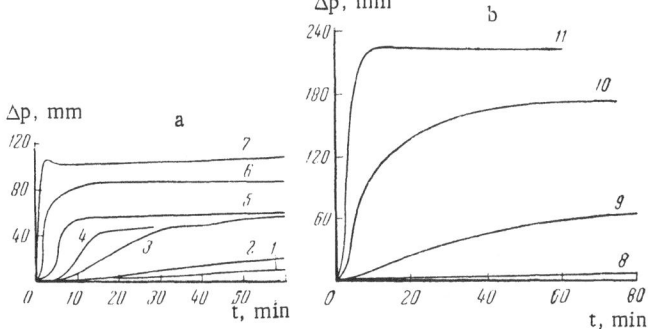

Fig. 140. Kinetics of the gas evolution in the thermal destruction of epoxide resin (a) and the resin hardened with 30% maleic anhydride (b) at various temperatures: 1) 243°; 2) 250°; 3) 277°; 4) 300°; 5) 326°; 6) 405°; 7) 503°; 8) 245°; 9) 301°; 10) 350°; 11) 450°C.

Low-molecular destruction products containing no epoxide groups in amounts sufficient for hardening predominate among the liquid destruction products of the hardened resins.

Figure 141 presents the degree of decomposition of a resin that was characterized by a loss in weight of the solid residue in the destruction of epoxide resins hardened with polyethylenepolyamine and maleic anhydride at various temperatures. We can see from the figure that the resin, hardened with polyethylenepolyamine, can withstand temperatures of 150-200°C without appreciable decomposition, while the resin hardened with maleic anhydride withstands higher heatings (~300°C).

A mass spectrometric analysis of the pyrolysis products of hardened epoxide resins based on polyphenols (Table 21) was conducted in [5]. The balance cited with respect to the amount of volatile and liquid pyrolysis products, as well as according to the magnitude of the solid residue, is in agreement with the data cited in [2].

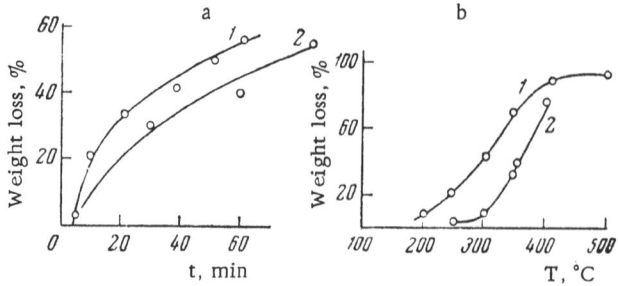

Fig. 141. Variation of the degree of decomposition of epoxide resin hardened with 7% polyethylenepolyamine [1] and with 30% maleic anhydride [2] with time at a temperature of 345°C (a) and as a function of the temperature when the destruction was conducted for 90 min (b).

As a result of a study of the kinetics of the process and the thermal decomposition products of epoxide resins, M. B. Neiman and B. M. Kovarskaya and associates have proposed a radical mechanism for the process [1-3].

Apparently the terminal groups of the resin are most readily cleaved, forming a radical

$$CH_2 \!-\! CH \!-\! CH_2O$$
$$\diagdown \ O \ \diagup$$

This radical can isomerize to the radical

$$\dot{C}H_2 \!-\! CH \!-\! C \overset{O}{\underset{H}{\diagup\diagdown}}$$
$$\quad\quad\ \ | \quad\quad\quad$$
$$\quad\quad\ OH \quad\ \ H$$

from which acrolein and hydroxyl are formed. In addition, the primary radical can decompose into formaldehyde CH_2O and the radical

$$CH_2 \!-\! \dot{C}H$$
$$\diagdown \ O \ \diagup$$

As a result of the isomerization of this radical, the acetyl radical $CH_3 \!-\! \dot{C}O$ can be obtained; by taking hydrogen from the epoxide resin, this radical forms acetaldehyde. Finally, the acetyl radical can decompose into CO and $\dot{C}H_3$; the latter, adding hydrogen, is converted to methane. Methane is also formed when methyl groups are stripped from the diphenylolpropane radical.

The proposed mechanism of the thermal decomposition of epoxide resins is confirmed by data [6] on the analysis of the destruction products

TABLE 21. Composition of Gaseous Pyrolysis Products of Hardened
Epoxide Resin

Composition	Pyrolysis temperature, °C			
	360	500	800	1200
	Amount of volatile products, %			
Hydrogen	–	–	0.8	2.1
CO	4.7	3.1	11.2	25.9
CO_2	16.2	6	3.7	1.8
Methane	1	0.8	1.6	4.3
Acetylene	0.3	–	–	2.5
Ethylene	–	–	–	3
Acetone	0.9	2.2	–	–
Propane	–	1.1	–	–
Propylene	6.5	2.3	2.2	–
Ethane	–	–	1.6	–
Cyclopentadienes	–	–	–	0.6
Pentane	–	0.5	–	–
Benzene	–	1.3	2.8	8.1
Methyl chloride	5.1	–	–	–
Ethyl chloride	1.7	–	–	–
Other gases	–	0.7	0	0.8
Heavy fraction volatile at the pyrolysis temperature and non-volatile at room temperature	63.6	82	73.1	50.9
Total amount	100	100	100	100
Volatiles in the sample, %	38	75	86	87

of an epoxide resin with a labelled central carbon in 4,4'-dihydroxy-
diphenyl-2,2'-propane and entirely explains the formation of all the decomposition products found in the gaseous phase. It was also shown in
this work that the predominant portion of the gaseous destruction products of the epoxide resin is obtained not from the radical of 4,4'-dihy-
droxydiphenyl-2,2'-propane, but from the aliphatic portion of the molecule.

The same gases and in the same amounts (~3% of the initial resin)
have been detected in the gaseous destruction products of hardened epoxide resins as in the nonhardened resin [2]. Apparently the gaseous
destruction products of hardened resins are also formed in the decomposition of free epoxide groups, which are always present in some amount
in the hardened products. Actually, the amount of carbon monoxide
formed in the decomposition of an epoxide resin hardened with polyethylenepolyamine corresponds to the decomposition of 5% of the total amount

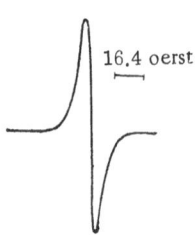

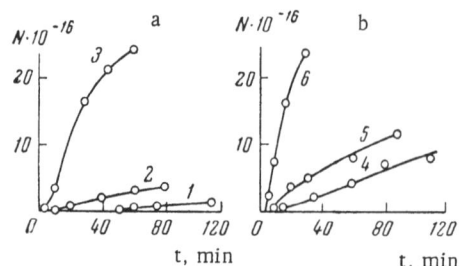

Fig. 142. EPR spectrum of epoxide resin hardened with polyethylenepolyamine and heated at 350°C for 30 min.

Fig. 143. Kinetics of the accumulation of systems with conjugated bonds in the decomposition of epoxide resins hardened with maleic anhydride (a) and polyethylenepolyamine (b). 1 and 4) 350°C; 2 and 5) 375°C; 3 and 6) 405°C.

of epoxy groups contained in the unhardened resin (in the decomposition of a resin hardened with maleic anhydride, larger amounts of CO and CO_2 are formed. It is quite evident that these gases are also formed in the decomposition of the maleic anhydride radical, which is included in the three-dimensional structure of the hardened resin).

In addition to the stripping of terminal groups, cleavage of bonds among carbon atoms in the aliphatic portions of the chains of the epoxide resin is also possible. In this case low-molecular-weight volatile products can be formed.

In all the cases considered, the reaction proceeds through active radicals, which cannot accumulate in large concentrations; hence they cannot be detected by the method of electron paramagnetic resonance.

The singlet signals detected at the more profound stages of thermal decomposition of epoxide resins are depicted in Fig. 142. They have the same appearance as the signals obtained by D. Ingram [7] in a study of the destruction of coals, and A. A. Berlin and L. A. Blyumenfel'd [8, 9], and A. V. Topchiev and V. V. Voevodskii [10] in investigations of polymers with a system of conjugated bonds as well as a number of carbonized materials.

Since the labelled central atom of 4,4'-dihydroxydiphenyl-2,2'-propane in the epoxide resin mainly remains in the solid destruction products, as was shown in [6], we might assume that it takes part in the creation of thermally stable condensed aromatic and other systems with conjugated double bonds.

Figure 143 shows the kinetics of the accumulation of systems with conjugated bonds in the residue after the destruction of epoxide resins hardened with maleic anhydride and polyethylenepolyamine.

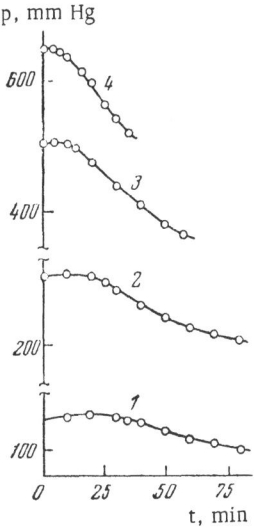

p, mm Hg

Fig. 144. Variation of the oxygen pressure as a function of the
duration of thermooxidative
destruction of an epoxide resin
hardened with 7.0% polyethylenepolyamine. Initial oxygen pressure: 1) 124; 2) 252; 3) 452, and
4) 525 mm Hg; at 200°C.

The activation energies were calculated for the processes of gas evolution, loss in weight of the residue, and rate of accumulation of systems with conjugated bonds in the thermal destruction of hardened epoxide resins.

For resins hardened with maleic anhydride, the energies of the first, second, and third processes proved equal to 30, 26, and 53 kcal/mole, respectively, while in the case of epoxide resins hardened with polyethylenepolyamine, the values were 25, 35, and 44 kcal/mole. Apparently the first two processes are related to cleavage of weaker bonds, while the third process is related to the cleavage of stronger bonds.

S. Madorsky and S. Straus [5] also found that the activation energy of the pyrolysis of hardened epoxide resins (pyrolysis temperature 360-1200°C) corresponds to 51 kcal/mole.

H. Anderson [11] used differential thermal and thermogravimetric analyses to confirm the free radical mechanism of the thermal decomposition of epoxide resins, proposed by M. B. Neiman and associates. It was shown in this work that an important stage in the pyrolysis of an epoxide resin is isomerization of the epoxide group to a carbonyl, which is accompanied by an exothermic effect and is observed in the temperature interval 350-400°C. The thermogravimetric curves confirmed that the rate of pyrolysis of samples of epoxide resins hardened with maleic anhydride is less than that of samples hardened with amines (metaphenylenediamine).

In a study of the thermooxidative destruction of epoxide resins (hardened with maleic anhydride and polyethylenepolyamine) [12], it was shown that the oxidation process is characterized by the presence of induction periods (Fig. 144), which decreased regularly with increases in the initial oxygen pressure (P) in the system and with increasing temperature (Fig. 145), in accord with the formula

$$\tau = \left(Q_1 + \frac{b}{P} \right) e^{\gamma/RT}$$

The quantity γ, characterizing the activation energy of the process of oxidation of the hardened epoxide resin, corresponds to 22 kcal/mole.

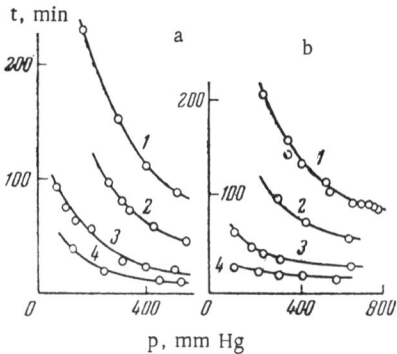

Fig. 145. Dependence of the induction period on the
oxygen pressure in thermooxidative destruction of an
epoxide resin. a) Hardened with polyethylenepoly-
amide: 1) 160°; 2) 170°; 3) 180°; 4) 200°C; b) hardened with maleic anhydride: 1) 200°; 2) 220°; 3) 230°;
4) 250°C.

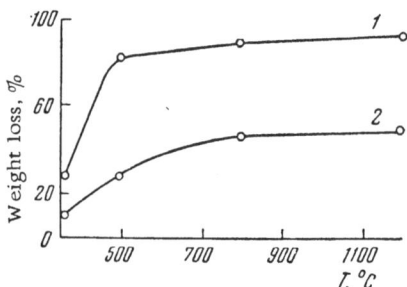

Fig. 146. Stability to thermal decomposition of hardened
epoxide (1) and phenolformaldehyde resins (2).

The presence of induction periods in the thermooxidative destruction of epoxide resins indicates an autocatalytic character of the reaction. By analogy with other polymers, the authors proposed that the
oxidation of epoxide resins is a chain reaction with degenerate branches,
which are related to decomposition of the hydroperoxides formed during
the oxidation process.

The experiments conducted showed that peroxide compounds are
formed in hardened epoxide resins oxidized under the same conditions,
the concentration of peroxides in the resin hardened by polyethylenepolyamine being considerably higher than that in the resin hardened with
maleic anhydride. The greater peroxide concentration in the resin hardened with polyethylenepolyamine also corresponds to a lower temperature (150°C) of the beginning of intensive development of the oxidation
process. The resin hardened with maleic anhydride, containing less peroxide compounds, begins to be intensively oxidized at a temperature of
~200°C.

The peroxide concentration increases exponentially with increasing duration of oxidation at the beginning of the process. The experiments described permit us to consider as probable the hypothesis that the oxidation of hardened epoxide resins proceeds according to a chain reaction through alkyl and peroxide radicals, forming hydroperoxides.

Below are cited the temperatures of oxidation of polymer materials, conducted under comparable conditions, for a comparison of their stability with the stability of epoxide resins (oxygen pressure 200 mm Hg):

	T, °C
Polypropylene	100
Polyformaldehyde	135
Aliphatic polyamides	130-140
Epoxide resin hardened with poly- ethylenepolyamine	150
Epoxide resin hardened with maleic anhydride	200
Polycarbonate	250
Polyarylates TD and ID	250

II. PHENOL-FORMALDEHYDE RESINS

Phenol-formaldehyde resins—condensation products of mono- and polyatomic phenols with aldehydes—are capable of being hardened upon heating, i.e., of forming infusible and insoluble products with a three-dimensional structure. G. S. Petrov [13] believed that hardened phenol-formaldehyde resins possess the structure of coal, since their heating above 1000°C without access of air leads to the formation of a large amount of coke – carbon residue. These data were confirmed in [5], where the pyrolysis of hardened phenol-formaldehyde resins of the novolac type under vacuum at temperatures of 360, 500, 800, and 1200°C was studied. Using mass spectrometry, the authors conducted an analysis of the volatile destruction products of the resin (Table 22). We can see from the data cited that the pyrolysis of phenol-formaldehyde resins at high temperatures (>800°C) leads to the formation of substantial amounts of coke (~50%).

Figure 146 [5] shows the relative stability of hardened phenol-formaldehyde and epoxide resins to thermal decomposition. In [14] the oxidative destruction and destruction in an inert medium of a hardened phenol-formaldehyde resin of the resol type were investigated by a thermogravimetric method. The volatile destruction products were studied by the method of direct chromatography. It was found that the initial activation energy of the process of oxidation of the resin within the temperature range 300-380°C corresponds to 15 kcal/mole (S. Madorsky and

TABLE 22. Composition of Gaseous Products of the Pyrolysis
of Hardened Phenol-Formaldehyde Resin

Composition	Pyrolysis temperature, °C			
	360	500	800	1200
	Amount of volatile products, %			
Hydrogen	–	–	3.6	5.6
CO	–	3.5	16.2	24.6
CO_2	0.5	5.5	2.7	2.1
Methane	–	4.3	12.6	9.0
Acetylene	–	–	–	2.8
Ethylene	–	–	1.5	2.4
Acetone	6.7	17.6	1	–
Propylene	4	–	1	3.4
Butanol	2.9	–	–	–
Propanol	10.9	11.1	–	–
Cyclopentadiene	–	–	–	3.1
Benzene	–	2.5	0.6	2.8
Toluene	–	4.7	0.5	–
Dimethylbenzene	–	0.9	–	–
Other gases	–	0	0.7	6.6
Heavy fractions volatile at the pyrolysis temperature and non-volatile at room temperature	75	49.9	59.6	37.6
Total amount	100	100	100	100
Volatiles in sample, %	11	28	44	48

S. Straus [5] determined that the value of the activation energy of the
oxidation of hardened phenol-formaldehyde resin is equal to 18 kcal/mole).

Oxidation at temperatures above 300°C is accompanied by complete
conversion of the resin to volatile products. However, heating to high
temperatures (900°C) in an inert atmosphere leads to the formation of a
stable carbon residue in amounts up to 75% of the initial weight of the
resin.

It was found that the basic volatile oxidation and pyrolysis products
of the resin are phenol and methylphenols, the oxidation products containing more phenol, as a result of the oxidation of the methylene group
to a carboxylic acid (through the stage of peroxide formation) followed
by decarboxylation. Water was also detected in the pyrolysis products
of the resin.

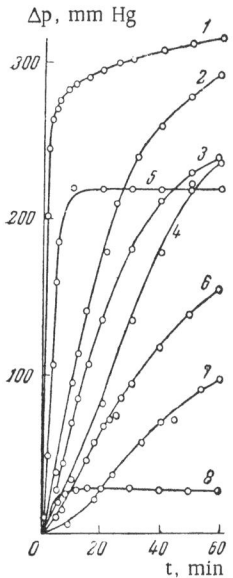

Fig. 147. Kinetics of the gas evolution in the thermal destruction of polyarylates (T = 450°C). 1) FD; 2) IPP; 3) TPP; 4) IH; 5) SD; 6) ID; 7) TD; 8) TH.

The formation of water and phenol agrees with the mechanism of pyrolysis of the resin proposed by Wolfs [15] and Ouchi and Honda [16]. Water is formed in the condensation of phenolic hydroxyl groups. Phenols and methyl-phenols are formed in the pyrolysis of methylene bridges. The free phenyl and benzyl radicals arising in this case remove hydrogen from the neighboring aromatic rings and are converted to volatile products. The carbon atoms lacking hydrogen atoms can become cross-linked. The formation of aromatic hydrocarbons is apparently the result of stripping of hydroxyl groups or cracking of the ethers formed.

The mechanism of the pyrolysis of phenol-formaldehyde novolac resins was studied in greater detail in [17]. A resin with labelled carbon C^{14} in the methylene bridges was synthesized to determine from what structural elements of the resin the coke and gases are formed.

Thermal destruction of the resin was conducted at temperatures of 550 and 800°C. An analysis of the destruction products obtained showed that most of the carbon from the methylene bridges of the resin participate in the formation of graphite (coke) and are not removed in the form of gases.

The carbon-containing gases formed in the pyrolysis of the resin derive partially from the bridge methylene groups and partially from the decomposition of some of the aromatic rings.

III. POLYARYLATES

New polymer materials based on hetero-chain polyesters of diatomic phenols — polyarylates — are of considerable interest thanks to their valuable properties: high thermal stability, mechanical strength, good dielectric properties, the ability to form strong elastic films, etc.

A large group of polyarylates based on diatomic phenols was synthesized by V. V. Korshak and S. V. Vinogradova, et al. [18, 19].

In [20], the thermal destruction of polyarylates based on 4,4'-dihydroxydiphenyl-2,2'-propane and hydroquinone with terephthalic, iso-

TABLE 23. Softening Points and Reduced Viscosities of Polyarylates of Various Structures

Polyarylate	Structural formula	Reduced viscosity	Softening point, °C
Based on 4,4'-dihydroxydiphenyl-2,2'-propane and terephthalic acid (TD)	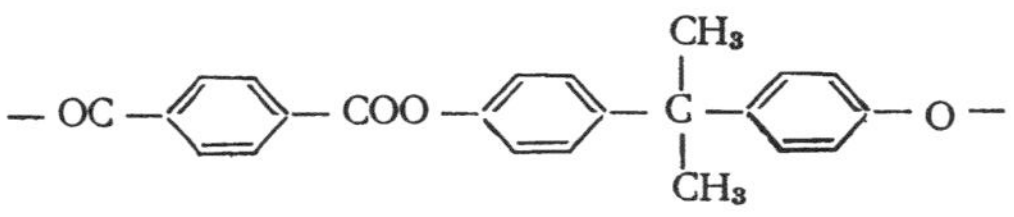	0.24	350
Based on 4,4'-dihydroxydiphenyl-2,2'-propane and isophthalic acid (ID)	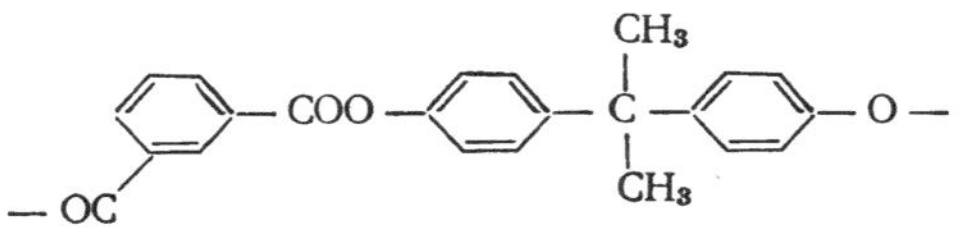	0.265	280
Based on hydroquinone and terephthalic acid (TH)		—	500
Based on hydroquinone and iso-phthalic acid (IH)	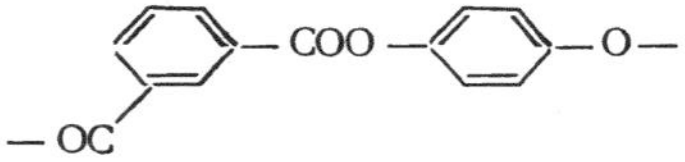	—	450
Based on 4,4'-dihydroxydiphenyl-2,2'-propane and sebacic acid (SD)	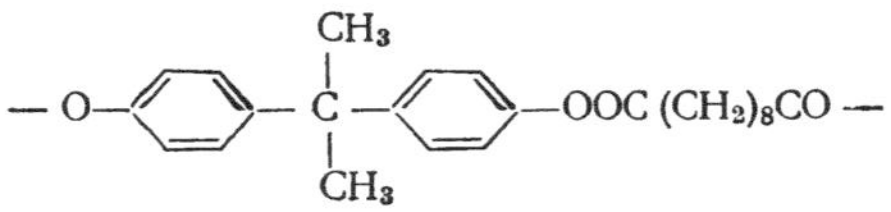	0.56	180

Table 23 (conclusion)

Polyarylate	Structural formula	Reduced viscosity	Softening point, °C
Based on 4,4'-dihydroxydiphenyl-2,2'-propane and fumaric acid (FD)	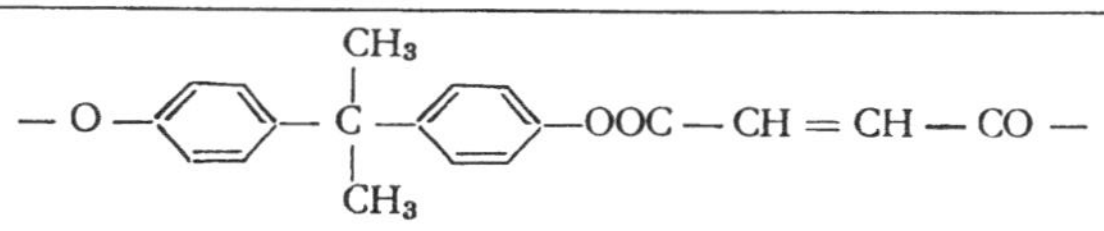	0.28	210
Based on phenolphthalein and terephthalic acid (TPP)	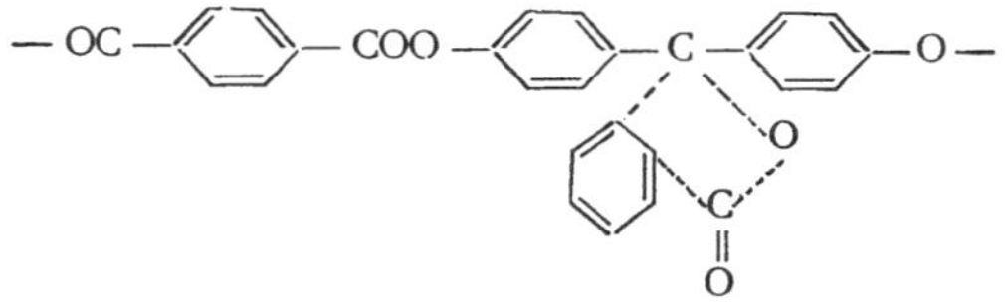	0.64	320
Based on phenolphthalein and isophthalic acid (IPP)	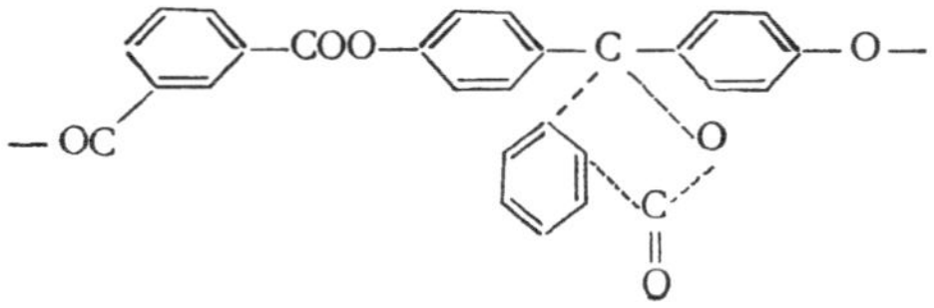	0.48	270

TABLE 24. Activation Energy of the Processes of Thermal Decomposition of Various Polyarylates

Polyarylates	Activation energy of decomposition, kcal/mole	Polyarylates	Activation energy of decomposition, kcal/mole
TH	84	FD	32
TD	60	SD	23
ID	51	TPP	41
IH	35	IPP	36

TABLE 25. Composition of Basic Gaseous Destruction Products of
Polyarylates at a Temperature of 450°C
(moles of gas/mole of polyarylate)

Polyarylates	CO	CO_2	CH_4
TD	0.35	0.3	0.13
ID	0.26	0.56	0.1
IH	0.23	0.64	–

phthalic, fumaric, and sebacic acids, as well as based on terephthalic
and isophthalic acids and phenolphthalein, was studied.

Table 23 presents the structural formulas and characteristics of
certain polyarylates, the thermal decomposition of which was investigated within a broad range of temperatures (250-550°C). The kinetics
of gas evolution in the decomposition of these polymers at 450°C is presented in Fig. 147. Appreciable gas evolution for polyarylates produced
on the basis of aromatic dibasic acids begins above 400°C and is sharply
intensified as the temperature is increased.

The amount of gases liberated from these polymers constitutes 20%
of the initial weight at 500°C. For polyarylates based on 4,4'-dihydroxy-
diphenyl-2,2'-propane with fumaric and sebacic acids, appreciable gas
evolution begins at only 300°C.

Table 24 presents values of the activation energy of the decomposition of polyarylates, calculated from the kinetic curves of gas evolution.

The composition of the gaseous products of the thermal destruction
of certain polyarylates is cited in Table 25.

In addition to the products cited in the table, traces of hydrogen,
ethane, and ethylene were detected among the gaseous decomposition
products of the polyarylates TD and ID. It is obvious that oxides of carbon are formed in the cleavage of ester bridges, while methane is formed

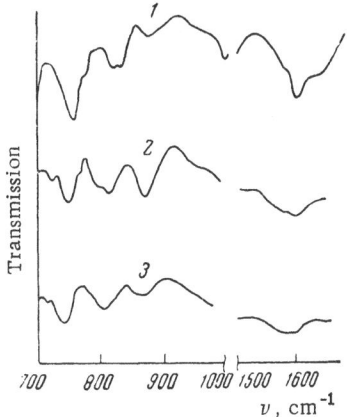

Fig. 148. Infrared absorption spectra. 1) Polyphenylene; 2) solid residue after destruction of the polyarylate TD (600°C, 60 min); 3) solid residue after the destruction of 4,4'-dihydroxy-diphenyl-2,2'-propane (500°C, 60 min).

in the stripping of methyl groups from the radical of 4,4'-dihydroxydiphenyl-2,2'-propane. The latter assumption is confirmed by the absence of methane in the destruction products of the polyarylates TH and IH.

In addition to gaseous products, liquid destruction products and a solid insoluble residue are formed in the thermal decomposition of polyarylates.

Fragments of the polyarylate chain, as well as free diphenylolpropane and terephthalic acid, formed in the decomposition of the polyarylate, have been detected in the infrared spectra of the liquid destruction products of the polyarylate TD, obtained at the temperature of 450°C.

There are no bands characteristic of the methyl group in the spectrum of the solid residue of the polyarylate TD after destruction at 450°C (conducted for 1 hr). The absorption bands characteristic of the ester bond, on the other hand, are absent only in the spectrum of the product subjected to destruction at 600°C for 20 min.

The absorption spectra characteristic of the benzene rings in the region of 700-900 and 1600 cm^{-1} remain rather intense when all the other bands decrease in intensity or disappear. This gives evidence that the residue after destruction is enriched in phenyl rings. The infrared spectra of the destruction products of the polyarylate TD obtained are definitely similar in the region of 700-900 and 1600 cm^{-1} to the infrared absorption spectrum of the polyphenylene obtained from dichlorobenzene (Fig. 148).

The electron paramagnetic resonance spectra, taken for the residues obtained after the destruction of the polyarylate TD at temperatures of 450-500 and 600°C, take the form of narrow singlet signals (width ~ 8 oersted), just as for the case of the polyphenylene structures with conjugated double bonds, described by A. A. Berlin and P. A. Blyumenfel'd [8]. Thus, an investigation of the decomposition products of the polyarylates TD and ID permitted the authors of [20] to hypothesize a possible mechanism of the process. In the thermal decomposition of these polymers, the ester bonds are cleaved, forming oxides of carbon, methyl groups of diphenylolpropane are stripped, forming methane, and the residue after destruction is enriched in phenyl nuclei.

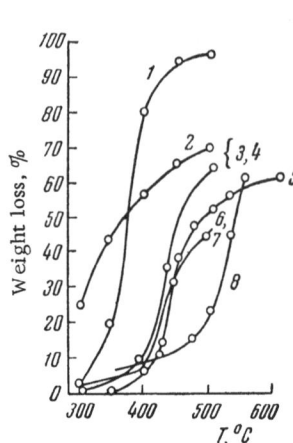

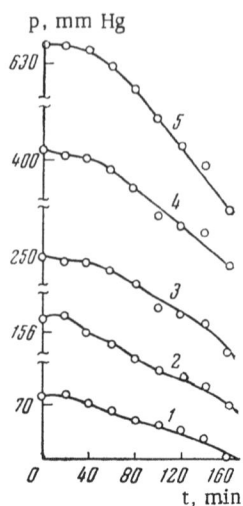

Fig. 149. Dependence of the
depth of decomposition of various polyarylates on the temperature in the case of heating for
60 min. 1) SD; 2) FD; 3 and
4) ID and TD; 5) IH; 6 and 7)
IPP, TPP; 8) TH.

Fig. 150. Variation of the oxygen pressure as a function of the
duration of thermooxidative destruction of the polyarylate TD
(T = 250°C). Initial oxygen
pressure: 1) 80; 2) 160; 3) 250;
4) 420; 5) 650 mm Hg.

The depth of decomposition of polyarylates, determined for various
temperatures according to the loss in weight during heating for 1 hr, is
cited in Fig. 149. We can see from the data obtained that the polyarylates TH, IH, ID, TD, IPP, and TPP are the most thermally stable.
In practice they begin to decompose appreciably at temperatures higher
than 400°C (polyarylate IH at temperatures above 500°C), and, consequently, are capable of prolonged operation at temperatures close to
400°C. The polyarylates SD and FD are less thermally stable and decompose appreciably at temperatures of only 300°C. The lower stability
to thermal decomposition of these polymers is apparently explained by
the presence of aliphatic portions in the polymer molecule, which more
readily undergo destruction than the rigid chains of the polyarylates IH,
TH, and others containing aromatic rings.

The process of oxidation of polyarylates (TD and ID) develops intensively at temperatures above 250°C and is characterized by autoacceleration (Fig. 150). The rate of this process increases sharply with the temperature and substantially exceeds the rate of oxidation of polycarbonate.
The activation energy of the process of oxidation of various polyarylates
lies in the range 20-25 kcal/mole.

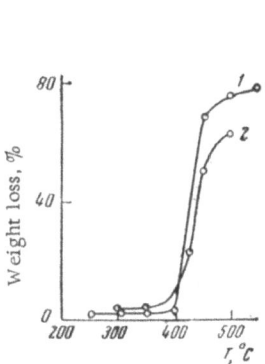

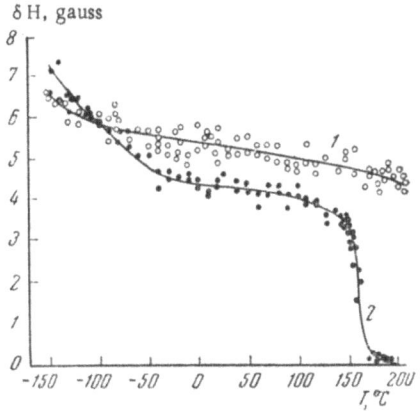

Fig. 151. Dependence of the depth
of decomposition of polycarbonate
(1) and polyarylates TD and ID (2)
on the temperature in the case of
heating for an hour.

Fig. 152. Dependence of the NMR line width (δ H)
on the temperature for polycarbonate destroyed at
500°C (1) and the initial polycarbonate (2).

The presence of hydroperoxides could not be established by an iodometric method at the early stages of oxidation; however, the presence of
products (apparently hydroperoxides) that decompose readily when polyarylates are oxidized was demonstrated by an indirect method [21].

The data obtained in the oxidation of polyarylates have made it possible to conclude that the mechanism of the process is a radical-chain
mechanism.

IV. POLYCARBONATE

Polycarbonate – the product of phosgenization or trans-esterification
of 4,4'-dihydroxydiphenyl-2,2'-propane:

$$\left[-O - \underset{}{\bigcirc} - \underset{\underset{CH_3}{|}}{\overset{\overset{CH_3}{|}}{C}} - \underset{}{\bigcirc} - OC - \right]_n \qquad \text{polycarbonate}$$

can also be included among the group of hetero-chain polyesters based
on diphenyls.

Polycarbonate possesses high mechanical strength, high thermal
stability, transparency, good dielectric properties, thanks to which it is
finding wide use in various branches of industry. It was shown in [22]
that the thermal decomposition of polycarbonate, analogous to certain
polyarylates, begins at temperatures above 400°C and is accompanied

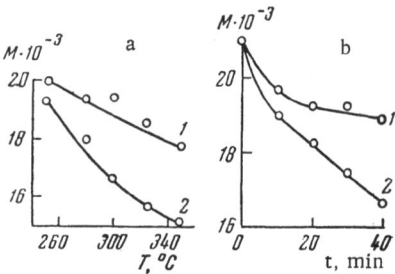

Fig. 153. Change in the molecular weight of polycarbonate as a function of the temperature (a) and
duration (b) of oxidation (1) and casting reprocessing (2).

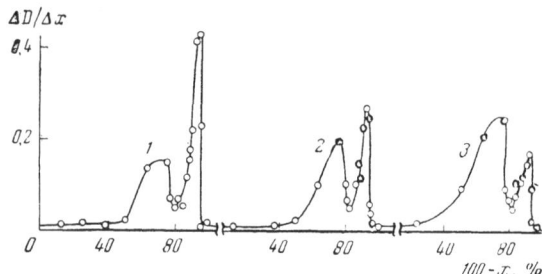

Fig. 154. Differential curves of the molecular weight distribution in samples of polycarbonate produced by casting reprocessing at 250° (1), 275° (2), and 350°C (3).

by appreciable gas evolution (Fig. 151). Considerable amounts of CO,
CO_2, and CH_4, ethane, ethylene, and traces of propylene have been detected among the gaseous destruction products.

The infrared spectrum of the residue of polycarbonates subjected to
pyrolysis at 500°C contains no bands characteristic of methyl groups; the
intensity of the absorption bands of the ester groups $-C = O$ and $-C-O-C-$
is significantly reduced, and the absorption bands corresponding to the
benzene ring are intensified.

The method of nuclear magnetic resonance (NMR) has been used to
obtain supplementary data on the change in structure and molecular motion in the destruction of polycarbonate. For the initial polycarbonate
there are two regions of decrease in the width of the NMR line (Fig. 152,
curve 2), from -10 to $+50°C$ and from 150 to 170°C. The constriction of
the NMR line in the low-temperature region is related to the motion of
the CH_3 groups in the polymer. An analogous decrease in the line width
in this temperature region has also been observed for a number of other
polymers containing CH_3 groups [23, 24]. The rapid drop in the line

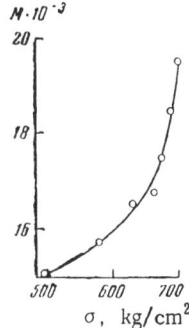

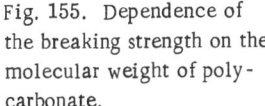

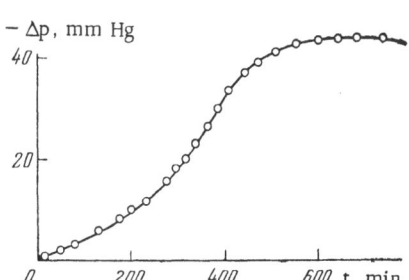

Fig. 155. Dependence of the breaking strength on the molecular weight of poly-carbonate.

Fig. 156. Kinetics of the oxidation of poly-carbonate at 300°C and an oxygen pressure of 200 mm Hg.

width at 150-170°C for polycarbonate is due to the beginning of motion of segments of the polymer chain during transition from the vitreous state to the highly elastic state. Essentially no decrease in the line width all the way up to 210°C (curve 1) is detected for the residue after destruction.

The change in the NMR spectrum after destruction shows that the number of CH_3 groups in the polymer is substantially reduced, and the rigidity of the structure increases considerably. Thus, analogously to polyarylates, thermal destruction of polycarbonate is accompanied by stripping of methyl groups from diphenylolpropane, forming methane, cleavage of ester bonds, forming oxides of carbon, and a significant increase in the rigidity of the structure as a result of enrichment of the residue in phenyl rings after destruction.

In establishing the systems of reprocessing of polymers, one must consider the processes of thermooxidative destruction and their influence on the molecular weight and strength characteristics of the material. This is especially important for such polymers as polycarbonate, poly-arylates, and other polymers that can be reprocessed at high temperatures, in which the oxidation processes develop at a considerable rate.

In [25] the change in the molecular weight and mechanical properties of polycarbonate ("diflon" USSR) during oxidation under laboratory conditions, as well as under the conditions of casting reprocessing, was studied.

Increasing the time of stay of the polymer in the casting machine and increasing the casting temperature leads to a substantial drop in the molecular weight of polycarbonate (Fig. 153a and 153b). An analogous picture is also observed under conditions of pure oxidation on a laboratory setup; however, the greater drop in the molecular weight under

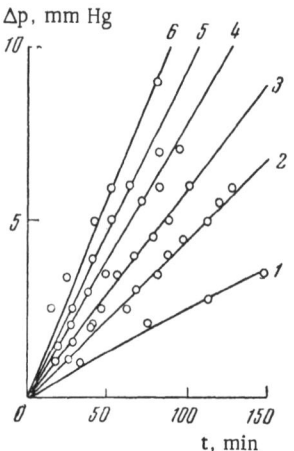

Fig. 157. Kinetics of the thermal destruction of samples of poly-carbonate, subjected to oxidation of varying duration. 1) Without preliminary oxidation; 2) oxidation for 50 min; 3) 90 min; 4) 120 min; 5) 150 min; 6) 190 min.

the reprocessing conditions indicates that a significant role under these conditions is played by mechanical destruction. Both these factors lead to an increase in the content of low-molecular fractions in the polymer, as a result of the cleavage of polymer bonds, which is distinctly evident from the differential curves of the molecular weight distribution in samples of polycarbonate produced according to various systems of casting reprocessing (Fig. 154).

A comparison of the data on the change in chemical properties and molecular weight of polycarbonate subjected to reprocessing under the conditions of various systems (Fig. 155) showed that the drop in strength is due to a decrease in the molecular weight of the polymer.

The process of oxidation of polycarbonate, as was shown in [26], is characterized by autoacceleration (Fig. 156) and is a chain reaction with degenerate branches. The process is accompanied by considerable gas evolution and is appreciably accelerated in the presence of impurities when the starting materials are insufficiently thoroughly purified, as well as in the presence of an excess of unreacted 4,4'-dihydroxydi-phenyl-2,2'-propane. Considerable amounts of CO and CO_2, traces of H_2, as well as water and formaldehyde were detected among the gaseous oxidation products of polycarbonate.

An analysis of the data obtained permitted the authors to propose the following mechanism for the oxidation of polycarbonate: the oxidation process begins with stripping of hydrogen from a methyl group according to the reaction

$$R-\underset{\underset{CH_3}{|}}{\overset{\overset{CH_3}{|}}{C}}-R+O_2=R-\underset{\underset{CH_3}{|}}{\overset{\overset{\dot{C}H_2}{|}}{C}}-R+HOO^{\cdot}$$

The radical obtained adds a molecule of oxygen:

$$R-\underset{\underset{CH_3}{|}}{\overset{\overset{\dot{C}H_2}{|}}{C}}-R+O_2=R-\underset{\underset{CH_3}{|}}{\overset{\overset{\overset{O-O^{\cdot}}{|}}{CH_2}}{C}}-R$$

The peroxide radical isomerizes to the radical:

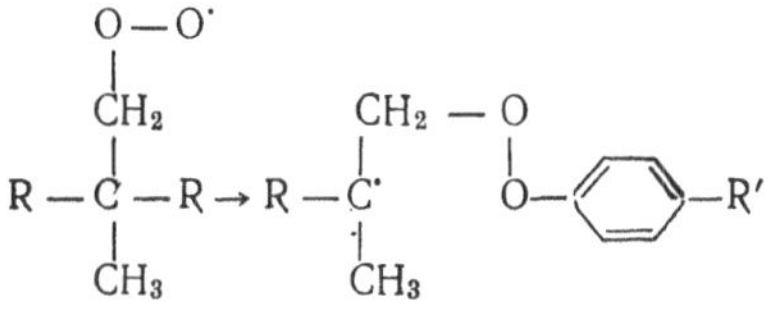

$$
\begin{array}{cccc}
O-O^{\cdot} & & & \\
| & & & \\
CH_2 & & CH_2-O & \\
| & & | & | \\
R-C-R \rightarrow R-C^{\cdot} & & O- \bigcirc -R' \\
| & & | & \\
CH_3 & & CH_3 &
\end{array}
$$

Then the weak $-O-O-$ bond is broken, forming radicals I and II:

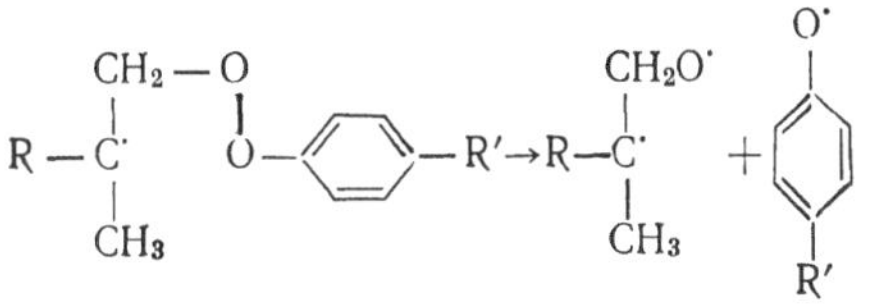

$$
\begin{array}{ccc}
CH_2-O & & CH_2O^{\cdot} \qquad O^{\cdot}\\
| \quad | & & | \qquad\quad | \\
R-C^{\cdot} \quad O- \bigcirc -R' \rightarrow R-C^{\cdot} \quad + \bigcirc \\
| & & | \qquad\qquad | \\
CH_3 & & CH_3 \qquad\quad R'
\end{array}
$$

$$\qquad\qquad\qquad\qquad\qquad\qquad\quad I \qquad\qquad II$$

The biradical I strips the hydrogen from the polycarbonate molecule, forming radical Ia:

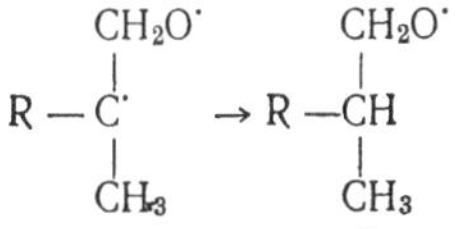

$$
\begin{array}{ccc}
CH_2O^{\cdot} & & CH_2O^{\cdot}\\
| & & | \\
R-C^{\cdot} & \rightarrow & R-CH \\
| & & | \\
CH_3 & & CH_3
\end{array}
$$

$$\qquad\qquad\qquad\qquad\qquad Ia$$

Then radical Ia decomposes into formaldehyde CH_2O and the radical $RC^{\cdot}HCH_3$. In thermal decomposition of formaldehyde, hydrogen and CO are formed, while a phenol derivative is apparently formed from radical II.

The radical $RC^{\cdot}HCH_3$, being oxidized, is converted to a peroxide radical

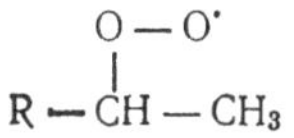

$$
\begin{array}{c}
O-O^{\cdot}\\
|\\
R-CH-CH_3
\end{array}
$$

which isomerizes to an aldehyde according to the scheme:

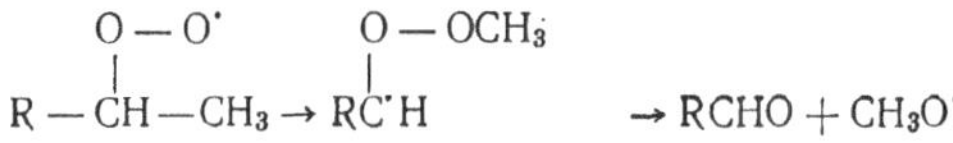

$$
\begin{array}{ccc}
O-O^{\cdot} & O-OCH_3 & \\
| & | & \\
R-CH-CH_3 \rightarrow RC^{\cdot}H & & \rightarrow RCHO + CH_3O^{\cdot}
\end{array}
$$

The methoxyl radical, adding hydrogen, is converted to a methyl radical.

The scheme cited shows that during oxidation not only are gaseous products liberated, but also, at the same time, the structure of the polymer chain is substantially changed; aldehyde and hydroxyl groups accumulate. If this assumption is correct, then the thermal stability of a polymer should decrease sharply as a result of oxidation. Actually, oxidized polycarbonate decomposes far more rapidly when heated under vacuum in the absence of oxygen than does unoxidized polycarbonate, the rate of thermal decomposition being proportional to the duration of preliminary oxidation and proceeding according to a linear law (Fig. 157).

The relative acceleration of the decomposition of oxidized polycarbonate is probably related to decomposition of the aldehyde groups formed in the oxidation. At high temperatures, in addition to the reactions indicated above, cleavage of ester bonds can also occur according to the scheme:

$$R-\bigcirc-O\overset{\overset{O}{\|}}{C}-O-\bigcirc-R \rightarrow R-\bigcirc-O\overset{\overset{O}{\|}}{C}\cdot+\cdot O-\bigcirc-R$$

$$\text{III} \qquad\qquad\qquad \text{IV}$$

Then the radical III splits out CO_2, forming the radical $R-\bigcirc-C\cdot$

$$\text{V}$$

The radicals IV and V capture hydrogen, forming benzene and phenol derivatives.

BIBLIOGRAPHY

1. M. B. Neiman, L. I. Golubenkova, B. M. Kovarskaya, A. S. Strizhkova, I. I. Levantovskaya, M. S. Akutin, and V. D. Moiseev, Vysokomolekulyarnye Soedineniya 1(10):1531, 1959.
2. M. B. Neiman, B. M. Kovarskaya, A. S. Strizhkova, I. I. Levantovskaya, and M. S. Akutin, Plast. Mass. No. 7:17, 1960.
3. M. B. Neiman, B. M. Kovarskaya, A. S. Strizhkova, I. I. Levantovskaya, and M. S. Akutin, Doklady Akad. Nauk SSSR 135(5):1147, 1960.
4. É. G. Gintsberg, B. M. Kovarskaya, and A. S. Strizhkova, Plast. Mass. No. 4:11, 1961.
5. S. L. Madorsky and S. Straus, Mod. Plastics 38(6):134, 1961.
6. V. D. Moiseev, M. B. Neiman, B. M. Kovarskaya, M. E. Zenova, and V. V. Gur'yanova, Plast. Mass. No. 6:11, 1962.
7. D. J. E. Ingram and T. G. Tapley, Chem. & Ind. 568, 1955.
8. A. A. Berlin, L. A. Blyumenfel'd, et al., Vysokomolekulyarnye Soedineniya 1:1361, 1959.
9. L. A. Blyumenfel'd, A. A. Berlin, et al., Strukt. Khim. 1(1):1960.
10. A. V. Topchiev, M. N. Genderikh, et al., Doklady Akad. Nauk SSSR, 128:3121, 1959.
11. H. C. Anderson, Polymer 2(4):451, 1961; J. Appl. Polymer Sci. 6:22, 484, 1962.

12. M. B. Neiman, B. M. Kovarskaya, M. P. Yazvikova, A. I. Sidnev, and M. S. Akutin, Vysokomolekulyarnye Soedineniya 3:602, 1961.

13. G. S. Petrov and S. N. Ustinov, POKh 3(7):393, 1937.

14. G. F. Heron, Thermal Degradation of Polymers, London, 1961.

15. P. M. J. Wolfs, D. W. Krevelin, and H. J. Waterman, Fuel. (London), 39, 15, 1960.

16. K. Ouchi and H. Honda, Fuel (London) 38:429, 1959.

17. V. D. Moiseev, M. B. Neiman, and E. N. Raspopova, Plast. Mass. No. 6:11, 1960.

18. V. V. Korshak and S. V. Vinogradova, Uspekhi Khim. 30:422, 1961.

19. V. V. Korshak, M. S. Akutin, S. V. Vinogradova, et al., Plast. Mass. No. 1:9, 1962.

20. B. M. Kovarskaya, A. S. Strizhkova, I. I. Levantovskaya, A. N. Shabadash, M. B. Neiman, V. V. Korshak, S. V. Vinogradova, and P. M. Valetskii, Vysokomolekulyarnye Soedineniya 4(3):433, 1962.

21. V. V. Ludorov, M. B. Neiman, and A. F. Lukovnikov, Plast. Mass. No. 12:3, 1961.

22. B. M. Kovarskaya, I. E. Zenova, I. Ya. Slonim, and M. B. Neiman, Vysokomolekulyarnye Soedineniya (in press).

23. A. Odajima, A. E. Woodward, and J. A. Sauer, J. Polymer Sci. 55:181, 1961.

24. H. S. Gutowsky and L. H. Meyer, J. Chem. Phys. 21:2122, 1953.

25. M. S. Akutin, V. N. Kotrelev, B. M. Kovarskaya, A. I. Sidnev, E. Podin, O. N. Nitche, and M. B. Neiman, Plast. Mass. No. 6:36, 1963.

26. B. M. Kovarskaya, A. I. Sidnev, M. P. Yazvikova, M. B. Neiman, and M. S. Akutin, Vysokomolekulyarnye Soedineniya 5(5):649, 1963.

Chapter IX

AGING OF POLYMERS WITH INORGANIC PRINCIPAL
CHAINS OF THE MOLECULES, FRAMED BY ORGANIC GROUPS

Among the polymers with inorganic chains of the molecules, framed by organic groups, polyorganosiloxanes have been most studied. These high-molecular substances, the principal molecular chains of which are constructed of silicon and oxygen atoms, differ from organic polymers in their composition and the structure of the polymer molecules. In a consideration of thermal and thermooxidative destruction, these theoretical differences must be taken into consideration.

The difference in the composition and structure of the polymers has its effect in the fact that in polymers with inorganic molecular chains, polyorganosiloxanes, the principal molecular chains are insensitive to oxidation reactions, in contrast to organic polymers. In polymers with inorganic principal molecular chains, the organic framing groups are most sensitive to thermooxidative destruction. Hence the stability of the Si – C bond acquires great significance [1].

In a consideration of the general properties of the bond of carbon to silicon, we must also consider the influence of the basic portion of the molecule. In a covalent bond between carbon and silicon there should be a definite charge distribution, since silicon more readily donates its electron than does carbon. Consequently, the electron cloud that forms the $-Si-C-$ bond is somewhat denser close to the carbon atom, because the charge of its nucleus is not shielded by a full L-shell, and hence exerts a stronger Coulomb attraction on the electrons responsible for the bond. As a result, the carbon atom is more electronegative than the silicon atom, with the same substituents on both atoms.

The general approximate method of theoretical calculation of the polarities of valence bonds has been developed only as applied to the simplest molecules and is based on a quantitative evaluation of the electron affinity of the atoms contained in the molecule, which can be defined as the energy of attraction by a given atom of the electrons bonded to it. The value of the electron affinity depends on the structure of the molecule; hence it varies for the same element.

Hydrogen is more electropositive than carbon.

Thus, the value of the polarity of the bond of silicon with carbon depends on the method by which the ions and molecules will be formed and directed in reactions.

Water is of definite significance in reactions of thermooxidative destruction. It is interesting to consider the reaction of alkali or water with hydroxyls. We should expect that the negative ion (OH) will approach the positive silicon more easily than the negative carbon, as a result of which, on the one hand, a product containing a $-Si-OH$ bond (or its condensation product) is formed, while, on the other hand, a product containing a $C-H$ bond is formed.

The character of the substituents on the carbon and silicon atoms can apparently increase or decrease the reactivity of the bond, depending on whether the substituents increase or decrease the polarity of the $C-Si$ bond. And yet, it is easy to show that it would be an error to overly simplify the problem by explaining the chemical behavior of the bond of carbon with silicon only by its partial ionic character. If we consider, in particular, the hydrides CH_4 (methane) and SiH_4 (silane), we can see that the difference between the electronegativity of silicon and that of hydrogen is somewhat smaller than that between the electronegativities of carbon and hydrogen. If this difference were the main guiding factor for the occurrence of the reaction, then we should expect that silane would be just as stable and inert as methane, which is not confirmed.

From the fact that silane is readily decomposed by aqueous alkalis, it follows that we should bring something else into our consideration, in particular, we should relate the values of the electronegativities of silicon and carbon to the ions or other agents entering into the reaction. Since silicon is more positive than hydrogen, it is especially receptive to the OH group and can split out a hydrogen atom together with its electron pair:

$$\Big\rangle Si : H + (OH)^- \to \left[\begin{matrix} OH \\ | \\ \Big\rangle Si : H \end{matrix} \right] \to \Big\rangle Si - OH + [: H]^- \atop \downarrow \atop H_2$$

The first step in this conversion can be better understood if the partially ionic formation is denoted in the following way:

$$Si^{\delta+} : H^{\delta-}$$

where δ^+ and δ^- denote fractional portions of electron charges. If we attempted to write an analogous scheme for methane, we would easily find an explanation for its observed reaction inertness, since carbon is more positive than hydrogen: $C^{\delta-}$: $H^{\delta+}$. Hence the attacking ion (OH) should go to hydrogen, leading to the formation of water, the H−OH bond. Thereupon the initial state would again be achieved, in which the free energy would not change; hence there are no grounds for the occurrence of such a reaction.

Hence it is essential to know whether a new bond arises before the old one is freed, or whether the old bond is liberated and the new one arises simultaneously. The maximum coordination number of carbon is equal to four, and as a result there is no bond capacity, on the basis of which a complex with five groups at the carbon atom might have an appreciable duration of existence. Hence the old bond is liberated and the new one arises simultaneously in hydrocarbons. However, in silicon such a complex is possible, since it possesses a greater "bonding capacity," i.e., as is frequently stated, it possesses the ability to expand its "valence sphere" to a maximum coordination number equal to six [1]. The possibility of the existence of a pentavalent intermediate complex in organosilicon compounds has been demonstrated:

$$\mathord{>}\!\mathrm{Si}^{\delta+}-\mathrm{C}^{\delta-}+(\mathrm{OH})^- \rightarrow \left[\mathord{>}\!\overset{\delta+\mathrm{OH}}{\ddot{\mathrm{Si}}} : \ \mathrm{C}^{\delta-}\right]^-$$

Thanks to this, the negative group containing carbon is stripped from the intermediate product and stabilized by capture of a proton. Products are thereby formed containing Si−OH, C−H groups, and a new ion (OH)⁻. The polarity of the C−Si bond can be reduced by the introduction, for example, of positive substituents (hydrogen atoms) on the carbon and negative substituents (oxygen atoms) on the silicon.

$$-\mathrm{O}-\overset{\delta\delta+}{\mathrm{Si}}-\mathrm{CH}_3^{\delta\delta-}$$

($\delta\,\delta$ indicates that here we are speaking of partial charges that existed on the silicon and carbon before the introduction of the substituent). As a result, organo-silicon compounds are formed, which are difficultly decomposed under the action of "nucleophilic attack," directed toward the silicon atom, so that it must be boiled with 10N KOH in an autoclave at 200°C to split out methane from it. The polarity of the Si−C bond can be increased by the addition of strongly negative substituents to the carbon at the silicon. Alkaline cleavage of the Si−C bond is thereby facilitated. Trichloromethyl groups are split out from silicon under the action of water, at room temperature, dichloromethyl groups at 45°C; chloromethyl groups are split out at 110°C, and only under the influence of a 10% KOH solution in butanol. Even such a weak base as an alcoholic solution of ammonia leads to the splitting out of chloromethyl groups.

The $-CHF_2$ and $-CF_3$ groups are split out from silicon under the action of cold water [2], which is an evidence of the sensitivity of such silicon-carbon-halogen structures to hydrolysis and to the action of nucleophilic reagents on account of the increased polarizability of the carbon-silicon bond. The stripping of a phenyl group from silicon under the action of acid is explained by the fact that the phenyl group repels electrons from the carbon atom of the Si–C bond and thereby increases the polarizability of the bond. As a result of this, hydrogen chloride attacks this bond, while silicon probably is displaced by hydrogen, splitting out carbon (or possibly a simultaneous attack of Cl^- on $Si^{\delta+}$ and H^+ on $C^{\delta-}$). Under these conditions the methyl group is not split out from the silicon atom.

Polyorganosiloxanes possess great stability to oxidation and stability to the action of high temperatures. Polyorganosiloxanes, for example, polydimethylsiloxanes with a linear molecular structure, are depolymerized with cleavage of the $\geqslant$Si$-$O$-$Si$\leqslant$ bonds, without cleavage of the Si$-CH_3$ bonds. In this case not only the bond energy, but also the structure of the polymer chain determines the temperature stability of the molecule. This is discussed below in greater detail.

The distinguishing feature of the process of oxidation of polyorganosiloxanes is the fact that the organic group is entirely stripped from silicon. Ethyl and methyl groups at the silicon atom are oxidized more readily than phenyl groups. The presence of phenyl groups increases the stability of polyorganosiloxanes, and in polymers when there are two organic radicals on the silicon atom – phenyl and methyl – only the methyl groups are oxidized. In the hydrocarbon chains of organic polymers, the stability to the action of oxygen is somewhat smaller among the same groups, which is explained by the influence of the ionic character of the siloxane bonds on the hydrocarbon portion of the molecule. This has been demonstrated by a study of the infrared spectra [3].

In an investigation of organosilicon compounds, a decrease in the absorption of the C–H bond of methylsiloxane was detected in the region of 3.38, 7.0, and 7.94 μ with increasing number of oxygen bonds on the silicon atom from 0 to 2. Weakening of the C–H bond in the methyl group of methylsiloxane in comparison with this bond in hydrocarbons was probably related to the inductive influence of the dipole of the Si^+–C^- bond, which reduces the dipole moment of the C–H bond. The large value of the Si–C dipole moment is due to the large value of the dipole moment, equal to 2.8 D for the Si^+–O^- bond. In addition, we should mention that silicon, in contrast to carbon, forms oxygen-containing compounds upon oxidation, which are nonvolatile, refractory, and represent polymer substances of complex structure. A typical representative of these substances is silica $(SiO_2)_n$, where each silicon atom is bonded to four atoms. Complete oxidation of any monomer or polymer compound of silicon leads to the formation of $(SiO_2)_n$, i.e., to a complex polymer. When carbon is oxidized, primarily volatile gaseous products are formed, in the form of aldehydes, carbon monoxide, or carbon dioxide. Profound oxidation of any monomer and polymer compounds of carbon leads to the formation of volatile substances.

Thus, silicon and polyorganosiloxanes form complex polymer compounds upon oxidation, with molecular chains $-\overset{|}{\underset{|}{Si}}-O-\overset{|}{\underset{|}{Si}}-O-\overset{|}{\underset{|}{Si}}-$ [4],

while the oxidation of carbon results not in the formation of polymers with chains $-C-O-C-O-C-$, but in the formation of CO_2.

It is known that specially synthesized polymer compounds, polyoxymethylenes, with a similar molecular structure $-C-O-C-O-$, are thermally unstable, being readily destroyed at 170°C.

The thermal stability of the siloxane with a $Si-O-Si$ bond in quartz is extremely high: quartz melts at a temperature of about 1800°C, while heating it at temperatures below the melting point involves no sharp destructure decomposition of the quartz molecules.

In polyorganosiloxanes, the temperature stability of the siloxane bond is considerably lower than that in quartz. A great influence on the thermal stability of the $-Si-O-Si-$ bond is exerted by what atoms or groups of atoms compensate for the other valences of the silicon atom, and especially the number of groups or atoms combined with the silicon atom in addition to oxygen. Organic radicals combined with the silicon atom in polyorganosiloxanes reduce the thermal stability of the siloxane bond, and the stability is reduced with increasing number of organic radicals on the silicon atom [5].

A study of the chemical nature of the radical or group in polyorganosiloxanes makes it possible to establish the influence of the organic part of the polymer on the thermal stability of a molecule with the same chain structure. A change in the number of radicals or groups in polyorganosiloxanes is related not only to a quantitative change in their content of the organic portion, but also to the fact that transition from two radicals at the silicon to one results in a change in the structure of the polymer molecules from linear to three-dimensional, which has an effect on the thermal stability of polyorganosiloxanes.

The literature contains fragmentary information on the thermal stability of certain liquid polyorganosiloxanes; it is indicated that polydimethylsiloxane liquids are stable up to 175°C [6], while polyphenyl-

TABLE 26. Stability of Polymers to Thermooxidative Destruction

Polymer	Chemical composition	Weight loss (in %) in 24 hr at a temperature, °C				
		250	300	350	400	500
Polydimethylphenylsiloxane	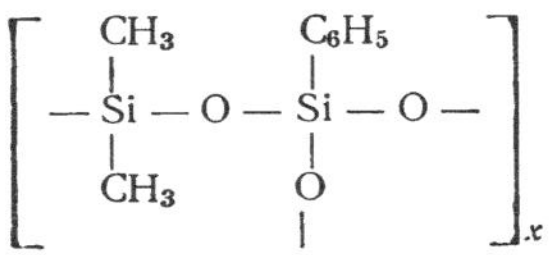 $\left[\begin{array}{c} CH_3 \\ -Si-O-Si-O- \\ CH_3 \end{array} \begin{array}{c} C_6H_5 \\ \\ O \end{array}\right]_x$	7.2	12.0	22.8	36.0	44.7
Polydiethylphenylsiloxane	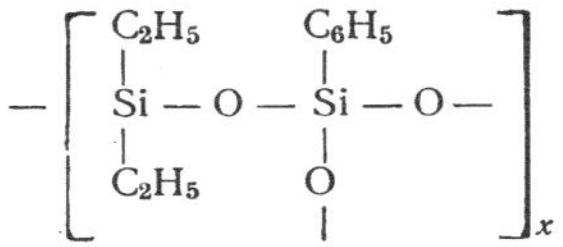 $\left[\begin{array}{c} C_2H_5 \\ -Si-O-Si-O- \\ C_2H_5 \end{array} \begin{array}{c} C_6H_5 \\ \\ O \end{array}\right]_x$	8.3	—	30.2	38.0	—
Polydimethylphenyl aluminosiloxane	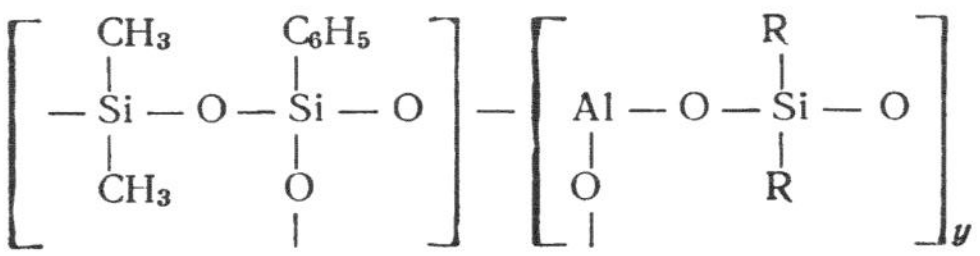 $\left[\begin{array}{c} CH_3 \\ -Si-O-Si-O- \\ CH_3 \end{array} \begin{array}{c} C_6H_5 \\ \\ O \end{array} \begin{array}{c} R \\ -Al-O-Si-O- \\ R \end{array}\right]_y$	—	5.0	8.8	13.0	—
Polytetrafluoroethylene	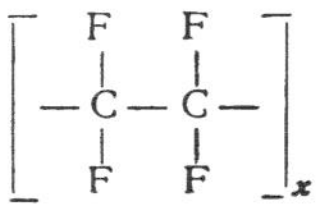 $\left[\begin{array}{c} F\ \ F \\ -C-C- \\ F\ \ F \end{array}\right]_x$	1.3	—	2.1	2.5	45.7

Table 26 (conclusion)

Polymer	Chemical composition	Weight loss (in %) in 24 hr at a temperature, °C				
		250	300	350	400	500
Polytrifluorochloroethylene	$\left[\begin{array}{c} F\ F \\ -C-C- \\ Cl\ F \end{array}\right]_x$	4.6	—	98.9	—	—
Polyamide (capron)	$\left[\begin{array}{c} O \\ -C-(CH_2)_5-NH- \end{array}\right]_x$	55.5	—	94.3	—	—
Epoxide resin	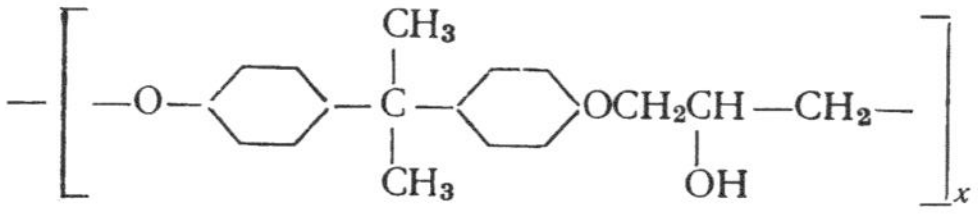 $\left[-O-\bigcirc-\overset{CH_3}{\underset{CH_3}{C}}-\bigcirc-OCH_2\overset{OH}{CH}-CH_2-\right]_x$	22.7	—	93.1	—	—
Polyester glycol maleate	$\left[-OCH_2CH_2O\overset{O}{C}CH=CHO\overset{O}{C}O-\right]_x$	20.5	—	88.7	—	—
Polyethyleneterephthalate	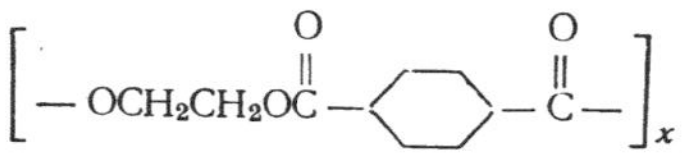 $\left[-OCH_2CH_2O\overset{O}{C}-\bigcirc-\overset{O}{C}-\right]_x$	7.5	—	91.2	—	—
Phenyl-formaldehyde resin	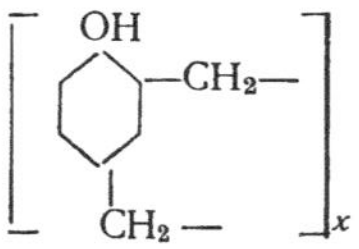 $\left[\overset{OH}{\underset{CH_2}{\bigcirc}}\underset{CH_2}{\overset{CH_2}{}}\right]_x$	5.3	—	68.0	—	—
Nitrile rubber	$\left[CH_2-\underset{CN}{CH}_2\right]_x$	9.38	—	72.0	—	—

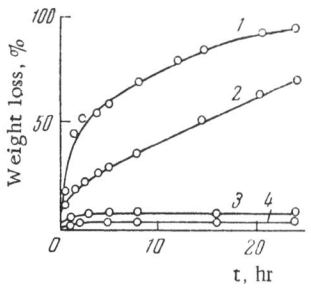

Fig. 158. Kinetics of the destruction of polymers at 350°C. 1) Viniflex; 2) phenol-formaldehyde resin; 3) polymethylsiloxane; 4) polyphenylsiloxane.

methylsiloxane liquid polymers are stable up to 250° [7]. With respect to solid polyorganosiloxanes with a three-dimensional molecular structure, it is known only that polydimethylsiloxanes decompose at 300 °C at a considerable rate [8]. The polydimethylsiloxane polymer decomposes at 400 °C under vacuum, forming low-molecular hexamethylcyclotrisiloxane and octamethylcyclotetrasiloxane [9].

The thermal stability of organosilicon dielectrics [10], liquids treated with ultrasound [11], has been studied. There are indications of an increase in stability to thermooxidative destruction for polydiethylsiloxane liquid polymers after the introduction of phenylnaphthylamine [12].

Table 26 presents comparative data on the stability to thermooxidative destruction of certain classes of organic polymers and polymers with inorganic principal molecular chains, and considers the influence of the structure and chemical composition of polymers with inorganic molecular chains on their thermooxidative stability. Thermooxidative destruction was determined on the pure polymers without fillers, and in certain cases in the presence of fillers. The following were taken as the criteria for evaluation: weight loss of the polymer during the process of heating at various temperatures in the presence of the oxygen of the air, determination of the thermal elasticity of films of the polymers on metallic plates, and change in chemical composition.

Table 30 shows the change in weight of various polymers after heating at temperatures of 250, 300, 350, 400, and 450°C for 24 hr [13]. As we can see a definite difference in the thermooxidative stability exists among organic polymers and polymers with inorganic molecular chains: in the organic polymers the loss in weight after heating is substantially greater than among the polymers with inorganic principal molecular chains.

Polymers with inorganic molecular chains undergo a considerable weight change at the beginning of heating (Fig. 158), but then the process slows down greatly [14]. In contrast to this, organic polymers, being greatly destroyed at the beginning of heating, also continue to undergo uninterrupted destruction subsequently, liberating volatile products at approximately the same rate.

TABLE 27. Change in the Elementary Composition of Polymers after
Thermal Destruction

Condition of destruction	Weight loss, %	C, %	Si, %	C : Si
Polymethylsiloxane (C − 17.8%, Si − 40.9%)				
24 hr at 250°C	2.76	17.12	40.73	1.07
2 hr at 350°C	6.02	8.67	42.68	0.47
Polyphenylsiloxane (C − 55.8%, Si − 21.7%)				
24 hr at 450°C	52.0	3.7	40.8	
6 hr at 550°C	57.5	0.33	46.32	
Polydimethylsiloxane (C −32.7%, Si−38.0%)				
5 hr at 300°C	29.5	28.86	38.78	1.74
5 hr at 350°C	35.0	5.89	42.05	0.32

The investigated organic polymers undergo great decomposition during thermooxidative destruction. Moreover, the thermooxidative
reactions take place not only in the groups framing the principal molecular chain, but also in the main chain itself. The destruction process is
accompanied by the formation of volatile oxidation products.

In polymers with inorganic principal molecular chains, carbon,
which gives volatile oxygen-containing compounds upon destruction, is
not included in the main chain of the polymer molecule, but is included
only among the groups framing the principal chain.

During the process of thermooxidative destruction in polymers with
inorganic principal molecular chains, the reactions develop mainly in
the organic portion of the molecule. In this case oxidation of the organic
groups and further structuring of the polymer occur, leading to steric
hindrance of the oxidation reaction. As a result of the process of thermooxidative destruction, the molecular chains become cross-linked, and
the inorganic portion of the molecule increases, which is confirmed by
the data of Table 27.

An analysis of the volatile products formed in the process of thermooxidative destruction of polymethylsiloxane showed that they consist mainly of carbon monoxide (25. 0%) and water (17%); there are also negligible
amounts of carbon dioxide (2%), formaldehyde (3.7%), methanol, and
traces of formic acid.

The mechanism of the thermooxidative destruction of polyorganosiloxanes probably consists of attack by the oxygen of the air upon a carbon atom on the silicon in the polymer molecule; hydroperoxides are
thereby formed initially, decomposing rapidly:

$$-O-\underset{\underset{|}{\overset{|}{O}}}{\overset{\overset{CH_3}{|}}{Si}}-O-+\dot{O}-O\cdot \rightarrow -O-\underset{\underset{|}{\overset{|}{O}}}{\overset{\overset{CH_2OOH}{|}}{Si}}-O- \tag{I}$$

Product I decomposes to form formaldehyde and a OH radical:

$$-O-\underset{\underset{|}{\overset{|}{O}}}{\overset{\overset{CH_2OOH}{|}}{Si}}-O- \rightarrow -O-\underset{\underset{|}{\overset{|}{O}}}{\overset{|}{\dot{Si}}}-O-+CH_2O+\cdot OH \tag{II}$$

The OH radical reacts with the positively charged silicon atom in the polymer molecule, being added to it:

$$-O-\underset{\underset{|}{\overset{|}{O}}}{\overset{|}{\dot{Si}}}-O-+OH\cdot \rightarrow -O-\underset{\underset{|}{\overset{|}{O}}}{\overset{\overset{OH}{|}}{Si}}-O- \tag{III}$$

$$-O-\underset{\underset{|}{\overset{|}{O}}}{\overset{\overset{|}{}}{Si}}-OH+HO-\underset{\underset{|}{\overset{|}{O}}}{\overset{\overset{|}{}}{Si}}-O- \rightarrow -O-\underset{\underset{|}{\overset{|}{O}}}{\overset{\overset{|}{}}{Si}}-O-\underset{\underset{|}{\overset{|}{O}}}{\overset{\overset{|}{}}{Si}}-O-+H_2O \tag{IV}$$

Formaldehyde undergoes thermal decomposition to form mainly carbon monoxide and hydrogen:

$$CH_2O \rightarrow CO + H_2 \tag{V}$$

Hence carbon monoxide is the basic gaseous destruction product. Under the conditions of our experiments, part of the formaldehyde is oxidized to carbon dioxide according to the scheme:

$$CH_2O + O_2 \rightarrow CO_2 + H_2O \tag{VI}$$

and part is oxidized to formic acid.

The presence of carbon dioxide in the products indicates that formaldehyde is partially oxidized to carbon dioxide. Such a mechanism is confirmed:

1) by the formation of formaldehyde, CO, and H_2O in the thermo-oxidative destruction of polymethylsiloxanes and by the good quantitative coincidence of the decomposition products obtained and calculated;

2) by the preservation of hydroxyl groups in large amounts at the temperatures at which intensive cleavage of the Si—C bonds in the poly-

TABLE 28. Influence of the Framing Groups on the Thermooxidative Stability of Polymers

Polymer	Chemical composition	Weight loss (in %) in 24 hr at a temperature, °C				
		250	300	350	400	450
Polymethylsiloxane	$\begin{bmatrix} CH_3 \\ | \\ -Si-O- \\ | \\ O \\ | \end{bmatrix}_x$	2.8	–	7	–	13
Polyphenylsiloxane	$\begin{bmatrix} C_6H_5 \\ | \\ -Si-O- \\ | \\ O \\ | \end{bmatrix}_x$	2	–	3	8.5	51.5
Polydimethylphenyl-siloxane	$\begin{bmatrix} CH_3 \quad C_6H_5 \\ | \quad\quad | \\ -Si-O-Si-O- \\ | \quad\quad | \\ CH_3 \quad O \\ \quad\quad | \end{bmatrix}_x$	7.2	12	22.8	36	44.7
Polydimethylsiloxane	$\begin{bmatrix} CH_3 \\ | \\ -Si-O- \\ | \\ CH_3 \end{bmatrix}_x$	4	–	34	–	–
Polydiethylphenyl-siloxane	$\begin{bmatrix} C_2H_5 \quad C_6H_5 \\ | \quad\quad | \\ -Si-O-Si-O- \\ | \quad\quad | \\ C_2H_5 \quad O \\ \quad\quad | \end{bmatrix}_x$	8.3	–	30.2	38	–

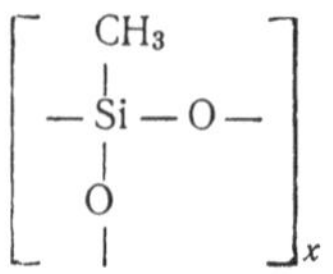

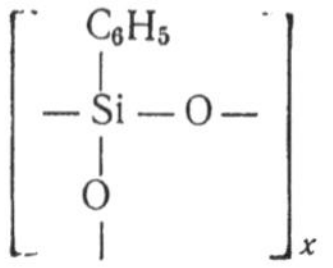

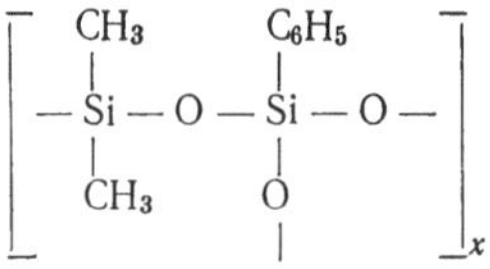

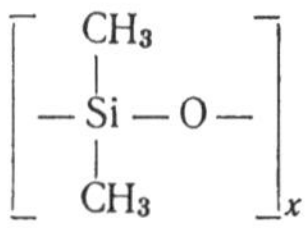

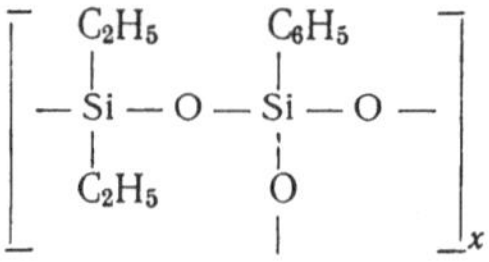

mers occurs; moreover, the HO groups difficultly enter into condensation at high temperatures on account of the great difficulty of their contact with one another.

According to the mechanism cited, it is easy to understand why polyethylsiloxane is more easily destroyed than polymethylsiloxane: the $-CH_2-$ group of the ethyl radical and the silicon atom is oxidized more readily than the CH_3- group. This reaction mechanism permits us to predict the thermal stability of polyorganosiloxanes on the basis of the structure of the organic radical or group at the silicon atom.

We should mention that polyorganosiloxanes with linear molecular chains are more sensitive to thermooxidative destruction than polyorganosiloxanes consisting of cross-linked and three-dimensional molecules. In the presence of the same organic groups, this is shown as a negligible

TABLE 29. Stability of Polymers to Thermooxidative Destruction

Polymer	Chemical composition	Temperature, °C	Time of loss by polymer of half its organic groups, hr
Polymethylsiloxane	$$\begin{bmatrix} & \overset{\displaystyle O^{1/2}}{\underset{\displaystyle O^{1/2}}{|}} & \\ CH_3 - Si - O^{1/2} & \end{bmatrix}_x$$	250 350 450	24 2 0.8
Polyethylsiloxane	$$\begin{bmatrix} \overset{\displaystyle C_2H_5}{\underset{\displaystyle O}{|}} \\ - Si - O - \\ | \end{bmatrix}_x$$	250 350 450	0.8 0.7 0.5
polyphenylsiloxane	$$\begin{bmatrix} \overset{\displaystyle C_6H_5}{\underset{\displaystyle O}{|}} \\ - Si - O - \\ | \end{bmatrix}_x$$	350 400 450 550	24 10 3.2 0.8
Polydimethylphenyl-siloxane	$$\begin{bmatrix} CH_3 & C_6H_5 \\ | & | \\ - Si - O - Si - O - \\ | & | \\ CH_3 & O \\ & | \end{bmatrix}_x$$	350 400	12 5
Polydimethylphenyl aluminosiloxane	$$\begin{bmatrix} C_6H_5 & CH_3 & C_6H_5 \\ | & | & | \\ - Si - O - Si - O - Si - Al - \\ | & | & | & | \\ O & CH_3 & O & O \\ | & & | & | \end{bmatrix}_x$$	350 400	20 10

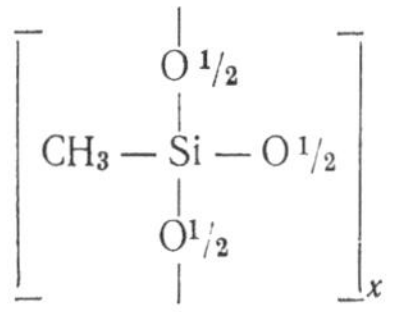

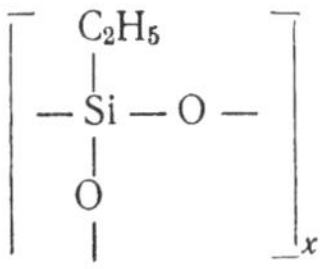

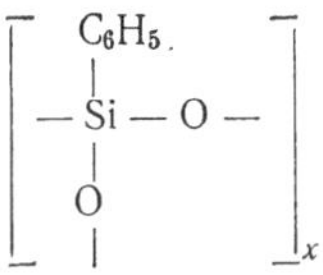

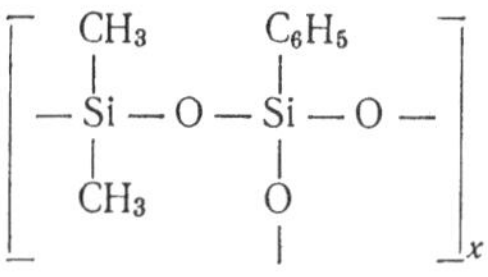

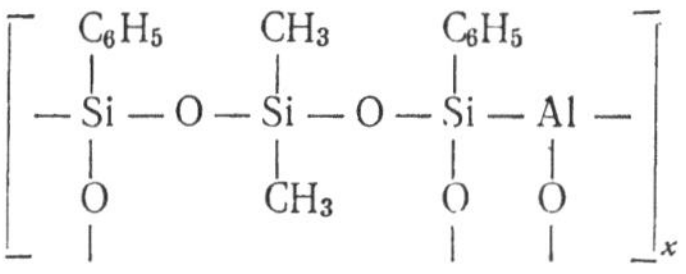

quantitative change in the chemical composition of the polymer, leading to a qualitative change in its structure, and permits the production of high-molecular substances with sharply differing properties.

Changes in the structure of polymer molecules (from linear to three-dimensional) lead to a substantial limitation of the mobility of the polymer chains under the action of temperature. In polymers with linear molecules, the chain mobility is limited by the rigidity of the Si−O bond, i.e., by one chemical bond. Cleavage of such a molecular chain involves cleavage of one Si−O bond. Hence, in polydiorganosiloxanes, thermooxidative destruction of the polymer molecules can proceed both at the less energetically stable Si−C bond, and at the stable Si−O bond. Cleavage of the Si−O bond requires the overcoming of a greater energy level

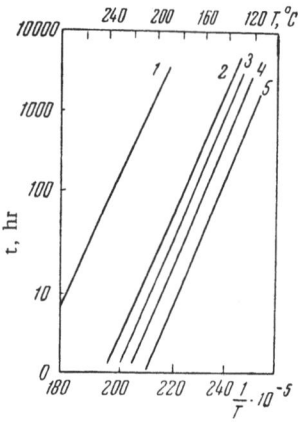

Fig. 159. Curves of the life of polymers as a function of the aging temperature. 1) Polydimethylphenylsiloxanes; 2) polyesterethylene glycol glycerin terephthalate; 3) epoxided polyester; 4) polyester ethylene glycol glycerin sebacate; 5) polyvinylformalethylal.

than for the Si—C bond, but here we must consider the structure of the linear molecules in polydiorganosiloxanes.

The newest data show that in polydimethylsiloxanes, the molecular chains possess a helical structure with three to six silicon atoms in the helix [16, 17]. Under the action of high temperatures, such a structure creates favorable conditions for the closing of rings with chain cleavage. This is also confirmed by the experimental observations. Polydimethylsiloxane decomposes into low-molecular cyclic polymers upon heating under vacuum at high temperatures, without appreciable cleavage of the Si—C bonds:

$$\left[-\ -O-\underset{R}{\overset{R}{Si}}\underset{O}{\overset{O}{\diamond}}\underset{R}{\overset{R}{Si}}-O- \right]_x$$

or

$$\left[-{}^{1\!/_2}O-\underset{\underset{O^{1\!/_2}}{|}}{\overset{\overset{R}{|}}{Si}}-O\underline{\overset{1\!/_2}{}} \right]_x$$

In polyorganosiloxanes with a three-dimensional molecular structure or a cyclo-three-dimensional structure, cleavage of the chain unit or a large portion of the chain of the polymer molecule involves the necessity of its cleavage at three "points," or at two in the case when there are cross-linked molecules in the composition of the polymer. This involves the cleavage of two or three energetically stable Si—O bonds. Hence, in polyorganosiloxanes with three-dimensional bonds, independent of the nature of the organic radical, destruction of the molecular chains at the Si—O bond is not observed even at 550°C.

The influence of the framing groups and structure of the molecular chain in the polymers on the change in weight is shown in Table 28. As we can see, the organic groups sharply influence the loss in weight of the polymer. Thus, polymers whose principal chains are framed by ethyl groups lose more weight than polymers containing phenyl and methyl groups, and only at temperatures above 400°C do polymers containing phenyl groups lose 52% of their weight. This shows that only at temper-

TABLE 30. Stability of Polymers Containing Various Atoms in the Principal Chain

Polymer	Chemical composition	Time, hr	Weight loss (in %) after heating at a temperature, °C				
			200	250	300	350	400
Polydiethylpolyphenyl-siloxane	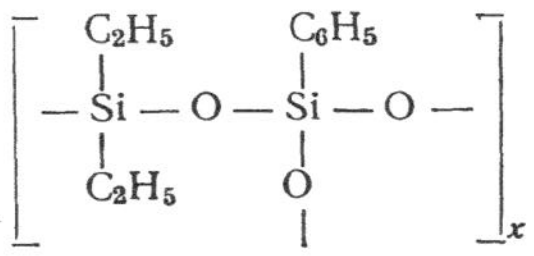 $\begin{bmatrix} C_2H_5 & C_6H_5 \\ -Si-O-Si-O- \\ C_2H_5 & O \end{bmatrix}_x$	24	10	16	20.5		
		72	13.5	28	33		
		360	21	30	45		
Polydimethylpolyphenyl-siloxane	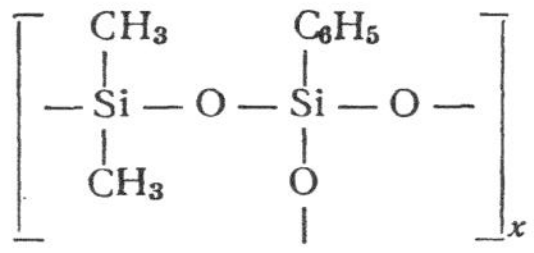 $\begin{bmatrix} CH_3 & C_6H_5 \\ -Si-O-Si-O- \\ CH_3 & O \end{bmatrix}_x$	24	3	4.5	12		
		72	3.75	5.8	17.2		
		360	4.5	6.3	22		
Polydimethylpolyphenyl aluminosiloxane	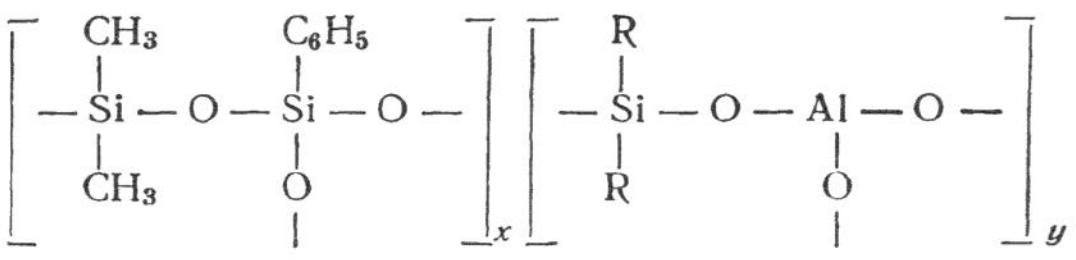 $\begin{bmatrix} CH_3 & C_6H_5 & R \\ -Si-O-Si-O- & -Si-O-Al-O- \\ CH_3 & O & R & O \end{bmatrix}_x \ \ _y$	24	—	—	5	8.8	13
		72	—	—	8.4	12	18
		360	—	—	10	15	29

TABLE 31. Loss of Elasticity of Polymers during Heating

Polymer	Chemical composition	Time of heating until loss of elasticity (hr) at a temperature, °C			
		180	200	210	220
Polydiethylpolyphenyl-siloxane	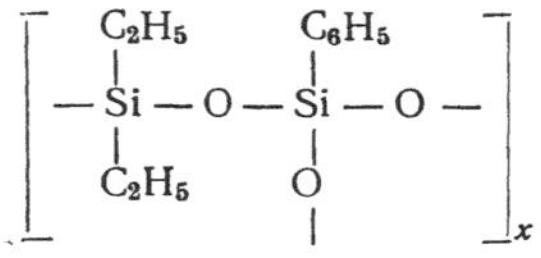 $\left[\begin{array}{cc} C_2H_5 & C_6H_5 \\ -Si-O-Si-O- \\ C_2H_5 & O \end{array}\right]_x$	—	90	—	18
Polydimethylpolyphenyl-siloxane	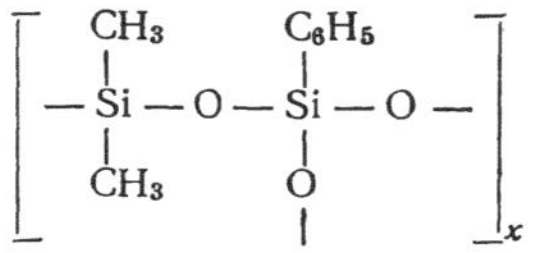 $\left[\begin{array}{cc} CH_3 & C_6H_5 \\ -Si-O-Si-O- \\ CH_3 & O \end{array}\right]_x$	—	700	—	150
Polydimethylpolyphenyl aluminosiloxane	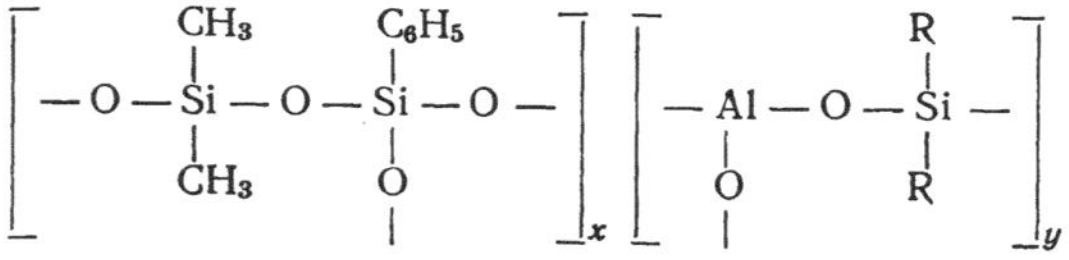 $\left[\begin{array}{cc} CH_3 & C_6H_5 \\ -O-Si-O-Si-O- \\ CH_3 & O \end{array}\right]_x \left[\begin{array}{c} R \\ -Al-O-Si- \\ O \quad R \end{array}\right]_y$	—	48	—	24

Table 31 (conclusion)

Polymer	Chemical composition	Time of heating until loss of elasticity (hr) at a temperature, °C			
		180	200	210	220
Polydimethylpolyphenyl-siloxane	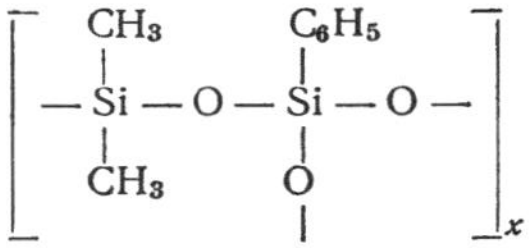 $\left[\begin{array}{cc} CH_3 & C_6H_5 \\ -Si-O-Si-O- \\ CH_3 \end{array}\right]_x$	2000	700	150	—
Polytrifluorochloro-ethylene	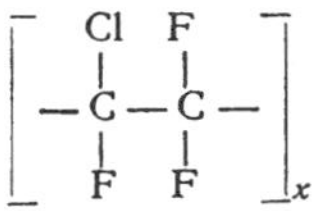 $\left[\begin{array}{cc} Cl & F \\ -C-C- \\ F & F \end{array}\right]_x$	120	70	30	—
Polyester glycol glycerin sebacate	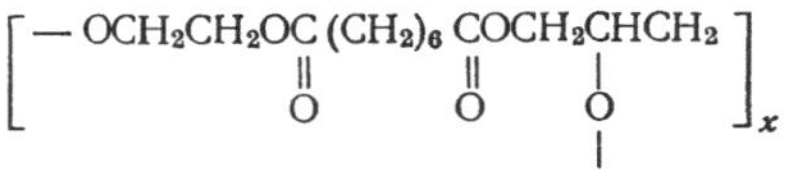 $\left[-OCH_2CH_2OC(CH_2)_8 COCH_2CHCH_2-O-\right]_x$	20	6	0	—
Polyester glycol glycerin terephthalate	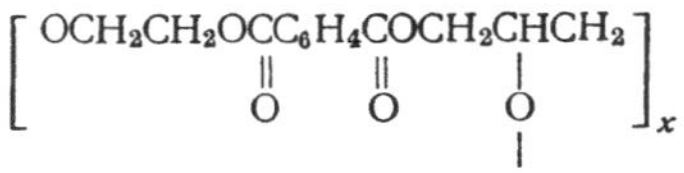 $\left[OCH_2CH_2OCC_6H_4COCH_2CHCH_2-O-\right]_x$	80	48	4	—
Polyvinylformalethylal	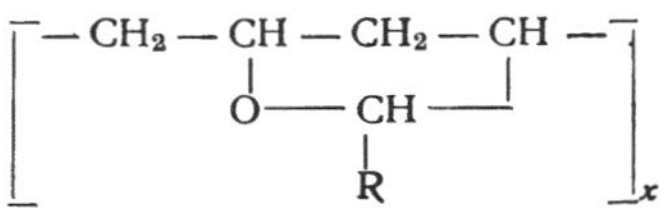 $\left[-CH_2-CH-CH_2-CH-\\ O-CH-R \right]_x$	8	2	0	—

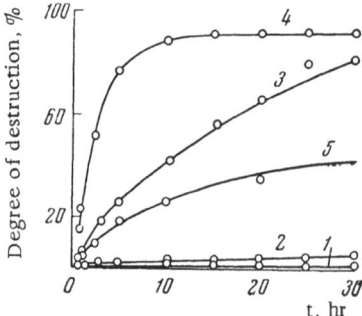

Fig. 160. Rate of thermal decomposition of polydimethylsiloxane in a stream of nitrogen. 1) 250°C; 2) 300°C; 3) 350°C; 4) 400°C; 5) thermooxidative decomposition of polydimethylsiloxane at 350°C.

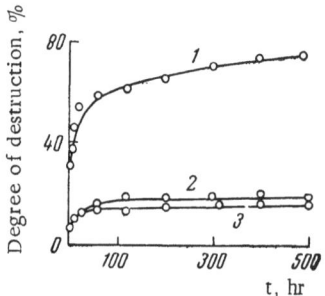

Fig. 161. Rate of thermooxidative decomposition at 300°C for polymethylsiloxanes produced under various conditions. 1) Conditions favorable for the formation of rings; 2 and 3) conditions promoting the formation of cross-linked chains.

atures above 400°C does rapid stripping of the phenyl groups occur, while the ethyl groups begin to be oxidized at a great rate at 250°C, while the methyl groups begin to be oxidized at 350°C.

Table 29 presents the time in which the polymer loses half of its organic groups. This time differs sharply for different polymers.

Various groups in the polymers can arranged in the following series with respect to increasing stability to heating in air:

$$C_2H_5- < CH_3- < C_6H_5-$$

It is interesting to follow the changes in the thermooxidative stability of polymers with inorganic molecular chains, when the principal chain of the molecule contains not only silicon, but also one or two other elements. Tables 29 and 30 show the stability to thermooxidative destruction of polymethylsiloxane containing links of

$$-\underset{\underset{R}{|}}{\overset{\overset{R}{|}}{Si}}-O-\underset{\underset{O}{|}}{\overset{}{Al}}-O-\quad\text{and}\quad-\underset{\underset{R}{|}}{\overset{\overset{R}{|}}{Si}}-O-\underset{\underset{R}{|}}{\overset{\overset{R}{|}}{Si}}-O-$$

As can be seen from Table 30, a significant reduction of the loss in weight during the process of heating is observed, but the elasticity is thereby substantially reduced in polymers containing aluminum.

In a study of the elasticity of polymers in the form of films 0.05 ± 0.005 mm thick at temperatures of 180, 200, 210, and 220°C (Table 31), a considerable difference was also observed in the properties of organic polymers and polymers with inorganic principal molecular chains, framed with various organic groups. All the tested films from organic polymers lose their elasticity considerably more rapidly during heating than do films from polymers with inorganic principal chains. Figure 159 shows the time of heating at various temperatures at which polymer films

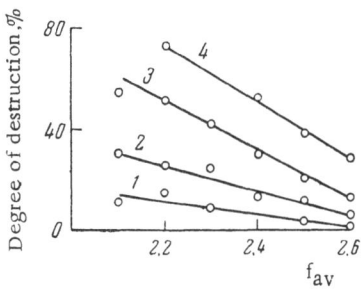

Fig. 162. Dependence of the degree of thermooxidative decomposition of polyphenylmethylphenylsiloxanes on the average functionality (degree of cross linking) of the polymer at various temperatures: 1) 250°; 2) 300°; 3) 350°; 4) 400°C.

reach a lengthening of less than 4%. As can be seen from the figure, the dependence of the time of loss of elasticity of various polymers on the temperature rather well obeys an Arrhenius equation.

By calculating the activation energy E according to the Arrhenius equation for a number of polymers, we find the following values (in kcal/mole): polydimethylphenylsiloxane 36, polydimethylphenylsiloxane modified 38, polyvinylformalethylal 26, polyester 25. Consequently, the activation energy of the process of thermooxidative destruction of polymers with inorganic molecular chains, calculated according to the thermoelasticity, is higher than the same energy for organic molecules. It is interesting that the calculation of the activation energy determined according to thermoelasticity rather closely coincides with the calculation of the activation energy according to the drop in the electric breakdown voltage during the process of aging. In this case we found E = 33 kcal/mole for polydimethylphenylsiloxane and E = 24 kcal/mole for polyethyleneterephthalate.

The character of the chemical processes that occur during oxidation is of great significance for an explanation of the high thermoelasticity of polyorganosiloxanes. In spite of the relative stability of the organic radicals on the silicon atom, they still become oxidized as time passes. Actually, for example, polymethylsiloxane with the structure unit $(CH_3SiO_{1.5})$ loses 2.8 and 7.0% of its weight in 24 hr at 250 and 350°C, respectively. It might seem that this loss at such temperatures is very small in comparison with organic polymers. But we should consider that a 7% weight loss corresponds to the stripping of approximately 75% of all the methyl radicals of the given polyorganosiloxane. The small weight loss is explained by the fact that an oxygen atom takes the place of two CH_3 radicals.

This also is one of the most important factors in the high thermoelasticity of polyorganosiloxanes. Oxygen takes the place of organic radicals in polyorganosiloxanes during thermooxidative destruction, thereby creating more and more new siloxane bonds, characterized by great thermal stability. As a result, the polymer is not destroyed at individual units of the principal chain, but only a branched siloxane structure of the molecule is produced. This process of "siloxane structuring" proceeds most intensively in the outer regions of the polymer. A layer that slows down the access of oxygen into the polymer is formed on its surface.

An analysis of the lacquer film of the dimethylphenyl polymer in the initial state and after heating at 250°C until loss of elasticity showed an increase in the Si content (in the form of SiO_2) from 25.49 to 25.92%, a decrease in the H content from 5.47 to 4.80%, and a decrease in the C content from 45.89 to 44.17%; the C : Si ratio was reduced from 4.47 to 3.68, while the H : C ratio remained approximately unchanged (before heating 1.35, after heating 1.30).

As is well known, two competing processes occur when polymers are heated: destruction and structuring. Polyorganosiloxanes are character-ized during thermal oxidation by a predominance of structuring, as a result of which the rate of decomposition of linear and three-dimensional polymers slows down as heating progresses. However, the rate of de-composition of the linear polymers remains higher than the rate of de-composition of the three-dimensional polymers with the same radical. This is explained by the greater mobility of the linear polymer chains in comparison with cross-linked chains, since cleavage of the principal molecular chain of cross-linked polymers requires the cleavage of more than one siloxane bond. The tendency toward cleavage of the siloxane bond is especially sharply manifested when polydimethylsiloxane is heated in the absence of oxygen (in a stream of nitrogen). The rate of decom-position under these conditions, judging by the weight loss of the sample, is almost twice as high as in the presence of oxygen (air) (Fig. 160). Ap-parently heating polydimethylsiloxane in a stream of nitrogen leads to a predominance of destruction, since there is no oxygen responsible for the oxidation of the radical and the formation of a transverse siloxane bond.

The rate of thermooxidative decomposition at 300°C for a three-dimensional polymethylsiloxane produced from a trifunctional monomer depends substantially on the method of its production. Figure 161 shows the rate of decomposition of such polymers with the same elementary composition, but prepared under different conditions. In one case (curve 1) the conditions were favorable for ring formation, while in the other two cases (curves 2 and 3) the conditions favored the formation of cross-linked chains.

A considerable influence on the rate of decomposition of polyorgano-siloxanes at increased temperatures is exerted by the catalysts that are usually used for hardening. Even a negligible amount of them (0.1%) quadruples the rate of decomposition of polymethylphenylsiloxane. These data are also of great practical significance and give evidence of the neces-sity for thorough purification of the polymer from polycondensation cata-lysts and a consideration of the catalytic action of certain fillers and pigments on the rate of decomposition of the polymers. Great influence on thermooxidative destruction is exerted by the average degree of cross-linking of the polymer (Fig. 162).

BIBLIOGRAPHY

1. K. A. Andrianov, Organosilicon Compounds, Moscow, State Press for Chemical Literature, 1955.
2. R. Krieble and J. Elliot, J. Am. Chem. Soc. 67:1811, 1945.
3. N. Wright and M. Hunter, J. Am. Chem. Soc. 69:803, 1947.
4. K. A. Andrianov and N. N. Sokolov, Khim. Prom. 6:329, 1955.
5. W. Patnode and D. Wilcock, J. Am. Chem. Soc. 68:358, 1946.
6. D. Atkins, C. Murphy, and C. Saunder, Ind. Eng. Chem. 39:1395, 1947.
7. C. Murphy, C. Saunder, and D. Smith, Ind. Eng. Chem. 42:2462, 1950.
8. E. Rochow and W. Silliam, J. Am. Chem. Soc. 63:798, 1941.
9. W. Patnode and D. Wilcock, J. Am. Chem. Soc. 68:358, 1946.
10. V. I. Kalitvyanskii, Elektrichestvo No. 3:55, 1955.
11. E. M. Oparina, G. S. Tubyanskaya, and A. S. Ermilova, Chemistry of Organosilicon Compounds, No. 2, Leningrad, Central Office of Technical Information, 1958, p. 50.
12. L. V. Tonets and T. Z. Lyazgunova, Ibid., p. 28.
13. K. A. Andrianov, Thermally Stable Organosilicon Dielectrics, Moscow, State Power Engineering Press, 1957.
14. K. A. Andrianov and N. N. Sokolov, Elektrichestvo No. 5:31, 1955.
15. K. A. Andrianov, High Temperature Resistance and Thermal Degradiation of Polymers, Soc. of Chemical Industry, Monograph N13, London, 1961.
16. A. Barry, J. Appl. Phys. 17:1020, 1946.
17. H. Fox and P. Taylor, Ind. Eng. Chem. 39:1401, 1947.

Chapter X

AGING AND STABILIZATION OF RAW AND CURED RUBBERS

In recent years a number of monographs and survey articles have been published on individual problems of aging [1-13], from which we can compile an idea of the modern state of the problem. In this article we shall attempt to generalize certain new results, the analysis of which may be of interest for the further development of work on the aging and stabilization of raw and cured rubbers.

The substantial difference in the mechanisms and rates of aging of purified and technical raw rubbers and cured rubbers prepared on the basis of them have many times been noted in the literature [12, 14, 15].

It is well known that the presence in technical raw rubbers of contaminations, introduced with the monomers, as well as residues of the catalysts, initiators, and regulators of polymerization exert a vital influence on the aging of rubbers and the behavior of inhibitors. It is also known that the introduction of a large number of ingredients into raw rubbers (sulfur, vulcanization accelerators, carbon blacks, plasticizers) radically changes the character of the processes of aging of rubber mixtures in comparison with technical rubbers.

Free impurities and ingredients in most cases weaken, and sometimes suppress the action of antioxidants. For example, sulfur substantially reduces the effectiveness of antioxidants belonging to the class of secondary aromatic amines [2]. Other cases are also known when the action of antioxidants is intensified by certain ingredients [16]. Thus, both a negative and a positive synergic effect may appear in rubber mixtures. Impurities contained in the monomers enter the structure of the polymer chains during polymerization, disturb their regularity, and frequently are the weak sites with which their thermal and thermooxidative decomposition begins. Hence the tendency to produce purer polymers from purified monomers, which will promote a substantial increase in their stability to aging, has now been outlined.

The development of a three-dimensional space lattice in raw rubbers during vulcanization creates a supplementary source for breakdown of the cured rubbers during thermal aging, since the transverse, especially polysulfide bonds themselves are capable of decomposing under thermal and

312

thermomechanical influences at an incomparably greater rate than the molecular chains of the raw rubbers. This problem will be discussed in greater detail below.

It is natural that the creation of strong transverse bonds by means of vulcanization will far increase the thermal and thermomechanical stability of the vulcanizates. Promising in this respect is rational vulcanization, promoting the formation of a space lattice of strong bonds.

I. THERMAL OXIDATION OF RAW RUBBERS
IN THE PRESENCE OF INHIBITORS
(At Moderate Temperatures)

Under the influence of small additions of amines, phenols, and certain other classes of compounds, the process of polymer oxidation is sharply decelerated; for example, 1% phenyl-β-naphthylamine reduces the rate of oxidation of polybutadiene during the steady-state period at 70°C more than six-fold, while the duration of this period increases sharply. By decelerating the oxidation process, the inhibitor naturally reduces the rate and even changes the character of the structural changes in the polymers.

Although structuring predominates from the very beginning in the free oxidation of polybutadiene, in the case of inhibited oxidation, a prolonged period arises during which destruction predominates. Thus, the inhibitor, while reducing the general rate of the processes leading to a change in the structure, influences the process of structuring of the chains to a greater degree. This is natural, since events of growth of the reaction chain lead to the structuring of the polymers. Decreasing the length of the reaction chain leads to a decrease in the rate of structuring.

Recently attempts were undertaken for the first time to study the nature of the oxygen-containing products that accumulate in the polymer during inhibited oxidation. In this study it was found that at temperatures above 100°C, a considerable portion of the oxygen is present in the form of aldehyde groups; the formation of alcohol and carboxyl groups is observed in smaller amounts. Peroxide groups are practically absent under these conditions.

The observed patterns contradict the generally accepted representations of the mechanism of polymer oxidation in the presence of amines and phenols, according to which the radical leading the chain is a peroxide radical, while inhibition of the process is accomplished as a result of the reaction of this radical with an inhibitor molecule. Although in the free autocatalytic oxidation of low-molecular hydrocarbons at moderate temperatures, hydroperoxides are practically the only reaction product of the early stages of the process, in the oxidation of polymers, only a small portion of the oxygen is present in the form of peroxides. N. N. Semenov [17] has proposed a mechanism of isomerization of the peroxide radical, forming aldehydes as the primary stable oxidation products, for low-molecular hydrocarbons oxidized at high temperatures.

New data on the inhibited oxidation of polymers, just like the results obtained in a study of a freely developing autocatalytic process, unambiguously indicate the correctness of the mechanism of isomerization of peroxide radicals of polymers even at moderate temperatures (80-110°C). This, of course, is specific only for polymers, as a result of the high viscosity of which the lifetime of the free radicals is increased, and favorable conditions are created for the occurrence of the reaction of their monomolecular decomposition.

Certain important principles have been detected in the behavior of inhibitors in the oxidation of rubbers. It has been established earlier that the oxidation of rubbers, as well as that of low-molecular hydrocarbons, is accompanied by consumption of the inhibitor, which is added to the molecules of the substance to be oxidized. In a number of cases it has been shown that the consumption of the inhibitor occurs at a constant rate. This phenomenon finds a kinetic explanation if we assume that the inhibitor is consumed only as a result of its reaction with the hydrocarbon being oxidized, leading to chain termination. In this case, if termination proceeds primarily on the inhibitor, in a steady-state process the rate of inhibitor consumption is equal to the rate of initiation. It has been established, however, that such a principle is manifested only at small inhibitor concentrations (up to 1%); at concentrations exceeding 1%, the consumption or addition of the inhibitor to the polymer already does not occur linearly, but the initial rate of consumption increases with increasing inhibitor concentration. Thus, the rate of inhibitor consumption does not always reflect the process of inhibited oxidation of polymers. In the case of inhibitors of the class of secondary amines, not only addition of the amine radical to the hydrocarbon being oxidized, but also partial recombination of its radicals occurs, forming dimers of the type

$$
\begin{array}{cc}
R_1 & R_1 \\
\diagdown & \diagup \\
N & - & N \\
\diagup & \diagdown \\
R_2 & R_2
\end{array}
$$

When the amine hydrogen atom is replaced by deuterium, the inhibiting action of the secondary amine is appreciably reduced. If this atom is replaced by a methyl group, the inhibiting action of the amine is entirely lost.

This gives evidence that the reaction of inhibition by secondary amines takes place primarily with cleavage of the N−H bond and formation of a relatively inactive diarylnitrogen radical

Dimers − recombination products of phenoxyl radicals − have also been detected in the inhibition of oxidation processes in polymers by phenols.

These facts indicate that the inhibition of the oxidative process by compounds of the phenol type is also based on a reaction of stripping of a labile hydrogen atom from the phenol molecule. Such a view of the mechanism of inhibition by phenols is maintained by many researchers. In the case of tertiary amines, it is assumed that the mechanism of the inhibition reaction is based on the transfer of an electron, forming a radical-ion, and ultimately stable products.

It has been shown that certain mercaptans of sulfonic acids and alkyl sulfides are capable of decomposing peroxides and other products of polymers, forming stable compounds. The same hypothesis is advanced with respect to derivatives of mercaptobenzimidazole − so-called deactivators of the process.

The material cited gives evidence that the mechanism of inhibition is varied in various classes of compounds. The possibility is not excluded that each of these classes can act in several ways. However, one of them is always predominant.

It is well known that as the temperature increases, the rate of inhibited oxidation of rubbers increases substantially, while the induction period decreases. The increase in the rate of the process with the temperature is related to an increase in the rate of chain initiation, and, partially, to a decrease in the effectiveness of the inhibitors. The latter follows from the fact that the inhibiting effect is determined by the ratio k_{inh}/k_{dev}.

Since E of the inhibition reaction is smaller than that of the reaction of chain development, the ratio of the constants drops with increasing temperature, and the decelerating action of the inhibitor is reduced. The rate of nonproductive consumption of the inhibitor, related to its direct oxidation by molecular oxygen, also increases with increasing temperature.

The effectiveness of the action of inhibitors depends to a great degree on their concentration.

millimoles of O_2/mole·sec

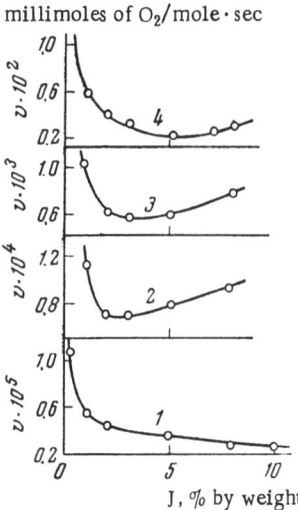

Fig. 163. Dependence of the rate of oxidation of divinylstyrene rubber on the concentration of phenyl-β-naphthylamine at various temperatures; 1) 70°; 2) 100°; 3) 130°; 4) 150°C.

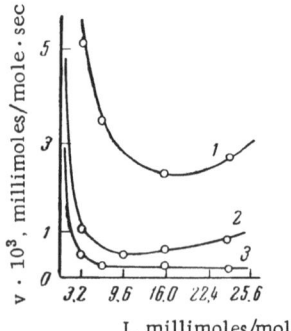

Fig. 164. Dependence of the rate of oxidation of divinylstyrene rubber on the concentration of various amines. 1) With diphenylamine; 2) with phenyl-β-naphthylamine; 3) with di-β,β'-naphthylamine at 100°C.

It has been shown that this dependence is described in a number of cases by a curve with a minimum [18, 19].

Figure 163 shows the dependence of the rate of inhibited oxidation of divinylstyrene rubber as a function of the concentration of phenyl-β-naphthylamine at various temperatures. This function is described by a curve with a minimum, and only at 70°C by a "saturation" curve.

A comparison of the experimental data, as well as a theoretical consideration of the problem has shown that the increase in the rate beyond the optimum concentration of the inhibitor is related to the ability of the inhibitor molecule to undergo direct oxidation by molecular oxygen, and thus to emerge partially in the role of an initiator of polymer oxidation.

As the inhibitor concentration is increased, the rate of its oxidation increases, and, consequently, so does its initiating action. From this follows the important conclusion that the more readily the inhibitor is oxidized, the smaller the amounts of it at which the minimum rate of oxidation of the polymer will be observed.

Thus, increasing the effectiveness of an inhibitor requires a reduction of its reactivity with respect to molecular oxygen.

In the case when the oxidation process is related to stripping of a labile hydrogen, which takes part in the inhibition reaction (as in amines), it is found that the requirements for an effective inhibitor are contradictory.

The smaller the energy of the N–H bond, the more readily the inhibition reaction proceeds, but, at the same time, the more the effect of initiation, which reduces its effectiveness, is manifested. Obviously,

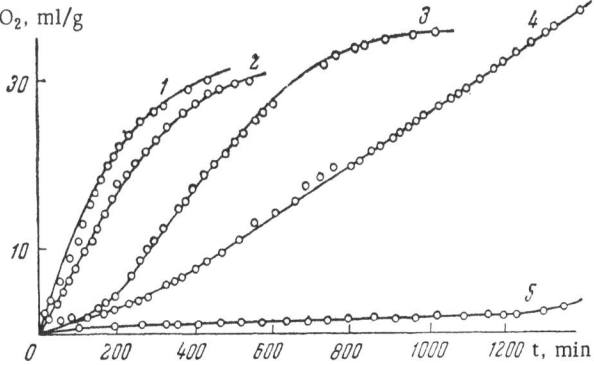

Fig. 165. Kinetics of the oxidation of purified methylvinylpyridine rubber MVP-15 at various temperatures (1-4) and that
of SKB (5). 1) 50°; 2) 60°; 3) 70°; 4) 80°; 5) 70°C.

in view of this a decrease in the energy of bond cleavage should have a
sensible limit, optimum for the given class of compounds.

Figure 164 presents the dependence of the rate of oxidation of rubber
on the concentration of inhibitors, distinguished by the effect of conjugation and, consequently, by the energy of the N−H bond in the molecule.
In the case of diphenylamine the minimum is displaced toward higher
concentrations in comparison with phenyl-β-naphthylamine. The curve
of di-β,β'-naphthylamine does not pass through a minimum, but approaches a constant value. The cause of this is the low activity of the
dinaphthylnitrogen radical, which is incapable of initiating oxidation of
the rubber.

Comparing rubbers differing considerably in reactivity, we should
note that the minimum oxidation rate for divinylstyrene rubber even at
100°C substantially exceeds the minimum oxidation rate of butadiene rubber at 120°C. Divinylstyrene rubber, just as we should have expected,
being more reactive, requires a larger inhibitor concentration to reach
the minimum rate. Moreover, v_{min} substantially exceeds the corresponding oxidation rate of butadiene rubber.

Below is cited a scheme of the mechanism of the oxidation of rubbers in the temperature region 100-150°C in the presence of inhibitors
of the class of secondary aromatic amines [20]:

1. $RH \xrightarrow{O_2} R^{\cdot}$
2. $JH \xrightarrow{O_2} J^{\cdot}$ } initiation reaction

3. $R^{\cdot} + O_2 \rightarrow RO_2^{\cdot}$
4. $ROO^{\cdot} \rightarrow R'CHO + R''O^{\cdot}$ } reaction of chain development
5. $R''O^{\cdot} + RH \rightarrow R''OH + R^{\cdot}$

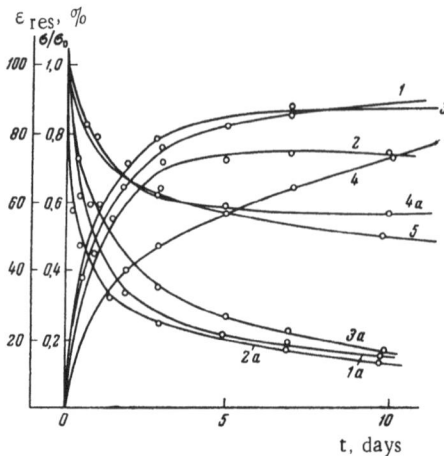

Fig. 166. Kinetics of chemical stress relaxation (1a,
2a, 3a, 4a) and the accumulation of residual deformation (1-4) in rubbers prepared on the basis of methylvinylpyridine rubber MVP-15. 1 and 1a) Rubber A,
vulcanizing group: sulfur + altax; 2 and 2a) rubber C,
vulcanizing agent: benzotrichloride; 3 and 3a) rubber
based on SKB; 4 and 4a) rubber B, vulcanizing group:
sulfur + altax and bonzotrichloride; 5) rubber D, vulcanizing agent: tetramethylthiuram disulfide.

6. $R''O^{\cdot} + -CH = CH - \rightarrow -CH - CH^{\cdot}-$
$\qquad\qquad\qquad\qquad\qquad\quad |$
$\qquad\qquad\qquad\qquad\qquad OR''$ $\qquad$ reactions of chain transfer

7. $RH + J^{\cdot} \rightarrow R^{\cdot}$

8. $R''O^{\cdot} + JH \rightarrow R''OH + J^{\cdot}$

9. $2J^{\cdot} \rightarrow J - J$
$\qquad\qquad\qquad\qquad$ termination on the inhibitor
10. $J^{\cdot} + R''O^{\cdot} \rightarrow R''OJ$

11. $R''OH \xrightarrow{O_2}$ aldehydes
$\qquad\qquad\qquad\qquad$ secondary reactions
12. $RCHO \xrightarrow{O_2}$ acids

Here RH represents the polymer molecule, JH the inhibitor molecule.

The scheme cited includes reactions characterizing the basic directions of the process.

An analysis of the theoretical material, as well as the experience in
the use and investigation of various antioxidants in the USSR and abroad
makes it possible to select classes of compounds for seeking new, effective, antiaging agents. They include:

1) secondary amines and diamines, as well as their derivatives,
which are little oxidized by molecular oxygen;

2) compounds of the quinoline series with various effects of conjugation in the molecule;

3) salts of derivatives of dithiocarbamic acid;

4) monoatomic and polyatomic alkylphenols, as well as

5) esters of phosphorous acid, used as noncoloring inhibiting compounds.

Recently various special-purpose rubbers have received wide circulation. Such rubbers should include methylvinylpyridine raw rubbers, cured rubbers made from which are distinguished by great ability to work under repeated deformations, high elasticity at low temperatures, and stability to the action of complex esters within a broad range of temperatures.

The rate of oxidation of purified methylvinylpyridine rubber (a copolymer containing 85% butadiene and 15% 2-methyl-5-vinylpyridine) within the interval 50-90°C substantially exceeds the rate of oxidation of butadiene rubber (SKB).

As is shown in Fig. 165 an induction period is observed only at temperatures not exceeding 60°C [21]. The increased reactivity of the rubber is apparently due to the greater content of 1,4-bonds in the principal molecular chains in comparison with SKB. During the process of oxidation, this rubber is intensively structured.

Cured rubbers based on methylvinylpyridine raw rubber, produced in the presence of sulfur, altax, and organohalogen compounds (first type) are structured considerably more intensively during the process of thermal aging than are cured rubbers vulcanized with only sulfur and altax (second type).

Chemical relaxation of stress in rubbers of the first type proceeds at a lower rate than that of rubber of the second type. In both types of rubbers, stress relaxation is determined by the breakdown of transverse bonds, formed during the process of vulcanization (Fig. 166). Moreover, the introduction not only of phenyl-β-naphthylamine, but also of such substances as the condensation products of acetone and aniline, mercaptobenzimidazole, or the condensation product of diphenylamine and acetone into rubbers vulcanized only with sulfur and altax substantially reduces the rate of decomposition of transverse bonds (Fig. 167).

For rubbers with stronger transverse bonds (thiuram rubbers), the introduction of the indicated products gives no appreciable effects.

The influence of thermal aging on the molecular weight distribution curves in polychloroprene (purified polymer of the type of neoprene-GN)

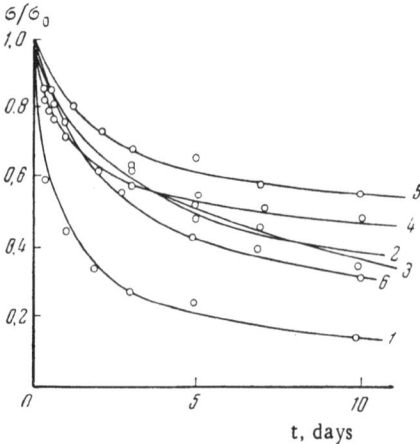

Fig. 167. Influence of antioxidants on the kinetics of chemical stress relaxation of rubber based on MVP-15, containing the vulcanizing group sulfur + altax (rubber A). 1) Rubber A; 2) rubber A + acetoneanil (3-methodi-hydroquinoline, polymerized); 3) rubber A + DPPD (diphenyl-p-phenylenediamine); 4) rubber A + BLE (condensation product of acetone and diphenyl); 5) rubber A + MB (mercaptobenzimidazole); 6) rubber A + oxynezone.

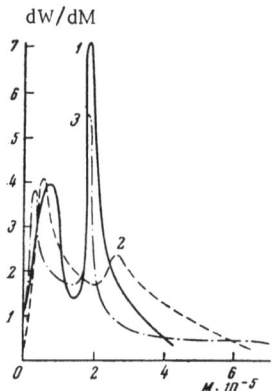

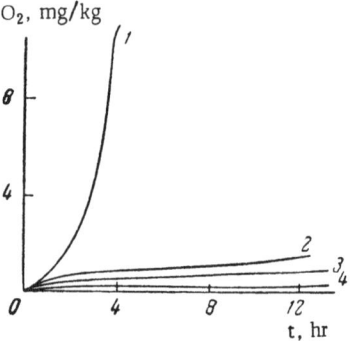

Fig. 168. Molecular weight distribution curves of polychloroprene. 1) Fresh polychloroprene; 2) aged in air; 3) heated in air for 2 hr at 80°C.

Fig. 169. Kinetics of the oxidation of technical natural rubber in the solid phase at 100°C. 1) 0.003 mole of copper per 100 g of rubber; 2) 0.003 mole of copper + 0.05 mole sodium diethyldithiocarbamate (mixed on rollers) per 100 g of rubber; 3) complex copper salt of diethyldithiocarbamic acid (synthesized) calculated for 0.03 mole of copper per 100 g of rubber; 4) rubber.

TABLE 32. Metal-Deactivator Ratio in Synthesized Compounds

Deactivator (D)	Gross formula of compound	
Phenyl-β-naphthylamine	Cu_2D	Fe_2D
Parahydroxyphenyl-β-naphthylamine	Cu_3D	Fe_2D
Dinaphthyl-p-phenylenediamine	Cu_4D	Fe_4D
Aldol-α-naphthylamine	Cu_3D	Fe_3D

was studied in [22]. Figure 168 presents the results of the investigation.
During aging the content of high-molecular fractions increases. The
total fraction of low-molecular fractions does not change in this case.

The oxidation of rubbers is substantially accelerated by salts of
metals of variable valence (iron, copper, manganese, cobalt, etc.).
Figure 169 presents the kinetic oxidation curves illustrating the premise
stated [23]. Homogeneous catalysis of the chain reaction of rubber oxidation occurs as a result of the reduction of the activation energy of initiation, as well as the decomposition of stable peroxide.

In view of the presence in rubbers of substances capable of reducing
Fe^{+3} to Fe^{+2} or Cu^{+2} to Cu^+, favorable conditions are created for the regeneration of active metal ions and the action of reversible oxidation-
reduction systems, which manifest activity even at low temperatures.
The inhibitors usually used can prove ineffective in the presence of metal salts.

Many substances capable of forming complexes with metal ions in
which transitions of the valence electrons are blocked have been studied.

Table 32 presents the composition of the complexes formed by metal
ions and certain antioxidants.

The composition of the complexes was established by the method of
quenching of luminescence [23].

The influence of complex compounds on the kinetics of the oxidation
of natural rubber is shown in Figs. 170 and 171. As can be seen, bonded
Cu and Fe ions not only do not accelerate the oxidation of rubbers, but
even decelerate this process.

In the inhibited oxidation of raw rubbers and in the presence of
catalytically inactive complexes, the process proceeds at a constant
rate. The activation energy of the oxidation in both cases is equal to
22-24 kcal/mole. The oxidation of the same rubbers with salts of copper and iron possesses a clearly pronounced autocatalytic character.
Thus, certain complex compounds represent strong inhibitors. When
the rubbers are mechanically mixed, complex formation proceeds practically spontaneously.

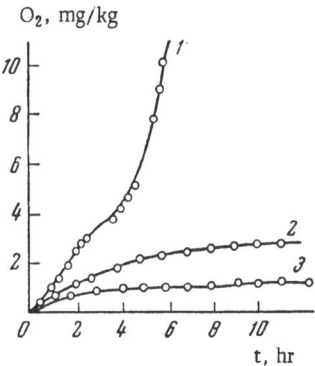

O₂, mg/kg

Fig. 170. Kinetics of the oxidation of technical natural rubber in the solid phase at 100°C. 1) 0.03 mole of iron per 100 g of rubber; 2) 0.03 mole of iron + 0.25 mole p-hydroxyphenyl-β-naphthylamine (mixed on rollers) per 100 g of rubber; 3) iron salt of p-hydroxyphenyl-β-naphthylamine (synthesized) in amounts of 0.03 mole of iron per 100 g of rubber.

Consequently, in cured rubber mixtures and cured rubbers, antioxidants and vulcanization accelerators can be partially or entirely bonded in catalytically inactive complexes. This should apparently explain the fact that salts of copper, iron, and manganese exert a harmful action on raw rubbers, but manifest no activity in many cured rubbers.

II. THERMAL OXIDATION OF CURED RUBBERS

The aging of cured rubbers, as has already been mentioned, differs substantially from the aging of the raw rubbers, on the basis of which they were prepared.

The method of vulcanization of cured rubbers using thiuram disulfides, in the absence of elementary sulfur, is widely used in the rubber industry. The basic advantage of thiuram rubbers is their high resistance to thermal aging. Thus, for example, cured rubbers based on butadiene nitrile rubbers, vulcanized with thiuram, manifest ability to work at 150-180°C, while the temperature limit of the operation of sulfur rubbers does not exceed 100-110°C.

It has been established that the cause of the high thermal stability of the indicated cured rubbers is the presence of zinc dithiocarbamates, which are formed from dithiocarbamic acid (one of the decomposition products of thiuram disulfide) and zinc oxide.

As is shown in Fig. 172 [24], vulcanizates obtained without zinc oxide exhibit no induction period in oxidation.

In a study of the inhibiting ability of dithiocarbamates containing various cations, as well as other radicals on the nitrogen atom, it was found that the most effective are the dithiocarbamates of Zn, Cu, and Bi (Fig. 173).

It was shown in [25] that in the thermal aging of massive rubber objects (for example, tire treads), only the outer surface of the tread and the boundary layer of rubber adjoining it are subjected to the influence of oxygen. The authors convincingly showed on extensive experimental ma-

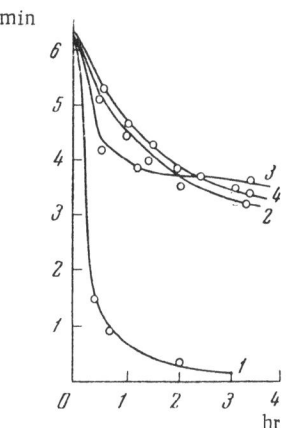

Fig. 171. Variation of the viscosity in the oxidation of natural rubber films (purified). Temperature 100°C. 1) 0.03 mole of iron per 100 g of rubber; 2) rubber; 3) 0.03 mole of iron + 0.15 mole p-hydroxyphenyl-β-naphthylamine per 100 g of rubber; 4) iron salt of p-hydroxyphenyl-β-naphthylamine (synthesized) in amounts of 0.03 mole of iron per 100 g of rubber. Along the horizontal axis: duration of oxidation of films; along the vertical axis: time of flow of the solution from the capillary.

terial that as a result of diffusion lags of the oxygen, the inner layers of rubber in the tire tread undergo practically no changes, while substantial changes in the mechanical properties are observed in the surface layer.

As has been noted, the ingredients exert a vital influence on the oxidation of cured rubbers and the character of the structural changes. Among other ingredients of rubber, a special place is occupied by carbon blacks.

A dual tendency is manifested in the influence of carbon blacks on the rate of oxidation of raw and cured rubbers. In the absence of antioxidants, carbon blacks play the role of weak inhibitors; in the presence of antioxidants, the initiating ability of carbon black, which substantially accelerates oxidations, is manifested. Evidently such behavior of the carbon black can be explained by the fact that its inhibiting action is negligibly small in comparison with the action of secondary aromatic amines, and hence is not manifested in the presence of the latter. In this case its initiating action becomes appreciable. The elucidation of which active centers on the surface of the carbon black particles so differently influence the character of the chemical processes in rubbers is of great interest; however, these problems have as yet received insufficient study.

III. LIGHT AGING

The photostability of raw and cured rubbers depends not so much on their nature as on the optical properties of the ingredients and impurities that they contain. Many ingredients of polymers, including the antioxidants used at the present time, are latent or patent photosensitizers.

Recent investigations have led to the creation of effective antiphotoaging agents. The best results are given by substances belonging to the

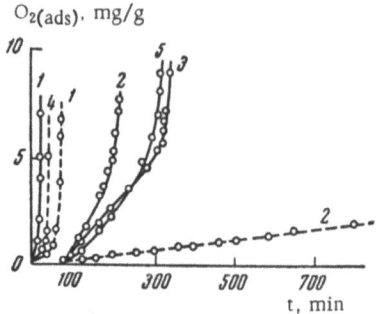

Fig. 172. Kinetics of the oxidation of thiuram vulcanizates. Solid lines, 150°; dotted lines, 130°C: 1) Vulcanizate II (with TMTD); 2) vulcanizate I (with TMTD); 3) vulcanizate I (with TEDT); 4) vulcanizate I (with TMTD), extracted; 5) vulcanizate I (STETD), extracted with an addition of 2% Zn DES; $O_{2(ads)}$ = adsorbed oxygen. t = duration of oxidation. I) With ZnO; II) without ZnO.

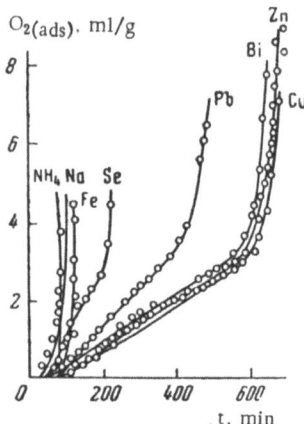

Fig. 173. Oxidation of sodium-butadiene rubber in the presence of various salts of diethyldithiocarbamic acid at 150°C (dose of salt 40 millimoles/liter). $O_{2(ads)}$ = adsorbed oxygen; t = duration of oxidation.

following classes: alkylated aromatic hydroxycompounds, phenylalkyls (ionol, antiaging agent 2256), substituted diphenylamines, derivatives of alkyldithiocarbamates (diethyl and dibutyldithiocarbamates of nickel).

For dark materials, a color change does not play a vital role; however, they also need photoprotection, since the absorption of light leads to their considerable heating, and as a result, to acceleration of photothermal aging.

IV. AGING OF RAW RUBBERS AT HIGH TEMPERATURES

The problem of the aging of raw rubbers and cured rubbers based on them, which manifest ability to work at high temperatures (200-400°C) has attracted the attention of many scientists in recent years [26-33]. Among the large number of thermally stable polymers synthesized in laboratories, copolymers of fluoroolefins and silicone rubbers have received the greatest industrial use abroad.

1. Copolymers of Fluoroolefins

Fluorine-containing elastomers combine high thermal stability with great stability to the action of various solvents, oils, fuels, and other aggressive media. The high stability of fluororubbers is due primarily to the high energy of the $C-F$ bond, equal to 124 kcal/mole.

In a study of the thermal decomposition of the copolymer of trifluorochloroethylene with vinylidine fluoride, it was established [34] that

TABLE 33. Values of the Activation Energy of the Accumulation of Residual Deformation of Vulcanizates of Raw Fluororubbers

Type of vulcanizate	Apparent activation energy (in kcal/mole) at a temperature, °C	
	90-150	150-280
Peroxide vulcanizate without filler	6.0	31.8
Peroxide vulcanizate with powdered silica gel	3.1	31.7
Radiation vulcanizate without filler	4.6	28.9
Radiation vulcanizate with powdered silica gel	3.9	36.5

HCl and HF are split out under vacuum at temperatures of 200-300°C. The activation energy of the splitting out of hydrogen halides from an elastomer of the type of Kel-F in the indicated temperature interval is equal to 29 kcal for HCl and 34 kcal for HF.

OH-groups in the form of alcohols and acids, as well as the presence of double bonds, have been detected in the liquid decomposition products by means of their infrared spectra. The liberation of volatile products is accompanied by the formation of a three-dimensional lattice.

Decomposition of the polymer chain is observed at 300°C. The color of the polymer changes from white to brown and even to black under thermal influence, which indicates the formation of conjugated double bonds. Above 300°C, in addition to HCl and HF, F is also split out. The rate of splitting out of hydrogen halides and the activation energy of their splitting out depend to a considerable degree on the nature of the lining. Thus, for example, the activation energy of the splitting out of halogens within the temperature range 340-380°C, is 50 kcal when the raw rubber is heated on a platinum lining and 57 kcal when it is heated on molybdenum glass. Monomers have not been detected in the decomposition products.

The initiation of thermal decomposition of the polymer begins with the oxygen-containing groups that are contained in the structure of the molecular chains and are "weak bridges," to a considerable degree reducing the thermal stability of the rubbers.

The introduction of various ingredients in the production of cured rubbers from raw fluororubbers sharply increases the rate of evolution of hydrogen halides. Organic products introduced into the rubber mixtures as vulcanizing agents, softeners, and stabilizers behave especially poorly. The indicated substances are hydrogen donors for the HCl and HF formed.

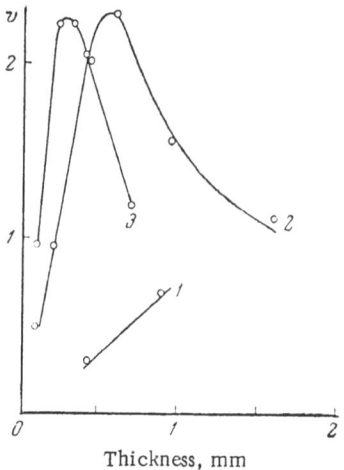

Fig. 174. Dependence of the rate of
splitting out of methyl groups on the
thickness of the sample at various temperatures. v is the number of methyl
groups split out per 100 monomer units
in 1 min. 1) 258° (v · 10); 2) 280° (v · 5);
3) 302°C.

Although raw fluororubbers withstand heating to 300°C, the cured rubbers based on them lose their valuable
properties at only 200°C.

Even if the mechanical properties of the cured rubbers undergo
small changes, but aging leads to the
formation of halogens and hydrogen
halides, this can prove harmful for
glass fabrics and metals, which are
construction elements in many objects.

Thus, the evolution of volatile
products should be taken into consideration in the use of cured rubbers
made from raw fluororubbers.

Of great interest is the behavior
of cured rubbers made from raw
fluororubbers subjected to thermomechanical influence. It has been
shown [35] that cured rubbers based
on SKF-32 and SKF-26 are characterized by an increased tendency toward
the accumulation of irreverislbe residual deformation. The residual deformation accumulated at 150°C is not restored at 25°C, but is reversible at 150°C. The authors explained this by an extremely slow occurrence of relaxation processes. The values of the apparent activation energy of the process of stress relaxation were studied in various temperature zones (Table 33).

At temperatures below 150°C, the activation energies correspond to
physical relaxation, related to displacement of segments of the molecular chains. At higher temperatures, chemical stress relaxation predominates. Hence, the activation energies cited are typical of chemical
processes.

2.   Raw Polydimethylsiloxane Rubbers

The literature contains few data on the principles of the thermooxidative transformations of polydimethylsiloxanes (see Chapter IX). It
is known that when these polymers are heated at temperatures above
200-250°C in the presence of oxygen, the methyl groups are oxidized,
leading to the splitting out and structuring of the polymer, as well as
breakdown of the main chains, accompanied by the liberation of cyclic
polydimethylsiloxanes [36-40]. It is assumed that the oxidation of methyl

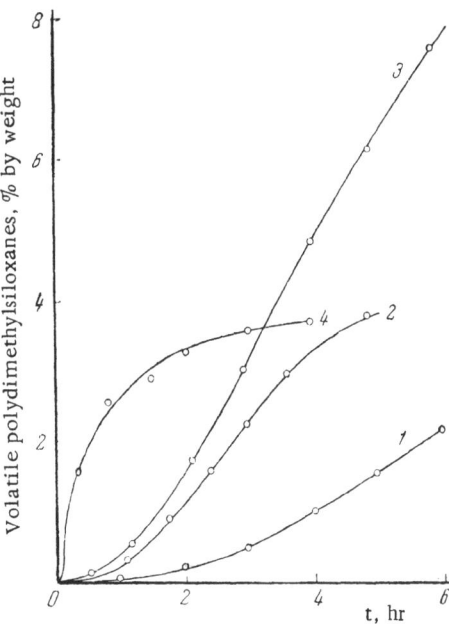

Fig. 175. Influence of oxygen on the process of splitting out of volatile organosilicon compounds during the heating of polydimethylsiloxane at various temperatures (film thickness 0.35 mm). 1) 278°C, in nitrogen; 2) 278°C, in oxygen; 3) 302°C, in nitrogen; 4) 302°C, in oxygen.

groups occurs according to a free radical mechanism through the formation of hydroperoxides; however, this viewpoint has not been substantiated experimentally. Cyclic polydimethylsiloxanes are formed not only in oxygen, but also in inert medium. These processes are highly accelerated in the presence of acid and alkaline additives [41–43].

The opinion is expressed that the process of chain breakdown under vacuum occurs according to an ionic mechanism [42]. The question of how oxygen influences this process has not been elucidated. In the oxidation of silicone rubbers, a very important role is played by the volatile decomposition products. It has been shown [44, 45] that in the oxidation of the polymer in a stream of oxygen a substantial portion of all the methyl groups split out are found in volatile products in the form of formaldehyde. Hence we might assume that the volatile compound that accelerates oxidation is formaldehyde, which, interacting with oxygen, gives rise to branching of the kinetic chain. This leads to an anomalous dependence of the rate of oxidation of the methyl groups on the thickness of the sample (Fig. 174). It is quite natural that the thicker the layer of polymer to be oxidized, the larger the amount of formaldehyde that has time to react in it before its emergence into the gaseous phase. As is shown in the figure, as the thickness of the sample is increased, up to a definite limit, the amount of methyl groups split out increases. The decrease in the number of methyl groups split out with increasing thickness above the "optimum" is explained by the substantial influence of diffusion lags of oxygen on the oxidation.

The molar ratio of hydrogen to carbon found in the volatile oxidation products (with organosilicon compounds deducted) as less than the ratio of hydrogen to carbon in the methyl groups and constitutes 2.5-2.9 (it increases with time).

The molar ratio between the oxygen added to the polymer and the carbon split out from it is higher than 0.5 and is equal to 0.55-0.75 (it decreases with the time of oxidation). On the basis of these data, the opinion has been expressed [44] that hydroxyl groups, bonded to a silicon atom, and siloxane bonds are formed at the site of splitting out of methyl groups.

The kinetics of the splitting out of low-molecular organosilicon compounds is characterized by an S-shaped curve (Fig. 175). A decrease in the oxygen concentration in the atmosphere where the polymer is heated (replacement of oxygen by nitrogen containing 50% O_2) leads to a decrease in the rate of splitting out of organosilicon compounds, especially in the initial period. Moreover, a decrease in the rate of chemical stress relaxation in a stretched film of the oxidized polymer is also observed: at 302°C, the rate of relaxation in oxygen is several times greater than that in nitrogen.

In the oxidation of films with thickness less than the "optimum," i.e., in the case when the diffusion lags are inconsequential, the relationship between the number of transverse bonds formed (n) and the number of methyl groups split out (x) practically does not depend on the conditions of oxidation (temperature, thickness; Fig. 176). This premise is equivalent to the assertion that the dependence of n on x is practically not influenced by the rate of splitting out of methyl groups and the relationship between the amounts of methyl groups and organosilicon compounds split out. Another important feature of this dependence is the fact that in the case of a small degree of oxidation, the ratio n : x is very small (<0.02); then it increases.

The indicated peculiarity of the dependence of n on x can be explained by the formation at the early stages of oxidation of predominantly intramolecular siloxane bonds and an increase in the subsequent fraction of transverse bonds. As is known from the literature, polyorganosiloxanes are extremely inclined to cyclize. Thus, for example, in the case

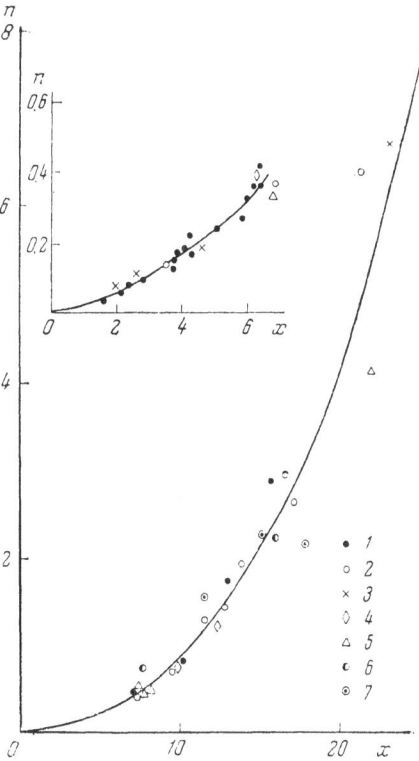

Fig. 176. Relationship between the number of transverse bonds formed and the number of methyl groups split out as a function of the temperature and thickness of the sample. 1) 258°, 0.45 mm; 2) 258°, 0.9 mm; 3) 280°, 0.25 mm; 4) 280°, 0.45 mm; 5) 280°, 0.6 mm; 6) 302°, 0.1 mm; 7) 302°, 0.25 mm; n is the number of transverse bonds per 100 monomer units; x is the number of methyl groups split out per 100 monomer units. The part of the curve near the origin is shown in detail in the insert.

of hydrolysis and joint condensation of di- and trifunctional organochlorosilanes, a polymer arises, containing cyclic portions in the macromolecular chains [39].

One of the important aspects of the problem is the behavior of antioxidants at high temperatures. Since the activation energy of the reaction of the antioxidant molecule with oxygen is always greater than that with the free radical of the oxidized polymer, the effectiveness of the antioxidants should decrease with increasing temperature. Their effectiveness also decreases because the low activity of the inhibitor free radicals at moderate temperatures becomes sufficient for the development of the chain process at high temperatures. Actually, it is known that the antioxidants usually used are relatively ineffective at temperatures above 150°C.

In addition to the factors indicated above, this phenomenon may also be due to the greater volatility of such compounds at these temperatures; for example, it has been shown that at a temperature of 130°C in a stream of nitrogen, about 50% of the phenyl-β-naphthylamine is volatilized from the polymer in 24 hr.

However, we cannot conclude from the above that the inhibition of aging processes at high temperatures is generally impossible: new types of inhibiting compounds are needed for this purpose. In connection with this, a new approach formulated by A. A. Berlin, for the selection of effective inhibitors of polymer oxidation at high temperatures [47], is extremely interesting. The author studied the action of polymer inhibitors with a system of conjugated bonds in the chain. It is important that free radicals in such systems are excited only at high temperatures, and hence manifest an inhibiting effect only in this region. It is natural that these inhibitors are nonvolatile. These substances manifest no activity at all at moderate temperatures.

V. CORROSION CRACKING OF CURED RUBBERS

The question of the interaction of deformed cured rubbers with various aggressive substances has long attracted the attention of researchers. The phenomenon of ozone cracking of such rubbers is well known. Yu. S. Zuev [48-54] has shown that this process is not specific only for rubbers and only for ozone.

The acceleration of static fatigue is observed for many high-molecular substances, as well as for various aggressive substances (solutions and vapors of acids, free radicals of various natures, hydrogen sulfide, etc.).

The rate of the cracking process varies within very broad limits, depending on the degree of deformation, adsorption capacity of the aggressor, on its concentration, temperature, and the dissociation constants of the acids. The presence of critical deformation in the cracking of cured rubbers is explained by the fact that, in addition to a decrease in the time until breaking with increasing stress, they are also oriented, and, consequently, strengthened. This effect is manifested especially at the apertures of growing cracks, where the deformation of the rubber is greater than its average value.

The process of decomposition in the presence of chemically active media, in contrast to the usual decomposition, does not proceed simultaneously in the entire volume, but at the openings of surface cracks as the aggressive medium penetrates into them, as a result of which the lifetime of rubbers increases with the thickness, in contrast to the analogous dependence in the case of their static fatigue.

Although the rate of interaction of undeformed rubber with an agressive medium is determined by diffusion, the decomposition of deformed rubber depends only on the rate of the chemical reaction of the rubber with the medium. The temperature coefficients of breaking depend on the type of bonds being broken and on the ability of the aggressive medium to be adsorbed on the rubber.

Corrosion cracking of cured rubbers during interaction with an aggressive medium is the result of destructive processes.

The temperature dependence of the breaking of rubbers in contact with aggressive media does not obey the principle found by S. N. Zhurkov for the breaking of metals and rigid polymers in air [55].

VI. AGING OF POLYMERS UNDER THE ACTION OF IONIZING RADIATIONS

An investigation of the variation of the properties of polymer materials under the action of ionizing radiations makes up one of the large sections of the general problem of aging of polymers. The first studies in this field contained mainly an empirical description of the observed phenomena, without any serious theoretical generalization of them. At the present time, as a result of extensive investigations, the behavior of the basic types of polymer compounds under the action of radiations has been described, and their radiation stability has been evaluated [56-71].

A number of works have been devoted to the study of the properties of irradiated vulcanizates of various rubbers. In the works of recent years, great attention has been paid to the investigation of radiation stability of new polymers and the analysis of the variation of the dynamic indices most sensitive to the changes that occur during irradiation.

An analysis of the accumulated material makes it possible to formulate the basic distinguishing features of this type of aging, determined by the specifics of the elementary processes. The chemical reactions initiated in rubbers by irradiation and occurring with the participation of free radicals in most cases do not lead to the development of chain processes. The presence of free radicals in irradiated polymers has been demonstrated directly by means of the method of electron paramagnetic resonance.

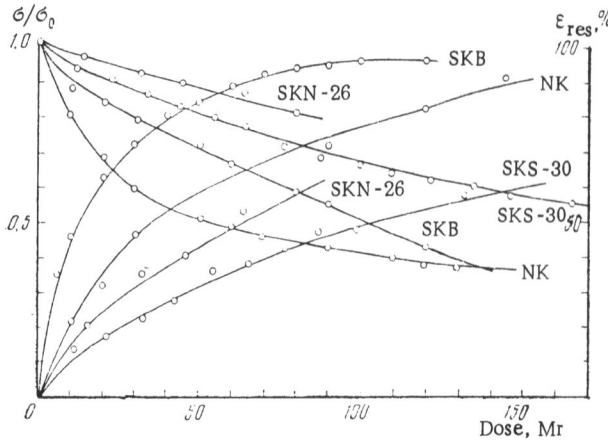

Fig. 177. Kinetics of the stress relaxation and accumulation
of residual deformation of thiuram vulcanizates prepared on
the basis of various raw rubbers. 1-4) Stress relaxation; 5-8)
accumulation of residual deformation.

The deciding factor determining radiation stability, in contrast to
light aging, is the chemical nature of the polymer. It has been shown
that the rate of recombination of the free radicals formed after the cessation of irradiation depends substantially on the temperature and the
presence of O_2.

It is known from the general principles of radiation chemistry that
aromatic compounds are the most radiation stable. Hence it is natural
that polymers and copolymers based on styrene are the most stable with
respect to radiation aging; moreover, the degree of their stability is
determined by the content of styrene units in the chain.

Good stability to the action of radiation is manifested by materials
prepared on the basis of vinylpyridine and urethane polymers.

A definite inverse relationship exists between the thermal stability
and radiation stability of rubbers. Thus, nitrile, polysiloxane, and
fluorine-containing raw rubbers are the most thermally stable and the
most unstable with respect to ionizing radiations. A significant influence
on the radiation stability of cured rubbers is exerted by various three-
dimensional structures formed during the process of vulcanization, as
well as by the ingredients (vulcanizing substances, fillers, softeners).
Thus, sulfur and thiuram (free and bound) decelerate radiation structuring [69, 70]. Carbon blacks participate in the formation of a space lattice under the action of γ-radiation [61, 71-76].

The introduction of mineral fillers (zinc oxide, lead oxide, litho-
pone, titanium dioxide) leads to an increase in the rate of structuring as

TABLE 34. Radiochemical Yield of Destruction and Structuring
Processes Developing Simultaneously in Vulcanizates
Under the Action of Radiations

Type of rubber	Yield of structuring (a)	Yield of destruction (b)	Ratio a : b
NK	7.37	6.85	1.16
SKN-26	11.0	0.55	20.0
SKB	12.7	2.49	5.12
SKS-30	4.5	2.72	1.64
SKN-26 + 4010•	3.6	0.65	5.6
NK + 4010	7.4	6.2	1.2

• 4010, i.e., N-phenyl-N'-cyclohexyl-p-phenylenediamine.

a result of an increase in the total amount of absorbed energy. The radiation stability of rubbers can be substantially increased by using softeners containing aromatic rings (phenol resins, naphthene oils, styrene,
epoxide, and epoxyamide resins). An increase in the radiation stability
can be achieved by introducing small amounts of various derivatives of
aromatic compounds, secondary aromatic amines, dinitrobenzenes, di-
nitrophenols, benzoquinone, naphthols, naphthaquinones, etc. In a combination of thermal and radiation aging, acroflex C (35% diphenyl- β -
phenylenediamine + 65% phenyl-β-naphthylamine) and quinhydrone have
been successfully used.

The mechanism of the protective action of such compounds has not
yet been studied. The effects that can be achieved from their introduction are far smaller than the action of inhibitors in polymerization or
antioxidants in oxidation. The cause of this difference in the effectiveness of inhibitors is apparently the absence of chain reactions during radiation aging.

The behavior of deformed cured rubbers under the action of radiation
is of great interest [77].

Figure 177 presents curves characterizing the kinetics of chemical
relaxation and the accumulation of residual deformation in vulcanizates
based on various raw rubbers. As we can see, the investigated vulcanizates are arranged in the following series according to rate of stress
relaxation: NK > SKB > SKS-30 > SKN-26, and according to rate of accumulation of residual deformation – in the series: SKB > NK > SKN-26
> SKS-30.

Table 34 presents the values of the radiochemical yields of the processes of structuring and destruction of the investigated vulcanizates for
the initial stages of irradiation (up to 10 Mrad), when various transverse
bonds formed during irradiation can still be neglected.

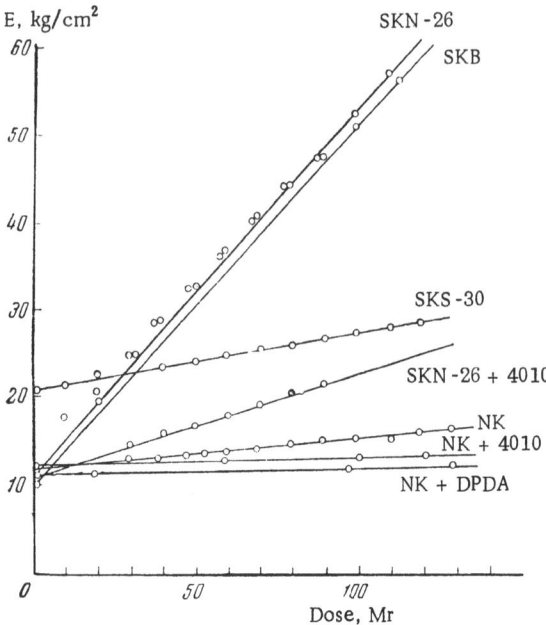

Fig. 178. Variation of the equilibrium modulus of thiuram
vulcanizates prepared on the basis of various raw rubbers.

The data on the variation of the equilibrium moduli, cited in Fig.
178, are in full agreement with the calculated ratios of the radiochemical yields, presented in Table 34.

For natural rubber the summary change in the density of the space
lattice occurs at the lowest rate, which indicates a small difference in
the rates of destruction and structuring. The rapid decrease in the
strength indices during irradiation of natural rubber vulcanizates is due
to the influence of the lattice formed on the ability to crystallize during
stretching.

In the case of the synthetic raw rubbers SKB, SKN-26, and SKS-30,
the gross effect during radiation aging is a substantial increase in the
density of the space lattice, which in turn is accompanied by a drop in
the relative lengthening.

The introduction of antiradiation agents selectively influences the
rate of structuring and has only a small effect on the rate of chemical
relaxation of the cured rubbers (Table 34, Fig. 179). A significant influence of antiradiation agents on the rate of chemical relaxation is manifested only in the case of irradiation of stretched cured rubbers in air.
In this case, however, the action of antiradiation agents should rather

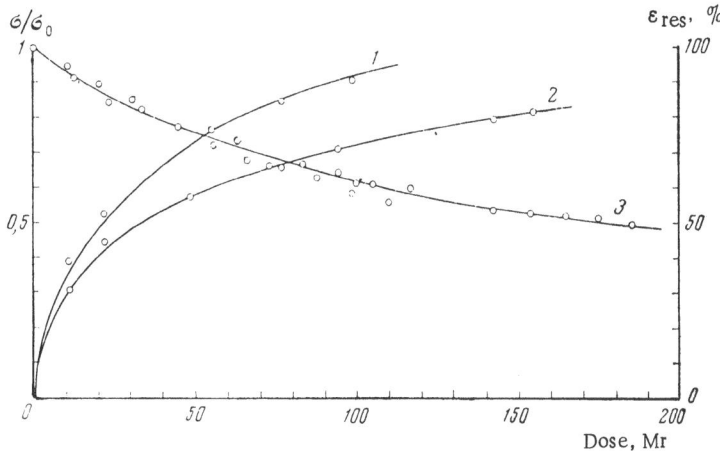

Fig. 179.   Influence of the antioxidant 4010 on the rate of chemical relaxation (3) and the rate of accumulation of residual deformation (1 and 2)
in vulcanizates of SKB. 1) Accumulation of residual deformation (without the antiradiation agent); 2) accumulation of residual deformation with
4010; 3) relaxation of stress of vulcanizates containing the antiradiation
agent (4010) and vulcanizates without the antiradiation agent.

be considered as that of antiozonants. It is well known that considerable
amounts of ozone, which destroys such rubbers (ozone cracking) are
formed in the case of irradiation of air or oxygen.

VII.   INFLUENCE OF MECHANICAL STRESSES
ON THE AGING OF CURED RUBBERS

Considerable attention is being paid to the aging of stressed cured
rubbers in the literature. This is natural, since all raw and cured rubbers are subjected to the action of deforming forces of various intensities and various characters during their reprocessing, storage, and
use. Let us consider the two most widespread systems: static and repeated stressing of the rubbers.

The question of whether fatigue of high-molecular materials is more
a mechanical breakdown or a chemical process is discussed in the literature [4]. The authors correctly indicate that the role of mechanical and
physical factors depends on the system of testing and on the medium.

We should think that since in most of the real cases, repeated deformation of cured rubbers is accomplished for prolonged periods in a
medium containing oxygen or other forms of still more aggressive substances, this process should be considered as a variety of aging. As is
well known, mechanical stresses activate chemical processes and inten-

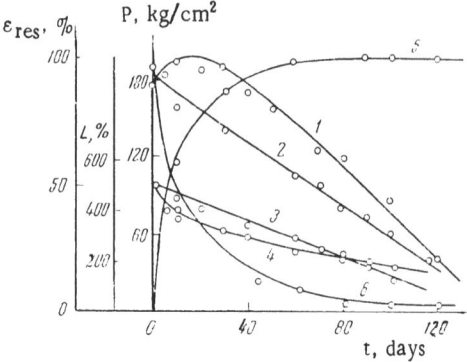

Fig. 180.   Influence of static deformations on the
aging of cured rubber No. 2959. Temperature, 70°C;
deformation, 50% stretching. 1 and 2) Breaking
strength (P); 3 and 4) relative lengthening (L); 1 and
3) aging in the undeformed state; 2 and 4) aging
under static deformations; 5) kinetics of the accumulation of residual deformation (ε_{res}); 6) kinetics of the
drop in stress.

sify the cleavage of chemical bonds in cured rubbers; hence the specific
feature of the aging of stressed rubbers is its mechanochemical character [2, 78].

Although a large number of works [79-83] have been devoted to the
problem of the fatigue of cured rubbers under repeated deformations, at
the present time such an important question as the influence of the nature
of the transverse bonds on the ability of cured rubbers to work is debatable.

It is indicated in [85,86] that the ability of cured rubbers to work increases with increasing energy of the transverse bonds, while the op-
posite dependence was observed in the investigations [84, 87]. The ability of cured rubbers to work increased substantially in the presence of
weak polysulfide bonds.

The indicated contradictions may be due to the fact that the authors
of the indicated works used different deformating systems in their investigations. It is also possible that cured rubbers whose space lat-
tices are formed by a set of bonds of different energies, for example,
of mono- to polysulfides, will be characterized by the greatest ability
to work. However, there are no data confirming such a hypothesis.

Of great importance is a study of the mechanism of antifatigue agents.
It is well known that the stabilizing action of antifatigue agents during repeated deformations is more effective than in the case of thermal aging

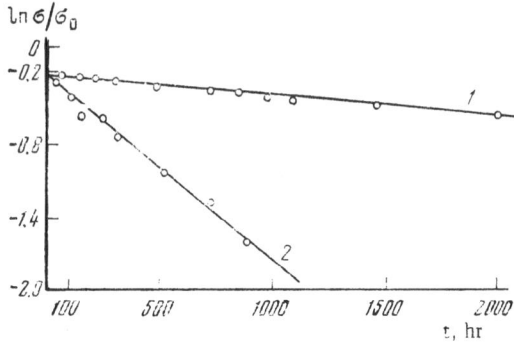

Fig. 181. Influence of the oxygen concentration on the kinetics of stress relaxation of monosulfide natural rubber vulcanizates at 100°C. 1) Vacuum (10^{-3} mm Hg); 2) air.

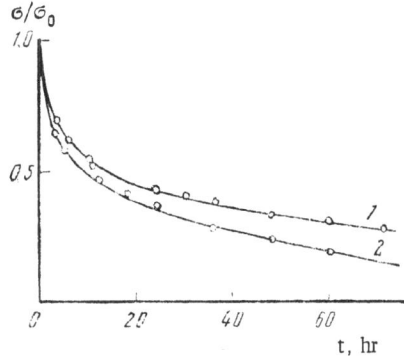

Fig. 182. Influence of the oxygen concentration on the kinetics of stress relaxation of polysulfide natural rubber vulcanizates at 90°C. 1) Vacuum (10^{-3} mm Hg); 2) air.

of cured rubbers. It was shown in the work of A. S. Kuz'minskii [88] that the action of secondary aromatic amines in vulcanizates is directed toward suppressing mechanically activated oxidation. Interesting results have been obtained using phenthiazine (thiodiphenylamine) as an antifatigue agent for vulcanizates from natural rubber, SKI, and SKS-30AM [89]. As the authors showed, phenthiazine reduces the rate of chemical relaxation in oxygen three- to seven-fold in vulcanizates of natural rubber, two-fold in vulcanizates of SKI, and two- to three-fold in vulcanizates of SKS-30AM, and increases the working ability by 20-50% (in one case it was even tripled). The authors indicate that in the absence of oxygen, phenthiazine stabilizes the decomposition of sulfur bonds under the action of static deformation.

Substantial results have been obtained on the influence of static deformations on the aging of cured rubbers. It has been shown by direct

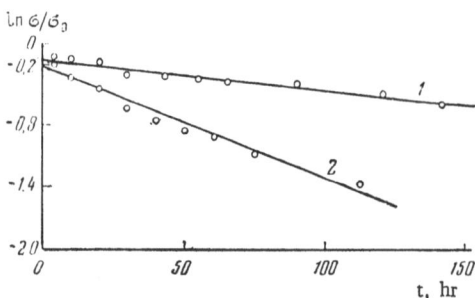

Fig. 183. Influence of iron naphthenate on the rate
of stress relaxation of monosulfide natural rubber
vulcanizate at 100°C. 1) Without iron naphthenate;
2) with 2% iron naphthenate.

experiments that, in contrast to repeated deformations, in the case of
static stressing in cured rubbers, the mechanical activation of oxidation
processes is extremely weakly pronounced. The rate of oxidation of vulcanizates does not increase with the application of a load, while the rate
of chemical relaxation is independent of the degree of deformation within
broad limits. However, the specific influence of static stresses is manifested in a change in the ratio of the rates of destruction and structuring
in the vulcanizates. The rate of structuring, as a rule, is reduced.

It has been established that standard physicomechanical indices,
which satisfactorily describe the aging of nondeformed cured rubbers,
are quite unsuitable for evaluating the aging of stressed rubbers (Fig.
180). The loss of the ability to work by statically deformed cured rubbers in many cases is due to a reduction of the stress to the critical
limit as a result of chemical relaxation. As can be seen from Fig. 180
a stress can be relaxed in cured rubber when the physicomechanical indices have still changed to a very small degree.

It is interesting to note that the equilibrium modulus also remains
practically unchanged during the period of total relaxation. This shows
that the breaking and formation of bonds occurs at the same rate, and,
consequently, the total number of bonds remains practically unchanged.
At the same time, the drop in the stress to a value close to zero indicates
that almost all the bonds situated under stress were rearranged into new
unstressed portions of the molecular structure of the vulcanizate.

In spite of the presence of a large number of investigations devoted
to chemical relaxation in rubbers, as yet no distinct representations
have been developed for a number of basic questions pertaining to the
mechanism of this process. At the present time the basic question of
the theory of chemical relaxation: whether the drop in stress occurs as

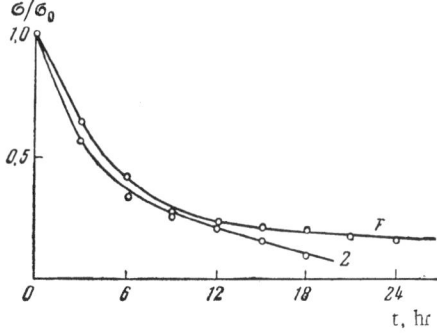

Fig. 184. Influence of iron naphthenate on the rate
of stress relaxation of polysulfide natural rubber vulcanizate at 100°C. 1) Without iron naphthenate; 2)
with 2% iron naphthenate.

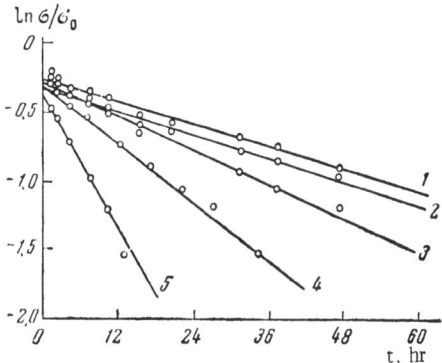

Fig. 185. Influence of fillers on the kinetics of
stress relaxation of monosulfide natural rubber vulcanizate at 90°C. 1) Unfilled vulcanizate; 2)
vulcanizate with 100 parts by weight chalk; 3) vulcanizate with 40 parts by weight thermal carbon
black; 4) vulcanizate with 40 parts by weight chimney soot; 5) vulcanizate with 40 parts by weight
channel black.

a result of the decomposition of transverse bonds, as is believed by Watson and his associates [90], or as a result of oxidative destruction of the
molecular chains in accord with the representations of Tobolsky and his
school [91], is being discussed in the literature.

The investigations we conducted of chemical stress relaxation of vulcanizates differing both in the nature of the transverse bonds and in the
reactivity of the polymer have made it possible to give an unambiguous
answer to the disputed questions [92]. It was shown that two competing
tendencies are observed in the process of aging of vulcanizates: oxida-

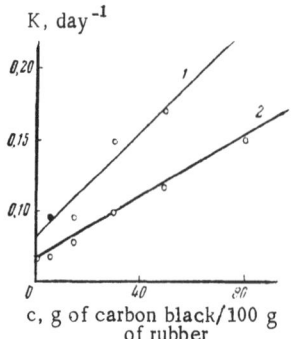

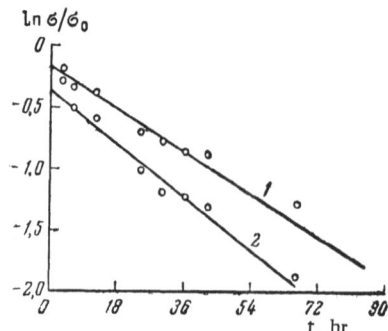

Fig. 186. Dependence of the rate constant of stress relaxation in thi-uram natural rubber vulcanizates and the degree of filling at various tem-peratures: 1) 90°; 2) 100°C.

Fig. 187. Influence of channel black on the rate of stress relaxation of polysulfide natural rubber vulcanizate at 90°C. 1) Unfilled vulcanizate; 2) vulcanizate with channel black.

tive destruction of the molecular chains of the polymer and thermal de-composition of the transverse sulfur bonds. If the three-dimensional lattice in the vulcanizate was formed from strong transverse bonds (thi-uram, thermal, radiation vulcanizates), then stress relaxation is deter-mined by oxidative destruction of the molecular chains. In this case, it is natural that the oxygen concentration, reactivity of the polymer to oxy-gen, and the introduction of antioxidants and substances that activate oxidation can substantially influence the rate of chemical relaxation.

When the lattice of the vulcanizate was formed by weak, especially poly-sulfide bonds, the rate of thermal decomposition of the latter ex-ceeds the rate of oxidative destruction of the molecular chains by an or-der of magnitude. In the indicated case, the rate of the chemical reaction is determined by the rate of decomposition of the transverse bonds, and all the factors that influence the oxidation of the polymer cannot essential-ly change the rate of chemical relaxation. Figures 181-184 illustrate the premise cited.

The literature contains only the work of Tobolsky [93] on the in-fluence of carbon blacks on the process of chemical stress relaxation; he arrives at the conclusion that carbon blacks do not influence the rate of chemical stress relaxation. However, the available data on the in-fluence of carbon blacks on the rate of oxidation of cured rubbers force us to treat Tobolsky's data with caution.

In [94] it was shown that carbon blacks substantially increase the rate and reduce the activation energy [95] of the oxidation of cured rub-bers.

v, cm³/kilowatt-hr

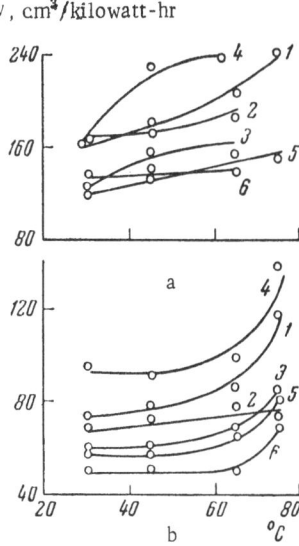

Fig. 188. Dependence of the index of abrasion (v) of tread rubbers made from natural rubber with various antiaging agents on the testing temperature. a) Air; b) nitrogen. 1) Without antiaging agents; 2) neozone D; 3) neozone D + flectol N; 4) neozone D + MV; 5) neozone D + 4010; 6) santoflex AW.

In the case of thiuram vulcanizates, stress relaxation, as was indicated above, is due to the oxidative destruction of the polymer chains. Hence we might assume that the presence of carbon black in thiuram cured rubbers produces an increase in the rate of chemical stress relaxation.

The influence of fillers on the rate of chemical stress relaxation has been investigated on thiuram natural rubber vulcanizates.

All the investigated carbon blacks increase the rate of chemical relaxation (Fig. 185). At the same degree of filling, vulcanizates with different carbon blacks are arranged in the following series with respect to the rate of chemical relaxation: with channel black > with chimney soot > with lamp black and thermal carbon black. Investigation of the influence of the degree of filling and the rate of chemical stress relaxation showed that the rate constant of chemical relaxation increases with increasing degree of filling (Fig. 186).

In the case of vulcanizates with polysulfide bonds, when the process of stress relaxation is due to the decomposition of thermally unstable transverse bonds, and not to thermooxidative destruction of the polymer chain, the difference in the rates of chemical stress relaxation of the filled and unfilled cured rubbers is comparatively small (Fig. 187).

The influence of carbon blacks on the rate of chemical relaxation cannot be explained by absorption of antioxidants by the carbon blacks. In the case of thiuram natural rubber vulcanizates, the difference in the rates of chemical relaxation of vulcanizates protected by various antioxidants and not protected by them is substantially smaller than in the case of vulcanizates filled with channel black, protected with an antioxidant. This also confirms the fact that, at least in the case of thiuram cured rubbers, the selection of the filler determines the stability of the rubbers to the joint action of heat and mechanical stresses to a greater degree than does the selection of the antioxidant.

The data cited permit us to formulate theoretically substantiated principles for the development of recipes for cured rubbers operating under conditions of static deformations.

The basic factor determining the behavior of cured rubbers during aging under conditions of static deformations is the nature of the transverse bonds. Since the character of the transverse bonds is determined by the method of vulcanization, proper selection of the vulcanizing agent and method of the vulcanization is a decisive condition, predetermining the properties of the cured rubbers obtained. Lack of fulfillment of this basic conditions cannot be compensated for by any other factors. If the first condition is fulfilled, then the proper selection of the polymer and filler acquire great significance.

Of considerable interest are new data on the role of thermooxidative destruction in the abrasion of cured rubbers [94]. The authors based their conclusions on the representation of the fatigue mechanism of abrasion [95, 96]. It was established that the rate of abrasion of cured rubbers based on natural rubber, SKS-30ARKM, and SKB is substantially greater in a medium of air than in nitrogen.

Cured rubbers from raw rubbers relatively unreactive to oxygen (low unsaturation) are insensitive to the medium in which abrasion testing is performed.

As Fig. 188 shows, antiaging agents exert a substantial influence on the abrasion stability of tread rubbers made from natural rubber, both in air and in nitrogen. The most effective combinations of antiaging agents proved to be santoflex AW, neozone D + 4010, and neozone D + flectol N.

We should keep in mind that high temperature is developed in the zone of contact of the working surfaces; hence the intensity of the chemical processes that develop will depend substantially on the thermal conductivity of the working materials.

BIBLIOGRAPHY

1. N. Grassi, Chemistry of Processes of Polymer Destruction [Russian Translation], Moscow, Foreign Literature Press, 1959.
2. A. S. Kuz'minskii, N. N. Lezhnev, and Yu. S. Zuev, Oxidation of Raw and Cured Rubbers, Moscow, State Press for Chemical Literature, 1957.
3. I. Kh. Dilon, Fatigue Phenomena in High Polymers, Moscow State Press for Chemical Literature, 1957.
4. M. M. Reznikovskii and L. S. Priss, Fatigue of High Polymers (Surveys and Translations), Moscow, State Press for Scientific and Technical Literature, 1957, p. 117.

5. Yu. S. Zuev and A. F. Postovskaya, Photoaging, Protection, and Recipes for Articles Made from Colored Rubbers, Moscow, State Press for Chemical Literature, 1959.

6. L. G. Angert and A. S. Kuz'minskii, The Role and Use of Antioxidants in Raw and Cured Rubbers, Moscow, State Press for Chemical Literature, 1957.

7. F. Bovei, The Action of Ionizing Radiations on Natural and Synthetic Polymers [Russian Translation], Moscow, Foreign Literature Press, 1959.

8. T. S. Nikitina, E. V. Zhuravskaya, and A. S. Kuz'minskii, The Action of Ionizing Radiations on Polymers, Moscow, State Press for Chemical Literature, 1959.

9. A. S. Kuz'minskii and E. V. Zhuravskaya, Khim. Nauka i Prom. 4:1, 1959.

10. J. R. Shelton, Rubber Chem. Technol. 30(5):1251, 1957.

11. P. Shneider, Angew. Chem. No. 2:61, 1955.

12. M. B. Neiman, A. S. Kuz'minskii, and L. G. Angert, Vestn. Akad. Nauk SSSR 11:36, 1960.

13. S. N. Borisov, Kauchuk i Rezina, No. 7:8, 1961; No. 8:16, 1961.

14. A. S. Kuz'minskii, Collection: Aging and Fatigue of Raw and Cured Rubbers and Increasing Their Stability, Moscow, State Press for Chemical Literature, 1955, pp. 6-7.

15. A. S. Kuz'minskii, Collection: Vulcanization of Rubbers, Moscow, State Press for Chemical Literature, 1954, p. 158.

16. L. G. Angert and A. S. Kuz'minskii, Kauchuk i Rezina No. 9:15, 1960.

17. N. N. Semenov, Some Problems of Chemical Kinetics and Reactivity, Moscow, Academy of Sciences USSR Press, 1959.

18. H. Winn and J. Shelton, Ind. Eng. Chem. 80:2081, 1948.

19. L. G. Angert and A. S. Kuz'minskii, J. Polymer Sci. 32:1, 1958.

20. L. G. Angert, Dissertation, M. V. Lomonosov Moscow Institute of Fine Chemical Technology, 1959.

21. L. G. Angert, A. I. Zenchenko, and A. S. Kuz'minskii, Summaries of Reports at the Conference on the Aging and Stabilization of Polymers, Moscow, Academy of Sciences USSR Press, 1961, p. 73.

22. I. Kessler, V. Matyska, and Ya. Polachek, Transactions of the International Symposium on Macromolecular Chemistry [Russian Translation], Section I, Moscow, Press, 1960, p. 328.

23. A. S. Kuz'minskii, V. D. Zaitseva, and N. N. Lezhnev, Doklady Akad. Nauk SSSR 125(5):1057, 1959.

24. L. G. Angert and A. S. Kuz'minskii, Kauchuk i Rezina, No. 9:15, 1960.

25. A. Antonova and M. V. Timofeeva, Kauchuk i Rezina, No. 5:12, 1960.

26. W. Postelnek, Rubber World 136(4):543, 1957.
27. W. Gruffin, Rubber World 136(5):687, 1957.
28. A. Wilson, C. Griffis, and Y. Montermoso, Rubber Age 83(4):647, 1958.
29. F. Eirich and G. Mark, Thermostable Polymer Symposium, London, 1960.
30. Madorsky, Ibid.
31. R. Show, Ibid.
32. R. Banks, J. Birchall, and R. Heszelchine, Ibid.
33. S. N. Borisov, Kauchuk i Rezina, No. 7:8, 1961; No. 8:16, 1961.
34. T. G. Degteva, Vysokomolekulyarnye Soedineniya 3:671, 1961.
35. A. N. Novikov, V. A. Kargin, and F. A. Galil-Ogly, Kauchuk i Rezina, No. 1:39, 1959.
36. E. Rochow, An Introduction to the Chemistry of the Silicones, J. Wiley and Sons, Inc., New York, 1951.
37. K. A. Andrianov, Organosilicon Compounds, Moscow, State Press for Chemical Literature, 1955.
38. D. Atkins, C. Murphy, and C. Saunders, Ind. Eng. Chem. 39:1335, 1947.
39. K. A. Andrianov and N. Sokolov, Khim. Prom. No. 6:9, 1955.
40. L. Scala and W. Hickam, Ind. Eng. Chem. 50:1583, 1958.
41. W. Patnode and D. Wilcock, J. Am. Chem. Soc. 68:358, 1946.
42. W. Sewis, J. Polymer Sci 33:153, 1958; 37:425, 1959.
43. M. Kuchera, I. Lanikova, and M. Elinek, International Symposium on Macromolecular Chemistry in Moscow, Section III, Academy of Sciences USSR Press, 1960.
44. A. S. Kuz'minskii and E. A. Goldovskii, Vysokomolekulyarnye Soedineniya 3:1054, 1961.
45. A. S. Kuz'minskii and E. A. Goldovskii, Doklady Akad Nauk SSSR 140(6):1324, 1961.
46. N. N. Sokolov, Methods of Synthesizing Polyorganosiloxanes, Moscow-Leningrad, State Power Engineering Press, 1959.

47. A. A. Berlin, Z. V. Popova, and D. M. Yanovskii, Summaries of Reports at the Conference on the Aging and Stabilization of Polymers, Moscow, Academy of Sciences USSR Press, 1961, p. 39.

48. Yu. S. Zuev and S. I. Pravednikova, Doklady Akad. Nauk SSSR, 116:813, 1957.
49. Yu. S. Zuev and S. I. Pravednikova, Zhur. Fiz. Khim. 31:2586, 1957.
50. Yu. S. Zuev and S. I. Pravednikova, Zhur. Fiz. Khim. 32:1457, 1958.
51. Yu. S. Zuev and S. I. Pravednikova, Trudy of the Scientific Research Institute of the Rubber Industry, No. 6:3, 1960.
52. Yu. S. Zuev and A. Z. Borshchevskaya, Doklady Akad. Nauk SSSR 124:613, 1959.

53. Yu. S. Zuev, N. N. Bukhanova, and T. I. Dorfman, Kauchuk i Rezina 10:44, 1960.

54. Yu. S. Zuev, A. Borshchevskaya, S. Pravednikova, and U. Yuétsin', Vysokomolekulyarnye Soedineniya 3(2):164, 1961.

55. Yu. S. Zuev and A. Z. Borshchevskaya, Summaries of Reports at the Conference on the Aging and Stabilization of Polymers, Moscow, Academy of Sciences USSR Press, 1961, p. 77.

56. R. Son, Mod. Plastics 32(1):141, 1954.

57. E. Collinson, Chem. Rev. 56(3):471, 1956.

58. A. Shapiro, Ind. Plastiques Mod. (Paris) 9(1):41, 1957; Collection: Khim. i Tekhnol. Polimer. No. 2, Moscow, Foreign Literature Press, 1958.

59. A. Miller, E. Lawton, and J. Balwit, J. Polymer Sci. 14(77):503, 1954.

60. J. Harrington, Rubber Age 81(9):971, 1957.

61. A. Charlesby and D. Groves, Rubber Chem. Technol. 30(1):27, 1957.

62. C. Bopp and I. Sisman, Radiation Stability of Plastics and Elastomers, ORNL-1373, 1953.

63. J. Ryan, Mod. Plastics 31(2):152, 1953.

64. A. Charlesby, Plastics 18(130):142, 1953.

65. R. Harrington, Rubber Age 82(3):461, 1957.

66. R. Harrington, Rubber Age 82(6):1003, 1958.

67. R. Harrington, Rubber Age 83(3):472, 1958.

68. M. Mezrobian, Transactions of the Second International Conference on the Peaceful Uses of Atomic Energy [Russian Translation], Moscow, Academy of Sciences USSR Press, 1958.

69. A. S. Kuz'minskii, T. Nikitina, and V. L. Karpov, At. Energ. 1(3):137, 1956.

70. T. Nikitina, A. S. Kuz'minskii, and V. L. Karpov, Collection: Action of Ionizing Radiations on Inorganic and Organic Systems, Moscow, Academy of Sciences USSR Press, 1958, p. 333.

71. B. Johnson, H. Adams, and M. Barzan, Rubber World 137(1):73, 1957.

72. Z. N. Tarasova, M. Kaplunov, B. A. Dogadkin, V. L. Karpov, and A. Breger, Kauchuk i Rezina, No. 5:14, 1958.

73. S. Gehman and J. Anderbach, Intern. J. Appl. Radiation Isotopes 1(1):102, 1956.

74. W. Jackson and D. Hall, Rubber Age 77:865, 1955.

75. S. Gehman and L. Hobbs, Rubber World 130(5):643, 1954.

76. P. Arnold, G. Kraus, and H. Anderson, Kautshuk Gummi 12(2):27, 1959.

77. A. S. Kuz'minskii, L. M. Fel'dshtein, E. Zhuravskaya, and L. I. Lyubchanskaya, Transactions of the Second All-Union Conference on Radiation Chemistry, Moscow, Academy of Sciences USSR Press, 1962, p. 576.

78. G. L. Slonimskii, V. A. Kargin, G. N. Buiko, E. V. Reztsova, and M. L'yuis-Riera, Collection: Aging and Fatigue of Raw and Cured Rubbers and Increasing Their Stability, Leningrad, State Press for Chemical Literature, 1955, p. 100.

79. G. L. Slonimskii and E. V. Reztsova, Zhur. Fiz. Khim. 33(2):480, 1959.

80. E. V. Reztsova, B. A. Lipkina, and G. L. Slonimskii, Zhur. Fiz. Khim. 33(3):656, 1959.

81. L. Begunovskaya, V. Zhakova, B. K. Karmin, and V. Epshtein, Aging and Fatigue of Raw and Cured Rubbers, Leningrad, State Press for Chemical Literature, 1955, p. 31.

82. B. A. Dogadkin and K. A. Pechkovskaya, Collection: Aging and Fatigue of Raw and Cured Rubbers, Leningrad, State Press for Chemical Literature, 1955, p. 53.

83. M. Reznikovskii, E. Vostroknutov, and L. Priss, Collection: Aging and Fatigue of Raw and Cured Rubbers, Leningrad, State Press for Chemical Literature, 1955, p. 76.

84. G. M. Bartenev and F. A. Galil-Ogly, Collection: Aging and Fatigue of Raw and Cured Rubbers, Leningrad, State Press for Chemical Literature, 1955, p. 119.

85. B. A. Dogadkin and Z. N. Tarasova, Kolloidn. Zhur. 15(5):347, 1953.

86. B. A. Dogadkin, Khim. Nauka i Prom. 4(1):55, 1959.

87. A. S. Kuz'minskii and L. I. Lyubchanskii, Doklady Akad. Nauk SSSR 93(3):519, 1953.

88. A. S. Kuz'minskii, Collection: Aging and Fatigue of Raw and Cured Rubbers, Leningrad, State Press for Chemical Literature, 1955.

89. Z. N. Tarasova, I. Éitingon, L. Senatorskaya, T. Fedorova, and B. Dogadkin, Kauchuk i Rezina, No. 9:15, 1961.

90. J. P. Berry and W. F. Watson, J. Polymer Sci. 18(88):201, 1955.

91. A. V. Tobolsky, J. Polymer Sci. 25(111):493, 1957; J. Appl. Phys. 27(7):672, 1956.

92. L. I. Lyubchanskaya, L. Fel'dshtein, and A. S. Kuz'minskii, Kauchuk i Rezina, No. 1:23, 1962.

93. A. Tobolsky, J. Prittyman, and J. Dillon, J. Appl. Phys. 15(4):380, 1944.

94. G. Brodskii, N. Sakhnovskii, M. Reznikovskii, and V. Evstratov, Kauchuk i Rezina, No. 8:22, 1960.

95. I. Kragel'skii, Kauchuk i Rezina, No. 11:20, 1959.

96. M. Reznikovskii, Nauchuk i Rezina, No. 5:34, 1960.

Chapter XI

MECHANOCHEMICAL PROCESSES IN HIGHLY ELASTIC POLYMERS

For a long time it was believed that the chemical bond can be broken only in some chemical reaction leading to a change in the structure of the molecule. Moreover, it was well known that the rate of chemical reactions can be substantially increased by means of physical influences (for example, heating). It was also known that chemical reactions themselves can be initiated by heating or irradiation by ultraviolet light.

With the creation of modern sources of powerful radiations, active physical intervention into chemical processes acquired an especially great scope and led to the rapid development of a new branch of chemistry – radiation chemistry. Investigations of the free radicals formed during the irradiation of substances, known earlier from a number of chemical reactions, in particular, from polymerization reactions, also developed in close relationship with the successes of radiation chemistry.

This young field of chemistry led to the creation of new concepts of the nature of the chemical bond and chemical reactivity, which permitted an understanding of the mechanism of many technically important, complex chemical and physicochemical processes (for example, the processes of synthesis and destruction of polymers) and made it possible to learn how to control them.

The possibility of breaking or changing a chemical bond through physical influences has usually been related only to such influences as heating and irradiation. The breaking of chemical bonds by means of mechanical influences seemed impossible for a long time.

However, the existence of mechanical cleavage of the chemical bond had already long ago been directly indicated by the well-known possibility, realized in jewelry work, of mechanical decomposition of diamond. More evidence of this possibility was the phenomenon of the breaking of textile fibers, representing highly oriented systems of long-chain macromolecules. Actually, the sum of even the weakest intermolecular interactions among long macromolecules should exceed the energy of the chemical bond for sufficiently long molecules. This actually occurs even at degrees of polymerization of the order of a hundred. Hence the breaking of fibers should not occur on account of the mutual sliding of straightened and oriented macromolecules, but should arise as a result of the energetically more profitable mechanical cleavages of the macromolecules at chemical bonds, i.e., as a result of mechanical destruction of the polymer substance.

There have also been other observations leading to the thought of mechanical breaking of chemical bonds in macromolecules [1]. For example, the widely known phenomenon of destruction of natural rubber during rolling has usually been explained by oxidative processes arising as a result of the formation of a large fresh rubber surface, local superheatings of the mass being rolled, and the action of ozone, which appears during the electric discharges that accompany the rolling process (triboelectricity). Moreover, the hypothesis of mechanical cleavage of the macromolecules during rolling was naturally also expressed, but it was believed that such cleavages, if they do occur, usually play a subordinate role.

Thus, the influence of mechanical action on the chemical bond was first denied, and then admitted, but considered as a secondary phenomenon.

However, during the last 10 years, it was found that mechanical cleavages of macromolecules in a number of cases play a very important role, since the free macroradicals formed in this case promote the beginning of complex chemical chain processes, leading ultimately to a profound change in the structure of the macromolecules and the properties of the polymer substances formed from them.

Since the primary event in these chemical processes is the mechanical cleavage of macromolecules, this new field of chemistry has received the name of mechanochemistry. *

I. CLEAVAGE OF THE MACROMOLECULE

First of all, let us consider the work of the destruction of one $C-C$ bond in a macromolecule. As is well known, the energy of formation of a $C-C$ bond is equal to approximately 83 kcal/mole, or 3.5×10^{12} erg/mole, which, converted to one bond, is equal to about 0.6×10^{-11} erg, i.e., it is equal to the work of a force of 0.006 mg over a path of 0.1 A. Thus, the work of destruction of an individual bond is negligible

---

* We should note the fact that processes of the transformation of chemical energy to mechanical energy during the work of a muscle and in the deformation of polymer electrolytes, due to a change in their charge or electrochemical potential, are also called mechanochemical. However, these processes would be more correctly called chemomechanical, since they are mechanical processes initiated by chemical reactions.

in comparison with the work of deformation of the entire body as a whole (for example, about 10^2 erg/cm^3 must be expended to stretch a sample of very soft rubber only 1%). At the same time, the work of deformation of the body itself is negligibly small in comparison with the total energy of all the chemical bonds contained in the body to be deformed; hence it is difficult to assume immediately that such a small quantity can be vital.

However, the picture is entirely changed if we consider the special character of the thermal motion of the flexible chain macromolecules and their mutual arrangement in the polymer body. Thanks to the poor order of the molecular structure (even in the presence of distinctly pronounced rods of macromolecules), an extremely inhomogeneous distribution of forces in volumes of molecular dimensions inevitably arises in a mechanically stressed polymer substance [2, 3]. This inhomogeneity is gradually reduced by rearrangement of the elements of the molecular structure, completing thermal motion. Processes of this type are well known as mechanical relaxation processes, manifested most plainly in highly elastic bodies.

It is quite clear that, as a result of the nonuniformity of the force distribution over the macromolecules, at any moment of the relaxation process there are individual, extremely stressed macromolecules. Hence there is nothing surprising in the fact that such macromolecules can be cleaved even under very small average mechanical stresses applied to the polymer body as a whole. This agrees with the possibility, noted long ago, of the destruction of polymers during flowing of their solutions through capillary tubes or during active mixing [4, 5].

Here we should emphasize that in addition to direct cleavage of the chemical bond, mechanical influences can also affect the chemical bond more mildly. If a chemical reaction leading to the destruction of a given chemical bond is possible, (for example, oxidative destruction of macromolecules) but requires an activation energy, then a mechanically stressed chemical bond requires less activation energy than an unstressed bond. Hence, chemical reactions of this type proceed at greater rates under conditions by mechanical influences [6-8]. Let us mention here that the acceleration of chemical reactions by mechanical influences is also possible in the case when the reaction rate is determined by the rate of diffusion of the reacting groups [9, 10]. An example is the reaction leading to the formation of intermolecular chemical bonds in a system of linear chain macromolecules at temperatures at which the viscosity of the system is rather high. Mechanical influences, producing mutual mixing of the macromolecules during deformation of the body, can substantially increase the probability of collision of the reacting groups and thereby increase the reaction rate. We might state that the deformation of a polymer body is a unique process of mixing.

TABLE 35. Dependence of the Molecular Weight of Polyisobutylene
on the Number of Periodic Deformations

Sample No.	No. of cycles of deformation, million	Molecular weight·10⁻⁶	Type of deformation	Amplitude of deformation	Frequency, cycles/min	Maximum temperature within the sample, °C
1	0	0.63	Shearing	0.15	500	45
	5.7	0.35	"	0.15	500	45
	8.6	0.22	"	0.15	500	45
2	0	7.0	Shearing	0.15	500	45
	7.2	1.6	"	0.15	500	45
3	0	0.63	Monoaxial compression	0.20	250	75
	4.0	0.11	The same	0.20	250	75

Thus, a mechanical field is capable of initiating chemical reactions by direct mechanical cleavage of chemical bonds, as well as of changing the rate of purely chemical reactions already begun.

II. MECHANOCHEMICAL PHENOMENA IN THE CASE OF REPEATED DEFORMATION OF HIGHLY ELASTIC POLYMERS

The possibility of cleavage of individual macromolecules even in the presence of small average mechanical stresses has been confirmed experimentally by a study of the change in the molecular weight of a chemically inert linear polymer – polyisobutylene, subjected to the influence of a large number of relatively small deformations [11, 12]. It was found that during such a prolonged experiment, the molecular weight of polyisobutylene is reduced several-fold (Table 35).

Such destruction can be produced only by direct mechanical influences on chemical bonds, since it is well known that even at higher temperatures and in the case of more prolonged heatings, purely chemical changes in polyisobutylene are only slightly noticeable.

Analogous results have also been obtained for butyl rubber and crude natural rubber [11, 12].

In the case of three-dimensionally structured polymers, the change in their structure under mechanical influences can be evaluated according to the change in the value of the equilibrium swelling in the corresponding liquid. Actually, a study of the swelling of unfilled vulcanizates of butyl rubber and natural rubber showed that the ability to swell is substantially changed as a result of the action of repeated deformations by shearing, monoaxial compression, or stretching. The butyl rubber vulcanizates substantially increase their maximum swelling in vaseline oil as a result of the mechanical influence. Vulcanizates of natural rubber increased their degree of swelling in polar liquids and reduced in it nonpolar liquids (for example, in vaseline oil) after prolonged repeated deformations. Moreover, transition from an increase in the swelling to a decrease in it occurred in the presence of greater polarity of the liquid, the longer the vulcanizate had been deformed. Thus, in contrast to butyl rubber vulcanizates, vulcanizates of natural rubber become more polar substances as a result of mechanical destruction, which indicates the development of reactions of oxidation of the rubber during the process of deformation.

In the case of vulcanizates of butadiene rubber, mechanical influences also led to its oxidation and related structuring.

All these data and a number of other known facts permit us to draw a conclusion on the nature of the important phenomenon, the so-called fatigue of a polymer body, consisting of breakdown of the body or a change in certain of its properties as a result of fatigue, i.e., as a result of a prolonged constant or repeated mechanical influence.

In the case of elastic polymer materials, it has been noted that, almost all the way up to decomposition, the strength of such bodies does not change, and that the strength of the fragments of the decomposed substance practically does not differ from the initial value. All this has given a basis for developing representations [11, 13] of the decomposition of polymer bodies as a result of fatigue (as well as the accompanying changes in a number of properties – solubility, swelling, modulus of elasticity, etc.) as a result of the development of secondary chemical processes (primarily chain oxidation processes) initiated by free macroradicals formed at the sites of mechanical cleavage of the macromolecules at chemical bonds.

According to these representations, the mechanism of the fatigue process is essentially analogous to the mechanisms of the processes of aging of polymers under the influence of light or heat, and hence can be considered as a variety of aging. The primary event in fatigue is cleavage of a chemical bond, forming free macroradicals. The secondary reactions that arise after this, depending on the concrete conditions, can lead both to destruction and to structuring of the polymer.

It is natural that as long as there are substances in the system that inhibit free radicals at the ends of the torn chain molecules, dangerous secondary chemical processes do not develop. Only after the inhibitors have been consumed does a rapid process of chemical change of the polymer arise, leading to its decomposition. Hence, appreciable secondary chemical processes begin only at the end of the induction period. Con-

sequently, the ability of a polymer body to work is determined by the induction period and should depend on all the factors determining its duration.

A characteristic feature of fatigue is the small number of free macroradicals arising in each deformation cycle. However, the heavier the
mechanical system of the influence, the greater the number of macromolecules that are broken in each deformation, and the shorter the induction period.

Thus, the concentration of the free macroradicals arising under
mechanical influences depends on the intensity of the influences, while
the possibility of the development of secondary chemical processes is
determined by the inhibitor concentration.

We should mention especially that there is no direct relationship between destruction (or structuring) and breakdown of a polymer body.

In the case of repeated mechanical influences with a constant deformation amplitude, destruction (or structuring) will lead to a reduction
(increase) in the stresses and, consequently, to a deceleration (acceleration) of the breakdown of the body. On the other hand, in the case of a
system of influence with constant stress amplitude, destruction (structuring) will already lead to an increase (decrease) in the value of the deformation amplitude and, consequently, to an acceleration (deceleration)
of breakdown.

It is easy to understand that if there are few radicals, but many inhibitor molecules, then no dangerous chemical reactions (primarily
chain oxidation processes) will have time to be initiated during the lifetime of the radical; hence, the few cleavages of the macromolecules do
not lead to any appreciable changes in the properties of the polymer substance (in the case of a very large number of deformations, these ac-
cumulating cleavages of the macromolecules will begin to be perceived
either by a change in the molecular weight or in the ability to swell).

However, not the gradual processes of change of the structure of the
entire body, but sharp changes in the structure in a few individual micro-
regions are vital for the breakdown of a polymer body. Since an elastic
polymer object is subjected to inhomogeneous deformation during the
fatigue process in almost all cases, various numbers of macromolecules
arise in different microvolumes of it. It is clear that the secondary
chemical processes arise earlier in those microvolumes in which an especially large number of free radicals was formed as a result of mechanical overloading.

Thus, appreciable changes in structure should occur only in individual microvolumes, which have become foci of decomposition. It is pre-

cisely there that the primary breakdown of the material occurs, forming a microcrack, which, expanding rapidly, breaks the body into parts, leaving the properties of these parts almost unchanged.

From all this it follows that in the case of the same work of deformation, the process of breakdown of a body during fatigue should develop differently depending on how this work was performed – uniformly over the entire volume or nonuniformly, for long periods in the presence of small stresses and deformations, or rapidly with large stresses. Hence the character of the structural changes is determined not by the integral value of the work of deformation of the body, but by the distribution of the influence with time and over the microvolumes of the body, i.e., by the differential characteristics of the mechanical fatigue system. This conclusion permits us to understand the well-known gross errors that arise in the use of accelerated methods of evaluating the fatigue strength of polymer materials. The cause of this is the fact that such methods are usually developed considering only the integral doses of the influence.

The high sensitivity of fatigue processes of polymers to the system of deformation and the possibility of the appearance of a multitude of different secondary chemical reactions result in an extreme variety of the forms of fatigue of polymer bodies, in spite of the unity of the initial elementary event – mechanical cleavage of the macromolecule. We might state that the fatigue of a polymer is a mechanicochemical process, in which the mechanical influences initiate, as well as accelerate and decelerate various chemical processes, leading ultimately to a breakdown of the deforming body.

The representations of the mechanism of fatigue breakdown of elastic polymer bodies presented are in good agreement with all the known characteristics of this phenomenon. However, one must also frequently encounter concepts of the deciding role of heat formation during repeated deformations. Actually, heat formation, producing an increase in the temperature of the deforming body, should accelerate the secondary chemical reactions (especially oxidation) and thereby sharply influence the process of decomposition of the material. Nonetheless, such an influence on fatigue is possible only in those cases when chemical reactions have already been initiated by free macroradicals of mechanical origin. * Hence, until the inhibitors present in the polymer permit the initiation of these reactions, heat formation cannot regulate the fatigue process, and, consequently, the lifetime of the object as well [14].

---

* If heating of the body as a result of heat formation produces thermal destruction, such a process will not be fatigue, but thermal aging under conditions when the source of heat is mechanical influence.

TABLE 36.   Lifetime of Filled Rubbers During Fatigue Under the Same
Conditions on an Instrument of the Martens-Shob Type [15]

Type of rubber	Without additive	Additives, % by weight					
		Azoisobutyrodinitrile				Di-tert-butyl-hydro-quinone	Benzo-quinone
		0.1	1	2	5		
		Lifetime, min					
Sodium-butadiene	1300	2	0.3	0.1	–	2500	2200
Butadiene-methyl-styrene	5400	5100	720	660-15	2	–	9200

   The concept of the process of fatigue of a polymer body as a mechanochemical process leads to clear ways to combat breakdown of elastic polymer objects under prolonged static or repeated deformations.   On
the one hand, any softening of the system of mechanical influences should
reduce the concentration of free macroradicals formed and increase the
service period of the object.   On the other hand, suitable inhibitors must
be introduced into the polymer.   Thus, increasing the ability to work of
elastic polymer objects working under conditions of prolonged mechanical
influences requires an optimum construction of the object (a guarantee of
the smallest mechanical stresses, most uniformly distributed over the
volume, under the conditions of operation), as well as the proper selection and measuring of the inhibiting substances (considering the chemical
peculiarities of the given polymer substance and the intensity of the mechanical influences).

   As an illustration of the correctness of mechanochemical representations of the fatigue of polymers, we might cite data on the action of substances active with respect to free radicals, introduced into rubber in
small amounts.

   Massive samples of filled rubber vulcanizates with various small
additives were fatigued under especially heavy dynamic conditions [14].
As can be seen from Table 36, the lifetime of the samples is extremely
sensitive to the active substances, being either increased or decreased.

   It is interesting to mention that the addition of azoisobutyrodinitrile,
which in this case is a supplementary source of free radicals and hence
sharply accelerates decomposition of the substance, leads to decomposition at considerably lower temperatures than the breakdown point of all
the remaining samples, which are greatly heated during prolonged deformation.   This clearly shows the subordinate role of heat formation in

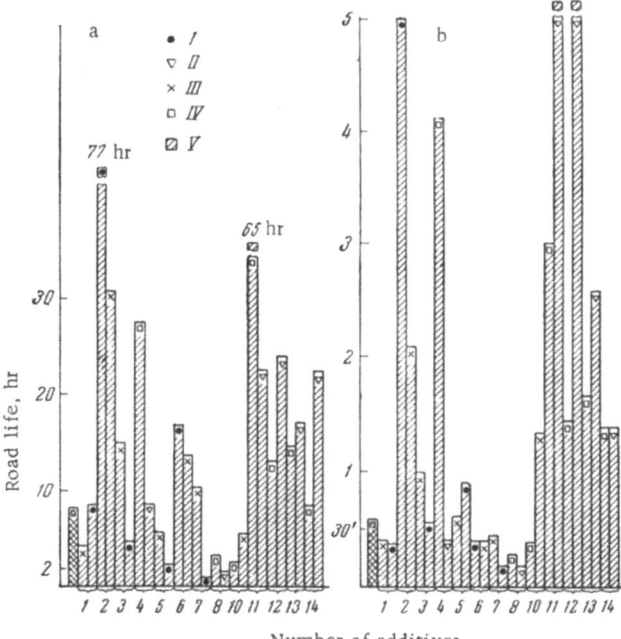

Number of additives

Fig. 189. Fatigue strength of rubbers made from SKS-30A after the introduction of various additives. a) At 20°; b) at 100°C. I) 10 parts by weight of the additive; II) three parts by weight; III) one part by weight; IV) 0.3 parts by weight; V) without additives. 1) Benzoyl peroxide; 2) dihydroxydiphenyl sulfide; 3) azoisobutyrodinitrile; 4) di-tert-butylhydroquinone; 5) tri-tert-butylphenol; 6) polyethylene-polyamine; 7) diproxide; 8) santovar 0; 9) chlorinated paraffin; 10) benzoquinone; 11) hydroquinone; 12) quinhydrone; 13) acetoneanil; 14) β,β'-dinaphthyl disulfide. System of fatigue: repeated stretching of samples in the form of blades; amplitude of deformation 150%; frequency 250 cycles per minute.

the process of fatigue. Other evidence of this is the distinctly greater heat formation in more long-lived rubbers made from butadiene-methylstyrene rubber in comparison with the rubbers of sodium-butadiene rubber, close to them in many mechanical properties. Analogous results have also been obtained for other types of repeated deformations of rubbers made from various crude rubbers after the introduction of active substances [14] (Fig. 189).

The development of concepts of the mechanochemical mechanism of fatigue has not yet been completed. The properties of the radicals formed and their transformations have received little study; scientific principles of selecting the type and amount of the inhibitor have not been developed; the quantitative relationships between the intensity of the me-

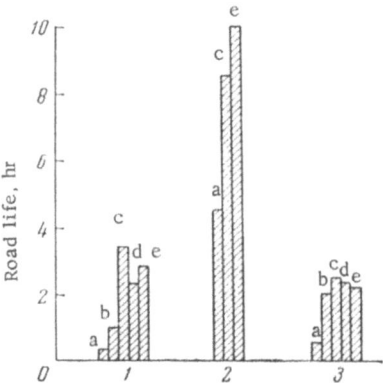

Fig. 190. Change in the strength of the bond of
two-ply samples based on the rubber SKS-30A
after the introduction of various additives in the
glueing layer. 1) Samples with a glueing layer
based on natural rubber with polyethylenepolyamine; 2) samples with a glueing layer based on
SKS-30A with di-tert-butylhydroquinone; 3) samples with a glueing layer based on natural rubber
with di-tert-butylhydroquinone. Content of additive in parts by weight per 100 parts by weight of
the rubber glue: a) without additive; b) 1; c) 3;
d) 5; e) 10.

chanical influences and the number of radicals formed have practically
not been studied. However, the presence of mechanochemical processes
and the possibility of chemical regulation of the lifetime of rubber objects by means of inhibitors have been reliably established, and this permits the assured use of the concepts outlined above on the mechanism of
fatigue.

The phenomenon of fatigue breakdown of homogeneous elastic bodies
is closely adjoined by the phenomenon of stratification of multilayered
elastic bodies during repeated deformations, i.e., the phenomenon of
breaking of bonds between layers of polymers differing in mechanical
properties.

In this case especially difficult systems of mechanical influences,
responsible for the difference in the mechanical properties of the layers,
are created in the transition region from one layer to another (close to
the boundary or at some distance from it, if the stress distribution is
complex). Hence, a more rapid mechanochemical process of fatigue,
leading to breakdown of the object at the boundary of the layers (or close
to it) develops in the boundary zones subjected to overloading.

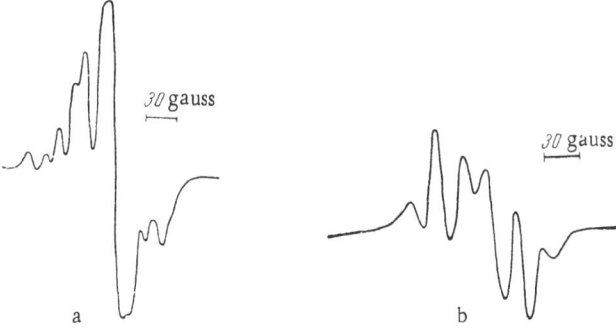

Fig. 191. Electron paramagnetic resonance spectra of samples of polymers mechanically broken down in liquid nitrogen [18]. a) Polyisobutylene; b) polyethylene.

Thus, if we neglect the elementary principles of stratification (for example, defects in the connection of various layers into a single body), then it is obvious that measures to combat stratification of multi-ply elastic objects should be the same as measures to combat fatigue breakdown. The difference in this case will be the necessity for introducing supplementary inhibiting substances in the region close to the boundary surfaces. This idea [16] is confirmed experimentally [17] by an increase in the bond strength between filled rubbers prepared on the basis of butadiene-styrene rubber (SKS-30A), as well as natural rubber (NK), glued by various rubber glues, after the introduction of inhibitors into the glue and into the rubber layers to be glued. For example (Fig. 190), the strength of the bond of two-ply samples based on the rubber SKS-30A in the case of the introduction of about 3% by weight polyethylenepolyamine into the glueing layer of natural rubber, as well as that of 3 to 10% by weight di-tert-butylhydroquinone into a glueing layer of SKS-30A, increased several-fold (the bond strength was evaluated according to the duration of work, the so-called "road life" of the two-ply sample before stratification under the conditions of periodic monoaxial compression of a cube; frequency 500 cycles/min, deformation amplitude 0.2) [17].

Apparently all the cases of breakdown of polymer bodies under mechanical influences are related to one degree or another to the development of mechanochemical reactions in these bodies. As was shown above, this is unquestionably observed in the breakdown of elastic polymer bodies under repeated mechanical influences. We can scarcely doubt that mechanochemical phenomenon also lie at the basis of the phenomena of mechanical wear of elastic polymer materials operating under conditions of friction.

In the case of the breakdown of solid (vitreous or crystalline) polymer bodies, in which the mobility of the long macromolecules is maxi-

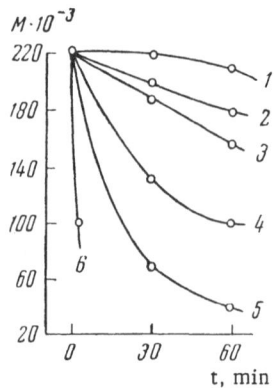

Fig. 192. Reduction of the molecular weight of polyisobutylene as a
result of rolling at various temperatures [26]: 1) 140°; 2) 100°; 3)
210°; 4) 60°; 5) 20°; 6) < −100°C.

mally limited, mechanical cleavages of
the macromolecules seem impossible.
Actually, in recent years it has been possible to show through the use of the phenomenon of electron paramagnetic resonance (Fig. 191) and other physical methods of investigating molecular structure,
that in the cutting and crushing of solid
polymers, as well as in a number of cases
of cracking of stressed polymer bodies,
mechanical cleavages of the macromolecules do occur [18-21]. Certain questions
of mechanical destruction in the pulveriza-
tion of polymers using various machines
have been investigated in greater detail
[22-26]. A consideration of the mechanochemical processes in solid polymer
bodies, as well as in solutions under the
action of ultrasound, goes beyond the
framework of our topic.

III. MECHANOCHEMICAL PHENOMENA IN THE REPROCESSING OF HIGHLY ELASTIC POLYMERS

In the reprocessing of highly elastic polymers (for example, in
rolling, extrusion, pressing), the polymer mass, flowing under the action of the applied stresses, experiences considerable deformations,
partially highly elastic, partially plastic. After all that was said above
with respect to mechanochemical reactions in deformations of elastic
polymer bodies, it is quite clear that these reactions also occur during
the reprocessing of the polymers.

Actually, it has long been known, for example, that the rolling of
isobutylene, natural rubber, and many other polymers reduces their
molecular weight (Fig. 192), while the rolling of sodium-butadiene rubber and certain other polymers leads to their three-dimensional structuring. As an example of how chemically more active mechanical influences during reprocessing are in comparison with deformations of
highly elastic bodies under conditions of operating systems, let us mention that the destruction of polyisobutylene under the action of repeated
deformations lasting for one to two weeks, described in Table 35, can
be achieved in 20-30 min by rolling. Thus, the intensity of the formation of free macroradicals during the reprocessing of polymers is far
higher than in processes of use of the finished objects.

It is natural that the appearance of free macroradicals should give rise (since this is not prevented by inhibitors) to complex secondary chemical processes, which change the structure and properties of the polymer mass to be reprocessed. At the same time, it is evident that the introduction of various substances active with respect to free radicals into the polymer during its reprocessing can regulate these secondary processes and, consequently, the properties of the mass to be reprocessed.

As an illustration of this, let us mention that rolling of sodium-butadiene rubber (containing no antioxidant) on cold rollers leads to the formation of a more rigid and poorly soluble material after only one to two hours. Continuation of the rolling leads to a further increase in the rigidity and complete loss of solubility. However, the introduction of several tenths of a percent by weight of di-tert-butyl-hydroquinone into the rubber before rolling holds back the complete loss of solubility for several hours [27]. The influence of the introduction of inhibitors on the change in the elastoplastic properties of crude rubber mixtures during rolling has also been demonstrated quite distinctly [28].

It is easy to understand that the intensity of mechanical cleavages of the macromolecules should depend sharply on the temperature [27, 29] at which the reprocessing is conducted. The lower the temperature, the higher the viscosity of the polymer mass, and the higher the stresses arising in it. Hence the intensity of the formation of free macroradicals increases sharply with decreasing temperature of the reprocessed mass and, of course, with increasing rigorousness of the mechanical system of reprocessing (for example, when the gap between the rollers is reduced). However, we should not forget that the rate of the most dangerous chemical reactions from the standpoint of structural changes (oxidation reactions) is reduced upon cooling.

The changes in the structure of the polymers can be especially substantial when they are reprocessed on powerful high-speed machines. The great mechanical stresses and high temperatures that arise under conditions should be extremely favorable to the initiation and development of the entire complex of mechanochemical reactions. Thus, the properties of polymer masses reprocessed in various temperature mechanical systems should not coincide in principle, and different amounts of additives regulating the mechanochemical processes should be introduced into the polymer for the production of products with the same properties under different systems.

Mechanical cleavages of chemical bonds in certain cases prove to be the cause of the appearance of the long since detected [30-32] phenomenon, encountered only in polymers, which has received the name of chemical flow. As is well known, the fluidity of a substance is always related to irreversible displacements of its structural elements. The flow of low-molecular liquids is accomplished by displacements of their molecules. In the case of polymers possessing long-chain molecules, the displacement of such molecules becomes possible thanks to flexibility and is accomplished by displacements of individual parts of the macromolecules — segments. It is natural that, as a result of the large number of segments in each molecule, the displacement of macromolecules occurs very slowly (the viscosity of polymers is enormous). In the case of a sufficiently rapid and powerful mechanical influence, the macromolecules do not have time to be displaced and breakdown into macroradicals. In view of this, the viscous flow of a polymer mass is always accompanied to some degree by cleavages of macromolecules and combination of the macroradicals arising into macromolecules.

In the case of three-dimensionally structured highly elastic polymers, the usual fluidity generally cannot occur; hence all the irreversible changes in the positions of the structural elements of the three-dimensional lattice can occur only in the case of mechanical cleavages of the chemical bond. In the presence of sufficiently rapidly applied high mechanical stresses, profound mechanical decomposition of the lattice into free biradicals, capable of being irreversibly displaced like the usual molecules, proves possible. A system of such "fragments" of the lattice flows like a liquid during a short period of time, and then, as a result of the enormous chemical activity of the free biradicals, again is converted to a chaotic three dimensional lattice, differing in no way from the initial lattice except for its external appearance. Thus, fluidity of a three-dimensionally structured polymer proves possible on account of profound mechanochemical transformation of its structure.

Chemical flow is encountered in the pure form in the case of impact pressing of objects made from three-dimensionally structured polymers, as well as in a number of cases.* As has already been mentioned, chemical flow is also accompanied to one degree or another by the usual flow of linear polymers.

It proves possible to convert three-dimensionally structured polymers to linear polymers (of course, highly branched) by means of chemical flow, with simultaneous inhibition of the free radicals formed [27, 34]. From this follows the possibility of creating a mechanochemical method of regenerating rubbers, as well as the necessity of considering

---

* We should not confuse chemical flow, based on mechanochemical reactions, with chemical relaxation and analogous phenomena [33], which are based on purely chemical decomposition processes (for example, oxidative destruction) of the stressed three-dimensionally structured highly elastic polymers.

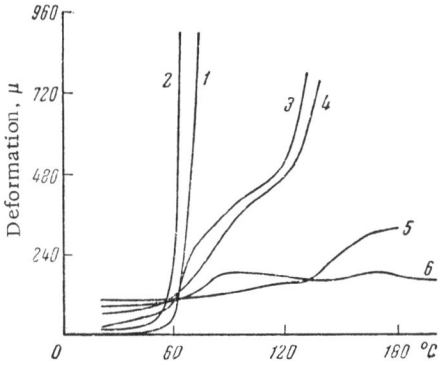

Fig. 193. Thermomechanical curves of block copolymers of epoxide resin and nitrile rubber [42].
1) Epoxide resin; 2) block copolymer with a 5 : 1
weight ratio of the resin and rubber; 3) the same,
2 : 1; 4) the same, 1 : 1; 5) the same, 1 : 2; 6)
nitrile rubber.

mechanochemical phenomena in the processes of rubber regeneration
presently carried out under conditions of active mechanical and chemical influences.

In view of all the questions of the reprocessing of linear and three-
dimensionally structured elastomers touched upon, we must not omit a
mention of the role of the medium in which the mechanochemical processes are carried out. It is clear that oxygen interacts actively with
free hydrocarbon radicals and changes their reactivity. The processes
of reprocessing in air and in an inert medium hence lead to various secondary mechanochemical processes, which exerts a substantial influence
on the structure and properties of the polymer. This very important
problem as yet has not been studied to the proper degree, but it seems
extremely real that the reprocessing of polymers will be accomplished
in the future considering the influence of the medium on the fate of the
mechanically formed macroradicals.

The influence of the medium is already being considered at the present time in conducting processes of mechanochemical synthesis and modification of polymers. Although a consideration of this rapidly developing direction of mechanochemistry (see, for example, [35-50, 55, 56])
goes beyond the limits of our topic, nonetheless let us briefly take it up.

Under mechanical influence on a polymer body, the free macroradicals that arise in it are capable, in the absence of oxygen and other
inhibitors, either of recombining with one another, or of reacting with
the macromolecules (forming new radicals), or, if a monomer is present

in the system, of initiating radical chain polymerization processes. All these possibilities are utilized for the synthesis and modification of polymer materials (see, for example, Fig. 193) by mechanical reprocessing in an inert medium (using specially designed machines) of mixtures of various polymers, as well as polymers swollen in low-molecular substances capable of polymerization.

In the first case, various types of macromolecules are cleaved into macroradicals, and the latter combine randomly into macromolecules. In this case so-called block copolymers arise, the chain molecules of which consist of different block pieces ("fragments" of the initial macromolecules). At the same time, the addition of various macroradicals to the middle portions of chain macromolecules, forming branches, is also possible. These branches can have side branches, coinciding with the principal chain or differing from it in composition and structure (depending on which free macroradical — of the same polymer or of another — reacted with the chain macromolecule). Thus, in addition to a block copolymer, grafted copolymers and branched polymers are also formed.

In the second case, the free macroradical initiates the polymerization process and grows thanks to the addition of a monomer to it. If this monomer is not the monomer of a mechanically treatable polymer in its composition and structure, then block copolymers and grafted copolymers are formed. If the treatable polymer contains its own monomer, then, depending on the concrete conditions, processes of increase in the molecular weight, increase or decrease in the degree of branching, etc., can occur.

At the present mechanochemical processes are attracting great attention, since they play an appreciable role in the processes of synthesis, modification, and reprocessing of polymers, as well as in the use of elastic objects under conditions of prolonged or transitory, but repeated mechanical influences.

Fatigue breakdown of highly elastic objects has proved to be a variety of aging, produced by mechanical stresses. In this case, the processes of fatigue of the object depend on the type of reprocessing to which the polymer was subjected in the preparation of the object, since in both processes mechanochemical reactions develop, and the content of inhibiting substances in the object, necessary for its protection from fatigue, depends on the system of mechanical reprocessing. The dependence of the lifetime of rubber objects on the duration and force system of reprocessing is confirmed experimentally [51]. Consequently, the protection of an object from fatigue under conditions of operation cannot be considered apart from the process of its preparation.

All this requires a systematic study of the mechanism of the mechanical cleavage of macromolecules, the chemical properties of free macroradicals, and their reactions with the inhibitor molecules and other macromolecules. A theory of mechanochemical reactions in the reprocessing of polymers and the use of objects made from them should be created on the basis of these investigations.

The selection of the most suitable types of inhibitors must be scientifically substantiated, and their proper dosing should be learned, depending on the intensity of the mechanical influences. The creation of such principles, as well as the elucidation of the interaction of photo-, thermo-, and mechanodestruction processes in polymers are undoubtedly primary problems in this important and very interesting field of phenomena.

Of no less interest is the elucidation of the similarities and differences between processes of fatigue and other types of aging.

BIBLIOGRAPHY

1. H. Staudinger, Kolloid Z. 51:71, 1930.
2. V. A. Kargin and G. L. Slonimskii, Collection: Conference on the Viscosity of Liquids and Colloidal Solutions, Vol. 1, Moscow-Leningrad, Academy of Sciences USSR Press, 1941, p. 117.
3. G. L. Slonimskii, Doctoral Dissertation, Moscow, L. Ya. Karpov Physicochemical Institute, 1947.
4. H. Staudinger, High-Molecular Organic Compounds [Russian Translation], Moscow-Leningrad, United Scientific and Technical Press, 1935, p. 211.
5. J. Frenkel, Acta Physicochim. USSR 19:51, 1944.
6. A. S. Kuz'minskii, M. G. Maizel's, and N. N. Lezhnev, Doklady Akad. Nauk SSSR 71:319, 1950.
7. A. S. Kuz'minskii, M. G. Maizel's, and N. N. Lezhnev, Collection: Chemistry and Physicochemistry of High-Molecular Compounds, Moscow, Academy of Sciences USSR Press, 1952, p. 99.
8. A. S. Kuz'minskii and L. I. Lyubchanskaya, Doklady Akad. Nauk SSSR 93:519, 1953.
9. G. L. Slonimskii, V. A. Kargin, and L. I. Golubenkova, Doklady Akad. Nauk SSSR 93:311, 1953.
10. V. A. Kargin, G. L. Slonimskii, and L. I. Golubenkova, Zhur. Fiz. Khim. 30:2436, 1956.
11. G. L. Slonimskii, V. A. Kargin, G. N. Buiko, E. V. Reztsova, and M. L'yuis-Riera, Doklady Akad. Nauk SSSR 93:523, 1953.
12. G. L. Slonimskii, V. A. Kargin, G. N. Buiko, E. V. Reztsova, and M. L'yuis-Riera, Collection: Aging and Fatigue of Raw and Cured Rubbers and Increasing Their Stability, Leningrad, State Press for Chemical Literature, 1955, p. 100.

13. V. A. Kargin and G. L. Slonimskii, Doklady Akad. Nauk SSSR 105:751, 1955.

14. E. V. Reztsova, B. G. Lipkina, and G. L. Slonimskii, Zhur. Fiz. Khim. 33:656, 1959.

15. G. Sh. Izraelit, Mechanical Testing of Raw and Cured Rubbers, Leningrad-Moscow, State Press for Chemical Literature, 1949.

16. G. L. Slonimskii, Collection: Strength of the Bond between Elements of Rubber-Fabric Multi-ply Objects in Production and Use, Moscow, State Press for Chemical Literature, 1956, p. 5.

17. G. L. Slonimskii and G. P. Drugova, Zhur. Fiz. Khim. 33:793, 1959.

18. S. E. Bresler, S. N. Zhurkov, E. N. Kazbekov, E. M. Saminskii, and E. E. Tomashevskii, Zhur. Tekh. Fiz. 29:358, 1959.

19. S. E. Bresler, É. N. Kazbekov, and E. M. Saminskii, Vysoko-molekulyarnye Soedineniya 1:136, 1959.

20. P. Yu. Butyagin, A. A. Berlin, A. É. Kalmanson, and L. A. Blyumenfel'd, Vysokomolekulyarnye Soedineniya 1:865, 1959.

21. M. P. Vershinina and E. V. Kuvshinskii, Vysokomolekulyarnye Soedineniya 2:1486, 1960.

22. N. K. Baramboim, Doklady Akad. Nauk SSSR, 114:568, 1957.

23. N. K. Baramboim, Zhur. Fiz. Khim. 32:433, 1958.

24. N. K. Baramboim, Zhur. Fiz. Khim. 32:806, 1958.

25. N. K. Baramboim, Zhur. Fiz. Khim. 32:1049, 1958.

26. N. K. Baramboim, Zhur. Fiz. Khim. 32:1248, 1958.

27. G. L. Slonimskii and E. V. Reztsova, Zhur. Fiz. Khim. 33:480, 1959.

28. N. V. Veresotskaya, K. D. Bebris, and G. L. Slonimskii, Kauchuk i Rezina No. 3:27, 1959.

29. A. A. Berlin, G. S. Petrov, and V. F. Prosvirkina, Zhur. Fiz. Khim. 32:2565, 1958.

30. V. A. Kargin and T. I. Sogolova, Doklady Akad. Nauk SSSR 108:662, 1956.

31. V. A. Kargin and T. I. Sogolova, Zhur. Fiz. Khim. 31:1328, 1957.

32. V. A. Kargin and T. I. Sogolova, Collection: Problems of Physico-chemistry, No. 1, Moscow, State Press for Chemical Literature, 1958, p. 18.

33. A. V. Tobolsky, Properties and Structure of Polymers, J. Wiley and Sons, New York, 1960, p. 223.

34. V. A. Kargin, T. I. Sogolova, G. L. Slonimskii, and E. V. Reztsova, Zhur. Fiz. Khim. 30:1903, 1956.

35. M. Pike and W. F. Watson, J. Polymer Sci. 9:229, 1952.

36. W. F. Watson and D. Wilson, J. Sci. Instr. 31:98, 1954.

37. D. J. Angier and W. F. Watson, J. Polymer Sci. 18:129, 1955.

38. D. J. Angier and W. F. Watson, J. Polymer Sci. 20:235, 1956.

39. D. J. Angier and W. F. Watson, Trans. JRJ 33:22, 1957.

40. G. Ayrey, C. C. Moore, and W. F. Watson, J. Polymer Sci. 19:1, 1956.

41. D. J. Angier, E. D. Farlie, and W. F. Watson, Trans. JRJ 34:8, 1958.

42. V. A. Kargin, B. M. Kovarskaya, L. I. Golubenkova, M. S. Akutin, and G. L. Slonimskii, Khim. Prom. No. 2:13, 1957.

43. V. A. Kargin, B. M. Kovarskaya, L. I. Golubenkova, M. S. Akutin, and G. L. Slonimskii, Doklady Akad. Nauk SSSR 112:485, 1957.

44. A. A. Berlin, Uspekhi Khim. 27:94, 1958.

45. A. A. Berlin, Collection: Successes of Chemistry and Polymer Technology, No. 2, Moscow, State Press for Chemical Literature, 1957, p. 33.

46. V. A. Kargin, N. A. Platé, and A. S. Dobrynina, Kolloidn. Zhur. 20:332, 1958.

47. N. A. Platé, V. P. Shibaev, and V. A. Kargin, Vysokomolekulyarnye Soedineniya 1:1853, 1959.

48. G. L. Slonimskii and E. V. Reztsova, Vysokomolekulyarnye Soedineniya 1:534, 1959.

49. G. L. Slonimskii, V. A. Kargin, and E. V. Reztsova, Zhur. Fiz. Khim. 33:988, 1959.

50. R. Zh. Sereza, International Symposium on Macromolecular Chemistry, Moscow, 1960, Reports and Authors' Abstracts, Section III, p. 148.

51. G. L. Slonimskii, E. V. Reztsova, Vysokomolekulyarnye Soedineniya 4:1571, 1962.

52. G. L. Slonimskii, Khim. Nauka i Prom. 4(1):73, 1959.

53. V. A. Kargin and G. L. Slonimskii, Brief Outlines of the Physicochemistry of Polymers, Moscow State University Press, 1960, p. 87.

54. A. A. Strepikheev and V. A. Derevitskaya, Fundamentals of the Chemistry of High-Molecular Compounds, Moscow, State Press for Chemical Literature, 1961, p. 281.

55. W. F. Watson, Makromol. Chem. 34, I. Sonderband, 240, 1959.

56. N. K. Baramboim, Mechanochemistry of Polymers, Moscow, State Press for Technical Literature, 1961.